[redacted]

ACS SYMPOSIUM SERIES **666**

Fuels and Chemicals from Biomass

Badal C. Saha, EDITOR
Agricultural Research Service,
U.S Department of Agriculture

Jonathan Woodward, EDITOR
Oak Ridge National Laboratory

Developed from a symposium sponsored
by the Division of Biochemical Technology

American Chemical Society, Washington, DC

Library of Congress Cataloging-in-Publication Data

Fuels and chemicals from biomass / Badal C. Saha, editor, Jonathan Woodward, editor.

p. cm.—(ACS symposium series, ISSN 0097–6156; 666)

"Developed from a symposium sponsored by the Division of Biochemical Technology at the 211th National Meeting of the American Chemical Society, New Orleans, Louisiana, March 24–28, 1996."

Includes bibliographical references and indexes.

ISBN 0–8412–3508–2

1. Biomass energy—Congresses. 2. Biomass chemicals—Congresses.

I. Saha, Badal C., 1949– . II. Woodward, J. (Jonathan) III. American Chemical Society. Division of Biochemical Technology. IV. American Chemical Society. Meeting (211th: 1996: New Orleans, La.) V. Series.

TP360.F816 1997
662′.88—dc21

97–6914
CIP

This book is printed on acid-free, recycled paper.

PRINTED IN THE UNITED STATES OF AMERICA

Advisory Board

ACS Symposium Series

Foreword

THE ACS SYMPOSIUM SERIES was first published in 1974 to provide a mechanism for publishing symposia quickly in book form. The purpose of this series is to publish comprehensive books developed from symposia, which are usually "snapshots in time" of the current research being done on a topic, plus some review material on the topic. For this reason, it is necessary that the papers be published as quickly as possible.

Before a symposium-based book is put under contract, the proposed table of contents is reviewed for appropriateness to the topic and for comprehensiveness of the collection. Some papers are excluded at this point, and others are added to round out the scope of the volume. In addition, a draft of each paper is peer-reviewed prior to final acceptance or rejection. This anonymous review process is supervised by the organizer(s) of the symposium, who become the editor(s) of the book. The authors then revise their papers according to the recommendations of both the reviewers and the editors, prepare camera-ready copy, and submit the final papers to the editors, who check that all necessary revisions have been made.

As a rule, only original research papers and original review papers are included in the volumes. Verbatim reproductions of previously published papers are not accepted.

ACS BOOKS DEPARTMENT

Contents

CHEMICALS

INDEXES

Preface

THE PRODUCTION OF FUELS AND CHEMICALS from biomass faces significant technical and economic challenges at present. Its success depends largely on the development of environmentally friendly pretreatment procedures, a highly effective multienzyme system for conversion of pretreated biomass to fermentable sugars, and efficient microorganisms to ferment mixed sugar substrates to fuels and chemicals. It is timely to provide a book that can guide further research and development in this area.

This volume was developed from a symposium presented at the 211th National Meeting of the American Chemical Society, titled "Fuels and Chemicals from Biomass", sponsored by the ACS Biotechnology Secretariat and the Division of Biochemical Technology, in New Orleans, Louisiana, March 24–28, 1996. This book presents a compilation of ten manuscripts from that symposium plus nine solicited manuscripts that represent recent advances in the production of fuels and chemicals from biomass. The chapters in this book are organized in two sections: fuels (ethanol, biodiesel, and hydrogen) and chemicals (lactic acid, succinic acid, 1,3-propanediol, 2,3-butanediol, polyhydroxybutyrate, and xylitol). A chapter on synthesis-gas fermentation is included in the chemicals section.

We hope that this book will serve as a valuable interdisciplinary contribution to the continually expanding field of the production of fuels and chemicals from biomass.

Acknowledgments

We are fortunate to have contributions from many leading authorities in their respective disciplines. We would like to take this opportunity to express our gratitude to the contributing authors, the reviewers who provided excellent comments to the editors, the ACS Biotechnology Secretariat and the Division of Biochemical Technology, and the ACS Books Department for making possible the symposium and the publication of this book. We also would like to acknowledge the Oak Ridge National Laboratory, which is managed by Lockheed Martin Energy Research Corporation for the U.S. Department of Energy under contract DE–AC05–96OR22464.

BADAL C. SAHA
Fermentation Biochemistry Research Unit
National Center for Agricultural Utilization Research
Agricultural Research Service
U.S. Department of Agriculture
1815 North University Street
Peoria, IL 61604

JONATHAN WOODWARD
Chemical Technology Division
Oak Ridge National Laboratory
Oak Ridge, TN 37831

January 24, 1997

FUELS

Chapter 1

Advanced Bioethanol Production Technologies: A Perspective

Michael E. Himmel, William S. Adney, John O. Baker, Richard Elander, James D. McMillan, Rafael A. Nieves, John J. Sheehan, Steven R. Thomas, Todd B. Vinzant, and Min Zhang

Biotechnology Center for Fuels and Chemicals, National Renewable Energy Laboratory, 1617 Cole Boulevard, Golden, CO 80401

Conversion of the fermentable sugars residing in lignocellulosic waste and energy crops can conservatively yield approximately 100 billion gallons of fuel-grade ethanol per year in the United States alone. However, the cellulosic biomass-to-alcohol bioconversion process must be proven economical before industry can commercialize this technology. The U.S. Department of Energy, Office of Fuels Development supports a program to develop a commercially viable process for producing ethanol transportation fuel from renewable biomass resources. Bioconversion technologies developed to date take advantage of a diverse array of pretreatments, followed by enzymatic or chemical saccharification, and fermentation. Progress during recent years in pretreatment and fermentation technologies promises to significantly improve overall process economics. Examples include the development of two-stage dilute acid, nitric acid/mechanical disruption, and countercurrent percolation pretreatments. Recently developed, genetically engineered ethanologens also promise to improve process economics and include *Escherichia coli*, *Saccharomyces* sp., and *Zymomonas mobilis*. These microorganisms reportedly ferment xylose and glucose mixtures with high efficiency. This review presents an up-to-date picture of the advanced technology aspects of unit operations key to successful biomass-to-ethanol processing plants with speculation about future focus in this field by NREL researchers.

Bioethanol: A National Strategic View

In the traditional jargon of strategic planners, organizations and programs are often called upon to do a "situation analysis" (*1*). For this type of analysis, planners scan the environment outside and within their organization to identify strengths, weaknesses, opportunities, and threats to it. This is the first step in identifying strategies for success. Strategic planning has clearly lost the allure it once had in this country as an approach to properly setting a course for the future of an organization, but even its most ardent critics recognize the value of a situation analysis in formalizing the process of setting strategies (*2*).

In the case of bioethanol, we look at the technology itself in identifying strategies for establishing this renewable energy industry. This perspective allows us to look at bioethanol as a new technology for the United States and what it will take to make it a success, while avoiding the more parochial concerns of a given organization. This section is a brief scan of some external issues related to the deployment of bioethanol technology in the United States; it is also an overview of the strategies for deployment currently in place at the U.S. Department of Energy (DOE). External factors that affect the deployment of bioethanol technology include (*3*):

- Environmental (ecological) issues
- Energy trends and national security
- Public opinion
- Public policy and legislative trends
- The market.

The Environment. Concerns about the health of our planet and the quality of life that we are leaving for future generations have become a major focus of our society during the past few decades, starting with Rachel Carson's landmark book *Silent Spring* (*4*), to the more recent book *Our Stolen Future*, which Vice President Al Gore touted as the unexpected sequel to *Silent Spring* (*5*). This literature focuses on the ecological and health effects of manmade chemicals, which travel through the air, water, and soil. A major theme of Coburn's new book is that even minute levels of many synthetic compounds have unpredictable effects on birth defects and basic human development (*5*). In this context, "natural" fuel products such as bioethanol look increasingly attractive. The more immediate and quantifiable environmental impacts of bioethanol focus on two key issues: global climate change and urban air pollution.

Global climate change is a surprisingly old issue. The principle of greenhouse gas effects was first proposed by the French mathematician, Fourier, early in the last century (*6*). In 1896, Svante Arrhenius identified the potential global warming effect of carbon dioxide produced from the burning of fossil fuels (*7*). But it was not until 1957 that the first definitive proof that carbon dioxide was indeed accumulating in the atmosphere was finally established (*8*). Many countries (including the United States) have made moves to reduce carbon dioxide emissions, but global climate change remains an issue plagued by political and scientific controversy. Global temperature data show trends of both increasing and decreasing temperature from 1880 to the present (*9*). Models being developed to predict the effects of increased carbon dioxide levels remain difficult to verify (*9*). Some members of the scientific community argue that, from a fundamental perspective, such models will never be reliable (*10*). Even more perplexing is how to predict the effects of a global temperature increase. Regional effects from flooding to droughts have been projected, but clearly we have no way of predicting these types of calamities regionally (*10-14*). And then there is the question of aerosols. Many have argued that aerosols have similar but opposite effects on climate change compared to greenhouse gases. The presence of anthropogenic aerosols may double the amount of sunlight scattered back into space. The uncertainty in our prediction of aerosol effects swamps any estimate of global warming potential associated with carbon dioxide accumulation.

But if the uncertainties that surround global warming are great, the potential risks to society if global warming is real are worse. Ice core data and other sources of paleoclimatic data have shown that our climate can abruptly and dramatically change (*15*). These changes have been assumed to occur only during ice ages, but there is now a growing body of evidence that dramatic climate shifts have occurred within the past 10,000 years. These changes are not as dramatic as those observed in glacial periods, but they would still be catastrophic if they occurred today (*16*). Thus, the comforting notion that periods of warm climate are relatively stable may be incorrect.

And so the controversy continues. Revelle and Seuss's conclusions on the question of global warming in 1957 still remain the best we can say about the risks we face (*8*):

> "Human beings are carrying out a large-scale geophysical experiment of a kind that could not have happened in the past nor be produced in the future. Within a few centuries, we are returning to the atmosphere and the oceans the concentrated organic carbon stored in sedimentary rocks over hundreds of millions of years."

Urban air pollution is another growing environmental problem, especially in terms of its impact on human health. Nowhere else in the United States has this problem been more evident then in Los Angeles, where 40 years of efforts to control smog and reduce health effects still leave this city with the worst air quality in the United States (*17*). A wide variety of pollutants may have effects on human health. These include carbon monoxide, sulfur dioxide, and heavy metals. Pollutants, such as nitrogen and sulfur oxides, particulates, and ozone are being scrutinized for their role in respiratory disease. Nitrogen dioxide is one of the most widely recognized respiratory irritants, and it often exceeds safety guidelines in urban areas. There is evidence that short-term spikes in nitrogen oxide are associated with increased hospital admissions for respiratory problems and asthma. Long-term exposure is linked to reduced lung function. In addition, nitrogen oxides contribute to ozone formation. Ozone, formed from the reaction of hydrocarbons, nitrogen oxides, and light, is an important pollutant. Individual responses to ozone vary, but people with respiratory disease or who exercise regularly are particularly prone to reduced lung function caused by ozone exposure (*18*).

During the past few decades, the U.S. Environmental Protection Agency's (EPA) regulation of pollutant emissions from stationary and mobile sources has resulted in major reductions in carbon monoxide, hydrocarbons, and sulfur oxides. Reduction of ozone and smog remains elusive (*19*). This is, in part, because of the complex atmospheric chemistry involved. EPA's current strategy for ozone focuses heavily on the reduction of nitrogen oxides. As with global climate change, the problems of dealing with urban ozone are fraught with uncertainty. Developing a more detailed picture of ozone chemistry is critical to developing the most effective strategies for ozone reduction, and to determining the role alternative fuels, such as bioethanol can play in these strategies.

The transportation sector has a major effect on environmental quality, and the use of bioethanol as an alternative to petroleum-based fuel is an important strategy in addressing environmental quality issues. Air pollution, global climate change, oil spills, and toxic waste generation are all results of petroleum-based transportation fuels. The transportation sector contributes almost 30% of the carbon dioxide produced in the United States (*20*). EPA estimates that transportation contributes 67% of carbon monoxide emissions, 41% of the nitrogen oxide emissions, 51% of reactive hydrocarbon emissions, and 23% of particulate matter emissions (*21*). DOE estimates that bioethanol use could reduce net carbon dioxide emissions from vehicles by 90% when used as 95% blend with gasoline in light-duty vehicles (*22*). This is due to the consumption of carbon dioxide by crops used as feedstock for the production of fuel ethanol. Net reductions in urban air pollutants also occur when a 95% ethanol fuel is used. Sulfur oxide emissions are 60% to 80% lower. Volatile organic compound emissions are 13% to 15% lower than those of reformulated gasoline. Net changes in carbon monoxide and nitrogen oxides are marginal (*22*).

Energy Trends and National Security. Concerns about energy security are among the greatest motivations for the DOE's Bioethanol Program. It is common sense that some day we will want to switch from a depletable resource such as petroleum to renewable and sustainable sources of energy. The tough questions are how and when. Should we pay more for renewable energy? Do we need to switch sooner rather than later? The answers to these questions have a dramatic impact on the near-term future of bioethanol technology and on how to deploy this technology. Projections for the depletion of domestic sources of conventional and unconventional petroleum suggest that we would run out of domestic oil within 70 years (*3*). The American Petroleum Institute (API) seems to recognize the legitimacy of these estimates (*23*), and has used the same U.S. Geological Survey estimates of reserves to show that, if we wished to be absurdly optimistic, there is a 5% probability that we will be able to sustain our petroleum production at current levels for the next 93 years! And there are other ways to extend this sense of optimism. Improvements in technology will increase supply from known resources. Ultimately, API relies on the argument that "unconventional" sources of fuel will triple our resource base. These "unconventional" sources, such as oil shale, are not cost competitive today.

Oil imports are on the rise. The Energy Information Administration estimates that, by the year 2010, we will be importing from 52% to 72% of the oil we consume. Even these estimates may be conservative; in 1995, we crossed the 50% threshold for imported oil. Rising imports not only increase our vulnerability to foreign control of energy supplies (*23*); they introduce a cost to our economy. DOE's 2010 projections for imports correspond to economic losses of $114 to $140 billion per year (*24*). This vulnerability is exacerbated by our transportation sector's reliance on petroleum for 97% of its fuel demand. These trends show that our energy outlook is clearly becoming a matter of national security.

Public Opinion. Public opinion is one of the most vexing aspects of establishing a strong renewable energy policy. In 1995, a poll asking for people's priorities in government funding of energy research showed an overwhelming preference for renewable energy as the top priority. But this same poll showed a great deal of ambivalence toward renewables when people use their votes or their pocketbooks to support renewable energy (*25*). A 1990 report summarizing public opinion polls on energy and the environment during the past 20 years shows erratic swings in our attitudes toward energy security (*26*). In times of crisis, concern over shortages is high, but it falls off dramatically between events like the Persian Gulf War or periods of long lines at the gas pumps. At the same time Farhar's report shows a reasonably steady (and more consistent) increase in concern over environmental quality.

Legislation and Policy Debates. Three major pieces of legislation affect the deployment of bioethanol technology:

- The Clean Air Act Amendments of 1990 (CAAA-90)
- The Energy Policy Act of 1992 (EPACT)
- The Alternative Motor Fuels Act of 1988 (AMFA)

CAAA-90 has brought about the increased use of ethanol as an oxygenate in reformulated gasoline for regions considered carbon monoxide and ozone nonattainment areas. EPACT places aggressive mandates on the use of alternative fueled vehicles. The law is intended to force a 10% displacement of fuel consumption with alternative domestic sources by the year 2000. This displacement is to reach 30% by 2010. AMFA complements the efforts of EPACT by putting specific mandates on alternative fuel vehicles for federal fleets. In addition to these three federal laws, there are tax incentives in various states and at the federal level for fuels with a renewable alcohol content of at least 10%. Blenders and sellers of renewable alcohols are eligible for income tax credits. All these incentives are scheduled to expire in the year 2000 (*27*).

These three pieces of federal legislation attempt to translate environmental and energy security issues into a direct cost of doing business. In other words, these legislative actions force the marketplace to recognize the cost of these societal issues while allowing the marketplace flexibility in finding the most cost-effective solutions to these problems. Vice President Al Gore has argued that the marketplace has long been blind to many types of "external" costs (*28*). Even more conservative business-oriented pundits such as Peter Drucker, have come to see this view lacking in the marketplace (*29*).

The continuing debate in policy circles is how the translation of externalities should be done. In 1992, for example, the Clinton Administration pushed for a "Btu tax" or a "carbon tax" that ultimately lost support and was dropped in favor of a motor fuels excise tax increase that actually penalized alternative fuels (*27*, *30*). Are there uniformly acceptable ways to calculate cost benefits for regulations that address hidden societal costs, such as energy security, the environment, health, and safety? Many economists argue that the answer to this question is yes (*31*). Such debates become even more difficult when surrounded by a high level of uncertainty. In the case of renewable energy, it is much easier to calculate the cost of mandating their use than to estimate their benefits, because there is no consensus about them. How valuable are reductions in greenhouse gases? Thus, policymakers need to temper their approach with a "risk management" analysis that allows for a balance of probabilities against the seriousness of the risk to society for these externalities (*32*). European political communities seem more successful at building consensus on these types of issues than their counterparts in the United States (*33*).

One hundred years after Arrhenius first raised concern about global warming from the burning of fossil fuels, the United Nations' Intergovernmental Panel on Climate Change (IPCC) concluded that "the balance of evidence suggests that there is a discernible human influence on climate." This may be one of the most important milestones in the policy debate on global warming. That report is now being assailed by an industry group in the United States which argues that no such influence has been proven. Despite the ongoing debate in the United States, the Clinton administration has reversed the U.S. position on limits for greenhouse gases. The United States has now agreed to mandatory limits. Representatives of the U.S. energy industry are reportedly upset by this new position. Specific limits have not been set, and details of how the mandates would be implemented remain unclear (*34*). Nevertheless, the policy discussion on global warming in the United States has taken one step closer to converting global climate change risk from an externality to real market cost.

The Market. The marketplace will, to the degree that government policy forces it to recognize externalities, ultimately decide the fate of bioethanol as a renewable fuel in the United States. One of the biggest factors that affect this decision is the price of oil. At the low end, DOE projects that oil prices will remain at a level of around $14 per barrel through the year 2010. At the high end, these projections suggest that oil prices could reach $28 per barrel by 2010 (*35*). DOE has consistently lowered the projected cost of petroleum during the past few years. The same perspective can be seen from annual oil price projections done by IEA (*36*). With each subsequent year, starting in 1981, IEA pushed its projections for oil prices lower and lower. Prices will arguably remain stable or even drop to levels of $10 per barrel. Stable pricing, so the argument goes, is based only on a psychological momentum that accepts current pricing. But a significant change in the oil market, such as the re-entry of Iraq as a supplier, might bring about a shift down in prices (*36*).

For the foreseeable future, the market for ethanol will be as an octane enhancer and as an oxygenate in reformulated gasoline (*37*). The former demand is purely market driven; the latter is driven by CAAA-90 discussed earlier. EPACT and AMFA will have little influence on bringing ethanol into the marketplace as a neat fuel, because its value to a refiner as an octane enhancer and as an oxygenate is much greater than as neat fuel (where it must compete with the continuing low cost of gasoline). A linear programming model has been used to estimate ethanol's value to refiners. Ironically, ethanol has more value as an octane enhancer than it does as an oxygenate (*37*). At the lower price projections for oil, ethanol may demand up to $0.70/gallon. At the high end for oil cost projections, ethanol can compete at around $0.90/gallon. This compares favorably with current estimates for ethanol production in the year 2000 from waste cellulosic materials of around $1.08/gallon (Glassner, D., personal communication, 1996). So there is clearly a gap to be filled between the price targets for bioethanol and its market value. The price projection for bioethanol is based on conventional financing of a grassroots (new) ethanol facility. Many other factors could come into play to reduce this cost, including the use of an existing facility to reduce capital cost and unique opportunities for financing and subsidies. Finally, as indicated earlier, the value of ethanol could be influenced by policies that bring externalities involving energy security and environmental quality to bear in the marketplace.

DOE's Strategy for Bioethanol. DOE outlined its strategies for biofuels in 1994 (*38*). Reductions in budgets since that time have caused some changes in these goals, but fundamentally, for bioethanol, the goals remain the same:

- Deploy commercial ethanol technology that utilizes waste cellulosic materials by the year 2000
- Introduce the first facility that utilizes a dedicated energy crop, switchgrass, as a feedstock for bioethanol by the year 2005
- Introduce hybrid poplar energy crops for use in bioethanol production after the year 2010

Waste feedstock technology has been chosen for the first deployment, because the feedstocks are cheaper, and such a scenario may be able to take advantage of unique environmental concerns to reduce the cost of ethanol and make it competitive in the market place. Though less predictable, the effect of policies on global warming and environmental quality are expected to improve the value of ethanol above that currently projected in the gasoline market. Clearly, we cannot count on these types of policies, so research and deployment efforts will focus on establishing ethanol production at prices that are supported by the fuel market. These feedstocks include:

- Softwood waste and materials collected from forest-thinning projects in the West aimed at reducing forest fires
- Sugarcane waste
- Hardwood sawdust waste

By 2005, switchgrass should be available for use in a facility that has the technology to use this dedicated energy crop, possibly in combination with other low-cost waste feedstocks. Ultimately, DOE plans to utilize short rotation woody crops, such as hybrid poplars, as a long-term, high-volume resource for producing ethanol. The specific technology approaches for meeting our strategic goals are discussed in the subsequent sections.

Advanced Pretreatment Technology at NREL

For more than 10 years, batch dilute acid pretreatment techniques have been extensively evaluated at NREL on biomass feedstocks, including several species of hardwoods, herbaceous crops, and agricultural residues (*39-42*). The pretreatment objectives have focused largely on a prehydrolysis approach, in which the hemicellulose component of biomass is hydrolyzed via dilute acid catalysis, leaving the cellulose fraction in an insoluble form for subsequent enzymatic hydrolysis by cellulase enzymes. A detailed process economic analysis of a base-case bioethanol production process, utilizing continuous cocurrent dilute acid prehydrolysis technology, has been conducted at NREL (*43*). Pretreatment conditions and performance parameters used in this evaluation were based on bench-scale batch, dilute acid pretreatment experimental data for hardwood poplar species. Since that time, pilot-scale batch (*44*) and cocurrent pretreatment reactors have been installed and operated to collect larger-scale reactor data for the pretreatment performance of selected feedstocks. For the most part, pretreatment performance data collected in these larger reactors have been comparable to results obtained in bench-scale batch dilute acid pretreatments (*44*).

Numerous of sensitivity analyses were performed as a part of this base case process engineering and economic analysis. The most significant parameter in improving the overall economics of bioethanol production was maximizing the yield of ethanol from a unit of biomass. Clearly, improving the yields of each individual unit operation will contribute to increasing the overall yield of biomass. The pretreatment step is very significant in this respect, as an improved pretreatment process would result in higher yields of soluble sugars from the hemicellulose fraction and a more digestible cellulose fraction, resulting in greater conversion of cellulose via enzymatic hydrolysis by cellulase. Because bench- and pilot-scale batch and cocurrent dilute acid pretreatment methods show definite limitations in the yields of soluble sugars from hemicellulose (about 80%-85% of theoretical) and in the yields of ethanol from cellulose in pretreated biomass solids from a simultaneous saccharification and fermentation (SSF) process (about 70%-75% of theoretical), new pretreatment approaches aimed at achieving higher yields are being developed and evaluated.

During the past few years, dilute acid prehydrolysis R&D at NREL has focused on the developing processes that exploit the biphasic kinetics of xylan hydrolysis from the hemicellulose fraction of biomass. In most species being considered for bioethanol conversion, xylan is the overwhelming majority of the hemicellulose component. The kinetics of xylan hydrolysis have been widely reported to be biphasic, with an "easy-to-remove" fraction that can be removed under relatively mild conditions and a "hard-to-remove" fraction that requires more severe conditions (*45-47*). The hydrolysis reaction mechanism for fast- and slow-reacting xylan fractions assumes conversion first to soluble xylose oligomers, followed by conversion to monomeric xylose and finally, to xylose degradation products. Three prehydrolysis reactor configurations (batch, cocurrent, and countercurrent) have been evaluated in a model of xylan hydrolysis kinetics to determine the best configuration for maximizing xylan hydrolysis yields, while minimizing degradation product formation (*48*). This work showed that a countercurrent configuration is the best design for achieving high recovery of xylose equivalents, especially at high conversion levels of xylan, which is critical to the economic viability of any bioethanol conversion process. This conclusion is somewhat intuitive, as in a countercurrent configuration, the residence time of the xylose from the "easy-to-remove" xylan fraction is short compared to a batch or cocurrent configuration, resulting in less degradation product formation.

Two-Stage Countercurrent Pretreatment. A large number of pretreatment data have been collected at NREL using hardwood yellow poplar sawdust in a system of percolation reactors operated in such a way as to simulate a two-stage countercurrent dilute acid process. The selection of the reaction conditions and the details of the operation of what is now referred to as the reverse-flow two-stage system have been described in detail (*49*). A factorial experimental design and subsequent reaction condition optimization was conducted in this study. The conditions selected as a result of this optimization were a 10-min residence time of the solids in each stage, an acid concentration of 0.07% (w/w) sulfuric acid, a first-stage solids temperature of 174°C, and a second-stage solids temperature of 204°C. A recovery of 97.0% of xylose equivalents in the liquor stream from the xylan content of the feedstock was achieved, with about 60% in the form of soluble oligomeric xylose and 40% as monomeric xylose. Less than 3% of the xylan degraded to furfural. About 10% of the glucan was solubilized, but with much lower levels of oligomeric glucose relative to monomeric glucose. About 35% of the lignin was solubilized. The structure of the soluble lignin compounds was not determined. The degree of lignin solubilization is significantly higher than with batch or cocurrent dilute acid pretreatments. It is believed that the reverse-flow mode of operation, in which the liquor is separated from the residual solids while temperatures are still high, prevents re-precipitation and/or re-condensation of solubilized lignin back on the pretreated solids particles.

As the reverse-flow procedure includes a hot water washing of the residual solids, with the wash liquor combined with the actual preydrolysis liquor, the resulting residual solids are free of soluble compounds, including sugars, oligomers, and soluble inhibitory compounds from lignin breakdown or sugar degradation. Thus, the residual solids can proceed to the SSF conversion process without any further washing or other detoxification procedure. Standard SSF conversion studies were performed using the conditions described by Torget and co-workers (*49*), including a cellulase loading of 25 FPU/g cellulose. For the reverse-flow two-stage optimization run described above, an ethanol yield from cellulose of 91% of theoretical was achieved in 55 h. Both the yield and the rate are substantially higher than those reported for similar feedstocks subjected to batch dilute acid pretreatments. This represents a substantial potential improvement to the overall process economics.

Because of the near quantitative yield of xylose equivalents, the partial solubilization of glucan to glucose equivalents, and the complete displacement of soluble compounds from the residual solids during the washing step, a significant fraction of the initial carbohydrates found in the feedstock end up in the prehydrolyzate liquor. Based on the initial composition of yellow poplar sawdust and the pretreatment performance achieved in the optimization run, more than 40% of the total available carbohydrates are found in this liquor. In the past, the evaluation of pretreatment performance has been based on the enzymatic digestibility and/or SSF production of ethanol from washed, pretreated solids only. It has largely been assumed that the soluble carbohydrates in the liquid fraction of the pretreated slurry or in a separate prehydrolyzate liquor, could be converted to ethanol at assumed yields and rates and would not require any type of post-treatment prior to either a separate or cofermentation.

Hydrolyzate Detoxification. The base-case process economic evaluation described above did not include any prehydrolyzate detoxification step. It has become recognized that some type of detoxification of at least the liquid fraction will likely be required. This is even more likely in flow-through, dilute acid pretreatments compared to batch or cocurrent pretreatments because of the higher lignin solubilization. Certain compounds that result from the solubilization of lignin, including certain organic acids, higher alcohols, and phenolic-based compounds, are known to be inhibitory to fermentative microorganisms. Also, feedstocks that have highly acetylated hemicellulose in their structure are likely to release inhibitory levels of acetic acid upon hemicellulose hydrolysis. Significant efforts are under way to develop cost-effective detoxification processes that can selectively remove or convert such compounds into non-toxic forms. At the same time, modified prehydrolysis conditions that result in similar pretreatment performance levels, but that produce fewer toxic compounds, are being evaluated.

Challenges to Sugar Yield. As stated above, the analysis of the carbohydrates in the optimized prehydrolyzate liquor shows that about 60% of the xylose equivalents in the liquor are in the form of oligomeric xylose. Previous process economic analyses have assumed that solubilized carbohydrates from pretreatment would be fermentable, i.e. monomeric sugars. Options for converting oligomeric xylose to monomeric xylose have been reported (*48*). Two options were investigated in work performed in the laboratory of Y.Y. Lee at Auburn University (*48*). The first was a mild-temperature hold of the prehydrolysis liquor at the pH (~2.2) of the liquor as it leaves the pretreatment reactor. A range of hold times and temperatures was examined. The best conditions were a temperature of 130°C and a hold time of 11 h, where the monomeric xylose content increases to 97% of total xylose equivalents and the oligomeric xylose content decreases to virtually zero. The furfural content increased only marginally, indicating little increased sugar degradation under these relatively mild conditions.

The other option evaluated for xylose oligomer hydrolysis was to determine whether sufficient xylanase activity exists in cellulase preparations to allow for enzymatically catalyzed hydrolysis of xylan oligomers to monomers under typical SSF conditions. In tests where Genencor Cytolase CL was added to prehydrolyzate liquor in amounts equivalent to 10 and 25 FPU/g cellulose as if 8% (w/v) cellulose was present, significant conversion to xylose monomer was achieved (Elander, R., unpublished results, 1996). Starting from an initial monomer level of 41% of the total xylose equivalents, the monomeric xylose level increased to 87% of total xylose equivalents with a 10 FPU/g assumed cellulose loading at 61 h, and 95% of total xylose equivalents with a 25 FPU/g assumed cellulose loading at 61 h. The 61-h time is similar to the 55-h SSF time required to achieve a 91% ethanol yield in SSF, as reported above for the reverse-flow, two-stage pretreated solids. The unknown of these experiments is whether these levels of conversion would be maintained when process concentrations of cellulose and lignin are present in the system.

In addition, the work of Elander and co-workers (*48*) included an analysis of potential process improvements that could result in higher concentrations of carbohydrates feeding into the fermentation steps, which would presumably result in a higher ethanol concentration exiting the fermentation step(s) and lower ethanol recovery costs. Using the volume of liquor generated from pretreatment, which was 2.2 reactor void volumes per biomass-packed reactor (*49*), and an assumed washed pretreated solids concentration of 40%, process analysis indicated that in a simultaneous saccharification and cofermentation (SSCF) process, in which the washed solids are combined with detoxified liquor and yields of ethanol from cellulose (*49*) and from xylose are known (*43*), an ethanol concentration exiting SSF of 3.05% (w/v) could be expected. This is somewhat lower that the base case value of ~4.5% (w/v) assumed by Hinman and co-workers (*43*), and indicates that the volumes of liquor generated could increase ethanol recovery costs. Also, the high liquor volumes result in increased fermentation tankage requirements and would require higher steam demands in pretreatment, which would be exacerbated by the higher temperature requirements that have been used in reverse-flow prehydrolysis. This situation will certainly have an impact on the energy balance of the process and will affect overall process economics.

Reactor Design. The ultimate ability to reduce prehydrolyzate liquor volumes, but still maintain prehydrolysis sugar yields and pretreated solids enzymatic digestibility, becomes a reactor design issue, not only at the bench-scale level but, more importantly, in an engineering-scale system that is representative of a commercial-scale countercurrent prehydrolysis reactor system. The ability to achieve the contacting and movement of liquid and solids effectively in a large-scale device is a major challenge and will likely determine the ultimate commercial success of this technology. Efforts are currently under way to work with equipment suppliers to design, develop, and test engineering-scale countercurrent dilute acid prehydrolysis reactor systems. Although biomass particle sizes, reaction conditions, and the ultimate goals of the reaction are somewhat different, an extensive body of knowledge in reactor design issues related to biomass solubilization exists within the pulp and paper industry. Collaborations with pulping equipment manufacturers are being investigated as a means to more rapidly develop commercial-scale countercurrent prehydrolysis reactor systems.

New Directions at NREL. To date, the only feedstock extensively investigated using the dilute acid countercurrent process is yellow poplar sawdust. This particular feedstock is representative of a near-term waste biomass source from sawmill operations, but also is similar to potential hardwood energy crop species; thus, it provides useful information for near- and long-term applications. Ultimately, feedstock options, including other waste biomass materials, such as agricultural residues and industrial lignocellulosic wastes, and energy crop feedstocks, such as herbaceous grasses and fast-growing hardwoods, will be investigated.

In addition to dilute acid, the feasibility of a pressurized hot water countercurrent prehydrolysis process that requires no added acid catalyst is being investigated on appropriate feedstocks. Eliminating the acid catalyst requirement has a number of potentially beneficial impacts on the process, including reduced chemicals costs, potentially less expensive metals for pretreatment reactor construction because of less corrosive conditions, and reduced formation of insoluble compounds, such as gypsum, upon neutralization. Such a process is also more environmentally benign than those that utilize harsh acids, bases, or solvents. This work is being conducted using a flow-through percolation system at NREL. In addition, the investigation of a neutral pH-controlled process via a NREL-sponsored subcontract at Purdue University is currently being pursued. Other workers (*50-52*) have recently reported promising findings with pressurized hot water prehydrolysis approaches, but the economic impacts of some reported process conditions, such as very small biomass particle size and relatively large volumes of pressurized hot water, have not been thoroughly evaluated.

A number of alternate pretreatment methods have also recently been investigated through a series of NREL-sponsored subcontracts. These include the use of various acid catalysts in a dilute acid prehydrolysis mode, such as nitric acid, phosphoric acid, and phosphoric acid/organosolv; alkaline-based methods, such as lime pretreatment and ammonia recycle percolation; and physicochemical methods, such as hydrogen peroxide extrusion and ammonia freeze explosion. Many of these methods have been widely discussed in the literature, but it is difficult to compare the relative effectiveness of these various approaches because of variation in feedstocks tested, analytical methods, and pretreatment performance measurement and reporting protocols. A common-basis evaluation of these approaches that will determine the relative effectiveness and process economic implications of these various pretreatment approaches is currently being conducted. This can serve as a basis for making rational choices in pretreatment technologies for given feedstock and process applications. Promising pretreatment technologies will likely be further developed from additional bench-scale work and initial integration with subsequent detoxification and fermentation steps, potentially leading to engineering scale equipment design and testing in the future. Another key goal of this evaluation is to determine which pretreatments are best suited to various feedstocks.

Complete Hydrolysis Studies. Very recently, the general concept of dilute acid countercurrent prehydrolysis developed at NREL has been extended to investigate the possibility of a full hydrolysis of both hemicellulose and cellulose to soluble sugars. This concept has been motivated by the fact that thus far, costs of commercial cellulase preparations from industrial enzyme suppliers have been prohibitive for use in bioethanol processes (*43*, *53*). Although ongoing process development and research plans at NREL show promise in substantially reducing cellulase production costs, the ultimate costs of cellulase production are still unclear. Thus, a full hydrolysis process that substantially reduces (or even eliminates) cellulase requirements is an option worthy of investigation.

The use of dilute sulfuric acid to totally hydrolyze the carbohydrates in lignocellulosic feedstocks for ethanol production was widely abandoned by 1990 due to the relatively low yields of glucose (50%-60%) as compared to potentially high yields (80%-95%) from enzymatic hydrolysis of cellulose (*54-56*). Even though kinetic modeling exercises conducted at NREL (*57*) suggested that yields as high as 88% could be obtained using a countercurrent reactor configuration, laboratory studies could only demonstrate total sugar yields (C_5 and C_6 sugars) of about 60% (*56*). However, several observations were made to explain the low yields. The material of construction of the experimental reactor (Carpenter Cb20-3), when subjected to 1% acid at elevated temperatures, leached chromium ions at a concentration that leads to degradation of glucose formed in the reactor. Additionally, as the biomass was hydrolyzed, the packed bed collapsed, which led to non-ideal fluid dynamics that caused increased degradation of the glucose.

The renewed interest in total hydrolysis dilute acid processes has been motivated by modeling exercises, which indicate that with a hydrolysis liquor residence time about one-seventh of the solid residence time, yields of glucose could approach those obtained with an enzymatic process (*57*). Such a process could result in high volumes of hydrolyzate liquor relative to countercurrent dilute acid prehydrolysis. The impact of producing liquor volumes even greater than those for countercurrent dilute acid prehydrolysis is to diminish the overall process efficiency. Although overall cost savings may be realized through reduced cellulase requirement, process improvements to reduce the quantity of the resulting liquor volumes is required. Again, innovative pretreatment reactor design aimed at construction feasibility and reducing required liquor volumes will play a key role in determining the viability of a dilute acid total hydrolysis process.

Advanced Cellulase Research at NREL

The promise of highly efficient conversion processes coupled to a "green" technology is now universally appealing to industry and governmental policymakers. In general, hydrolytic enzymes offer depolymerization of naturally occurring polymers in high yield, with few, if any, by-product disposal problems, unlike acid-based hydrolysis processes. More than a decade ago, advances in the production of fungal cellulase preparations rekindled interest in enzyme-based biomass conversion processes. However, there is some evidence that this technology has now reached its zenith. Also, cellulase biochemistry has reached an enabling phase of development, in which combined efforts in biochemistry and molecular biology may be able to deliver improved cellulase systems for industrial application. Many factors that govern cellulase component action on crystalline cellulose (cellulase synergism, crystallographic structures, cellulase active site structure/function relationships) have been established (or at least preliminarily elucidated), so design of engineered recombinant cellulase systems by means of advanced concepts, such as enzyme component selection and site-directed mutagenesis (SDM), is now at hand. To be commercially viable, engineered cellulase systems must ultimately produce highly active cellulases with improved specific activities, protein yield, and/or production cost relative to current submerged culture fungal preparations. Once these goals are achieved, the ability to "tune" recombinant systems should permit access to new process feedstocks and markets.

Cellulosic Biomass and Cellulase Action. Biomass feedstocks most commonly considered for conversion to bioethanol in the near term are waste wood, agricultural wastes, and the paper fraction of municipal solid waste. The fermentable fractions of these feedstocks include cellulose (β-1,4-linked glucose) and hemicellulose. Although it is an abundant biopolymer, cellulose is unique because it is highly crystalline, water insoluble, and highly resistant to depolymerization. Three features of the cellulose in pretreated biomass make it extremely resistant to enzymatic hydrolysis.

First, the β-1,4-glycosidic linkage renders cellulose oligomers (cellodextrins) with a chain length in excess of six glucose residues extremely insoluble in aqueous systems. Cellulose chain lengths in plant cell walls are very much longer, and therefore, completely insoluble in water.

Second, in plant tissues, cellulose is organized by extensive hydrogen bonding and hydrophobic interaction into semicrystalline bundles of parallel cellulose chains, called microfibrils. This structural feature contributes to the rigidity and strength exhibited by wood and simultaneously protects most of the glycosidic bonds from attack by cellulases.

Third, acid pretreatment alters the chemistry of biomass to yield a substrate totally unlike materials encountered in nature. Specifically, the acidic cooking of biomass at temperatures above the phase transition of lignin (about 160°C) leads to redistribution of lignin during cool-down, perhaps with the effect of "coating" residual polysaccharide fibers. It is reasonable to assume that this redistribution of lignin impedes enzyme action through strong, nonproductive binding. This hypothesis follows, because several studies have shown a strong tendency for lignin in wood to interfere with cellulase action (*58-59*). Pretreated hardwood pulp also harbors a weak net negatively charged surface not associated with native wood, because regions of hemicellulose adjacent to 4-*O*-Me-glucuronate branches are protected during dilute acid hydrolysis and probably remain in the fiber (*60-61*) and glucuronate content in cellulose increases as a function of normal wood oxidation (*62*). The result of the alteration of wood surface chemistry following pretreatment most certainly alters interactions between cellulases and the biomass surface.

As a consequence of these properties, converting cellulose to glucose requires substantially larger ratios of enzyme to substrate than for otherwise similar processes which convert starch to glucose. Despite this fact, it is remarkable that an amount of cellulose approximately equivalent to that synthesized during the yearly growing season is recycled on a global scale during the same period of time in the environment.

Cellulose is enzymatically degraded to glucose by the synergistic action of three distinct classes of enzymes: the "endo-1,4-β-glucanases" or 1,4-β-D-glucan 4-glucanohydrolases (EC 3.2.1.4), which act randomly on soluble and insoluble 1,4-β-glucan substrates, the "exo-1,4-β-D-glucanases," including both the 1,4-β-D-glucan glucohydrolases (EC 3.2.1.74), which liberate D-glucose from 1,4-β-D-glucans and hydrolyze D-cellobiose slowly, and 1,4-β-D-glucan cellobiohydrolase (EC 3.2.1.91), which liberates D-cellobiose from 1,4-β-glucans, and the "β-D-glucosidases" or β-D-glucoside glucohydrolases (EC 3.2.1.21), which act to release D-glucose units from cellobiose and soluble cellodextrins, as well as an array of glycosides. The concepts of exo-endo and exo-exo synergism are shown diagrammatically in Figure 1.

Synergism between exo- and endoglucanases is best explained in terms of providing new sites of attack for the exoglucanases. These enzymes normally find available chain ends at the reducing and nonreducing termini of cellulose microfibrils. Each random internal cleavage of surface cellulose chains by an endoglucanase provides two additional sites for attack by cellobiohydrolases. Therefore, each hydrolytic event by an endoglucanase yields both a new reducing and a new nonreducing chain end. Thus, logical consideration of catalyst efficiency dictates the presence of exoglucanases specific for reducing and nonreducing termini, which has now been confirmed for *Trichoderma reesei* CBH I and CBH II (*63*). Exo-exo synergism may be explained by considering the tendency of cellulose chains to "reanneal" or return to the crystallite surface following hydrolysis if unimpeded by the presence of an exoglucanase. It is possible that maximal initiation of the processive hydrolytic process catalyzed by exoglucanases occurs only if both exoglucanases (reducing and nonreducing specific) are present at the site of internal bond hydrolysis immediately following endoglucanase action. In order to reduce the effects of steric hinderance at this site, both exoglucanases must bind to their respective substrates with high precision.

As of early 1996, glycosyl hydrolases have been grouped into 56 families of related proteins, based on amino acid sequence homology (*64-66*) and hydrophobic cluster analysis (*67*) of catalytic domains. Cellulases are included in 11 of the 56 glycosyl hydrolase families so far elucidated by these methods. Some cellulase families include endo- and exoglucanases from either fungal or bacterial systems. The mechanism of cellulase action is a major feature common to all members of a family (e.g., single displacement, leading to inversion of configuration at the anomeric carbon, or double displacement, leading to retention of configuration at the anomeric carbon (*68*, *69*). Subsets of these glycosyl hydrolase families have been grouped into super-families, or clans, on the basis of conservation of enzyme mechanism and tertiary structure of the molecule (*70*, *71*).

Applications and Economics of Cellulase Production. The ultimate goal of advanced cellulase development research is to develop a robust and easily integratable cellulase production system that can produce adequate amounts of highly effective enzyme to satisfy process requirements at a cost compatible with overall ethanol process economics–probably in the range of \$0.05 to \$0.10/gallon ethanol produced. For comparison, starch-based ethanol processes currently consume approximately \$0.04 to 0.05 in enzyme per gallon of ethanol produced (Miller, C., personal communication, 1995). Current assumptions dictate that the enzyme production system must be able to produce vast quantities of enzyme. For example, at an enzyme loading of 25 FPU/g cellulose, a bioethanol process will require about 11 million FPU (19 kg, 42 lbs) of cellulase to process 1 ton of biomass (1000 lb cellulose) to 84 gallons of ethanol. This amount of enzyme is equivalent to about 143 L of a commercial preparation that contains 80,000 FPU/L. Therefore, a single 2,000 ton/day bioethanol plant would require a staggering 15,000 ton/year cellulase, which is conservatively one-fourth the entire 1994 U.S. market for <u>all</u> industrial enzymes (*72*). Put another way, at 25 FPU/g cellulose, 136,000 FPU (approximately 1.8 L of commercial enzyme) of cellulase will be required to produce 1 gallon of ethanol.

Most industrially useful enzymes are produced using fermentative processes, which involve capital- and labor-intensive production plants that use large stainless steel tanks, huge volumes of media, expensive energy, and material inputs (agitation, oxygenation, media, steam sterilization, etc.). Commercially available cellulase preparations are currently employed on a large scale almost entirely in non-biomass conversion applications, such as textile processing ("biostone"-washed jeans), detergents, and food processing. Each of these markets commands a much higher price for cellulase than can be afforded by any projected bioethanol process in the United States today, which is due, in large part, to the low cost of gasoline and the high cost of feedstock materials.

Given that cellulases are a critical component of lignocellulose conversion technology and that commercially available cellulase preparations are currently far too expensive for use in bioethanol processes, the alternatives are many and can be summarized briefly.

- Develop an on-site liquid or solid-state fermentative process for producing cellulase with optimized microbial sources (e.g., filamentous fungi, such as *T. reesei*, *Aspergillus niger*, or *Humicola insolens*).
- Develop a genetically engineered cellulase production system that uses a bacterial or fungal host to express and secrete effective cellulase preparation, or some other genetically engineered cellulase production system that uses a non-microbial host organism, such as insect cells, crop plants, or lactating mammalian systems.

Developing cost-effective cellulase production methods, designed for complete and efficient hydrolysis of cellulose in relevant feedstocks is therefore essential. Besides the cost that results from losses in overall ethanol yield when enzyme is produced from biomass, high cellulase production costs are also due to the intrinsic costs of fermentative processes, including capital-intensive tankage, agitation, and sterilization equipment, and costly media and contamination control chemicals.

Prospective for *T. reesei* Cellulase Use in Near-Term Bioethanol Plants. *T. reesei* mutants are generally recognized to be the best strains currently available for the industrial production of cellulases (*73*). Yet, the cost of bulk quantities of cellulase to the ethanol from biomass process has remained an area of uncertainty. Although few detailed economic studies are available, an estimate for the cost of cellulase production from lactose for an advanced bioethanol plant based on Iogen technology was proposed as $0.53/gallon ethanol (*74*). More recently, an estimate of $0.30 to $0.81 per pound cellulase protein was proposed for on-site cellulase production based on Army-Natick data (*75*). (These values may be cautiously converted to a cellulase cost of approximately $0.11 to $0.30/gallon ethanol assuming a specific activity of 600 FPU/g protein and that 100,000 FPU are required to produce 1 gallon ethanol). Cellulase cost data from the 1993 study by Hayn (*76*) proved to be in agreement with the earlier Iogen data and were based on actual separate hydrolysis and fermentation (SHF) pilot plant results, which provided a cellulase cost estimate of approximately $0.68/gallon ethanol. Because no detailed pilot-scale studies of cellulase production from pretreated woody biomass are readily available, production parameters based on small-scale *T. reesei* growth and induction studies are critical.

Enzyme Technology for Next-Generation Bioethanol Plants and Beyond. Because the cost of producing the enzymatic catalysts for the SSF process is a critical issue, the available enzymatic activity must be maximized. This requirement can be met by ensuring that the enzymes used are obtainable at minimal cost and have the highest specific activity and the highest possible stability at the pH and temperature of the intended application. Esterbauer and co-workers (*77*) caution that, "In retrospective, we and others feel that cellulase production by *Trichoderma* has its limitations and a significant further improvement cannot be expected. In the future, efforts should also be focused on other cellulolytic microorganisms, both bacteria and fungi."

We feel that cellulase systems capable of greater productivities and carbon conversion efficiencies than those possible from fungi are required for the success of bioethanol plants targeting objectives for the next decade (i.e., \$0.67/gallon ethanol) (*78, 3*). Preliminary technoeconomic analyses of the bioethanol process at NREL show that the cost of on-site cellulase production is keenly sensitive to delivered feedstock cost (\$/ton feedstock), moderately sensitive to the carbon conversion efficiency of cellulase production [FPU produced/gram carbon consumed from feedstock = (gram cellulase/gram carbon) x (FPU/gram cellulase)], and less sensitive to enzyme loading (FPU/gram cellulose content in SSCF) (Glassner, D., personal communication, 1996). Thus, the key to this strategy is to increase the specific activity, thus deriving more FPU activity per gram of protein produced and to increase both carbon conversion efficiency and effective enzyme loading. The degree to which the specific activity of the cellulase system can be increased is not known. A related issue of equal importance is the development of an expression system that can produce large quantities of recombinant enzymes at low cost, preferably from low-value processing plant streams.

Strategies for Improving Cellulase Systems. The first major research goal for advanced cellulase technology is to increase the specific activity of cellulase systems. We propose approaching this challenge through the use of at least four distinct strategies. First, the component enzymes constituting known cellulase systems can be isolated and recombined to create new, non-natural systems to be evaluated for possible improvements in activity–to a large degree, work initiated at NREL (*79*) and elsewhere (*80-83*) has shown that significant, but probably limited, increases in system efficiency can be expected from this approach. One important discovery from two-component enzyme mixing studies at NREL was that a significant improvement in degree of synergism (DSE) and reducing sugar (RS) release (i.e., as much as 40% DSE and 30% RS release) can be found when mixing enzymes produced from diverse organisms, in this case a hot-spring bacterium and a filamentous fungus (*79*). Our progress in demonstrating the potential for new cellulase system engineering by assessing the efficacy of mixed origin and native binary and ternary systems is shown in Table 1. Although admittedly not as active on Sigmacell as the native *T. reesei* ternary system, the new ternary system based on *Acidothermus cellulolyticus* EI, *Thermomonospora fusca* E_3, and *T. reesei* CBH I is competitive and offers the potential for another important advantage for bioethanol process cellulases, thermal tolerance and increased process half-life.

Engineered Cellulase Systems at NREL. *A. cellulolyticus* is a thermotolerant, cellulolytic bacterium which was originally isolated from a Yellowstone National Park hot spring (*84*). One component enzyme from this cellulase system, EI, has been purified, characterized, crystallized, and subjected to x-ray crystallographic analysis (*85*). EI is a 60 kD endoglucanase that is highly thermostable, demonstrates high specific activity on carboxymethylcellulose (*86, 87*), and is a member of the glycosyl hydrolase Family 5, Subfamily 1 (*88*). Because of its membership in this family, EI is expected to show retention of the stereochemistry at the anomeric hydroxyl following catalysis.

Cellobiohydrolase I (CBH I), a 52 kD exoglucanase produced by the filamentous fungus, *T. reesei*, is the most abundant component of that organism's cellulase system. *T. reesei* also produces another exoglucanase, CBH II, and at least three endoglucanases, referred to as EG I, EG II and EGV. CBH I is produced in extremely large amounts from a single genomic gene and may represent up to 60% of the protein secreted by strains of *T. reesei*that can produce extracellular protein at concentrations up to 40 grams/L (*89*).

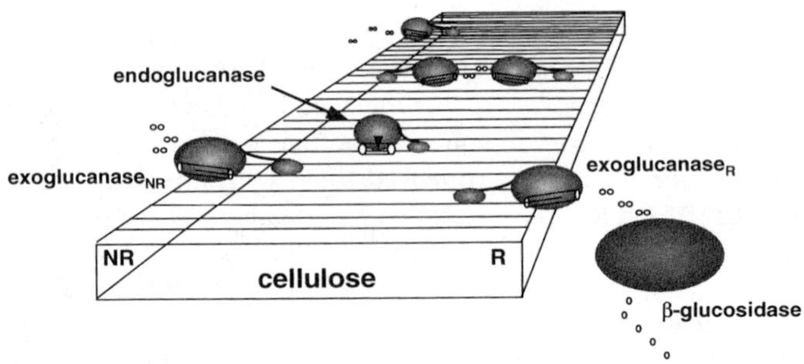

Figure 1. Depiction of the action of cellulase component enzymes on cellulose. This general fungal cellulase system consists of one or more endoglucanase types and cellulose reducing terminus-specific and cellulose nonreducing terminus-specific exoglucanases (cellobiohydrolases).

Table 1
Non-Native Endoglucanase/Exoglucanase Mixtures Tested at 10% Total Saccharification for Degree of Synergistic Effect (DSE) and Reducing Sugar (RS) Release

Enzyme Mixtures	DSE	%Max. DSE	RS Release	% Max. RS Release
Cellulase Binary Systems[a]				
A. cellulolyticus EI & *T. reesei* CBH I	2.75	100	40	100
T. fusca E5 & *T. reesei* CBH I	2.16	79	30	75
T. reesei EG I & *T. reesei* CBH II	1.72	63	31	75
T. reesei EG I & *T. reesei* CBH I	1.58	57	27	68
Cellulase Ternary Systems[b]				
T. reesei EG I & *T. reesei* CBH I & *T. reesei* CBH II	2.51	100	55	100
A. cellulolyticus EI & *T. reesei* CBH I & *T. reesei* CBH II	2.16	85	50	91
A. cellulolyticus EI & *T. reesei* CBH I & *T. fusca* E_3[c]	1.87	75	44	80
M. bispora endo A & *T. reesei* CBH I & *T. fusca* E_3	1.79	71	36	65

Reducing sugar (as mM glucose) from Sigmacell 20 measured after 120 h at 50°C and pH 5.0 according to Baker and co-workers (*79*). [a] Data collected at an endo/exo ratio of 20/80 (*79*). [b] Data collected at an endo/exoR/exoNR ratio of 20/20/60. [c]Data for this study collected at an endo/exoR/exoNR ratio of 20/30/50.

CBH I is not thermotolerant, acts processively from the reducing end of a cellulose substrate, demonstrates retaining-type product stereochemistry (*63*), is a member of glycosyl hydrolase Family 7 (*90*), and exhibits a synergistic activity in combination with every endoglucanase that has so far been tested, including EI (*79*).

E_3 is a nonthermotolerant exoglucanase secreted by the actinomycete *T. fusca* (*80*). This enzyme is about 60 kD in molecular weight, demonstrates an inverting-type product stereochemistry, and is a member of glycosyl hydrolase Family 6 (Wilson, D., personal communication, 1996).

We feel that when acting on pretreated biomass, engineered enzyme mixtures can be formulated that are more effective than the most active native mixtures, at least at the binary enzyme level. Based on the notion that an engineered mixture of activities is desirable, and that an effective cellulase system must have at least one highly active endoglucanase, one cellulose reducing terminus-specific exoglucanase, and one cellulose nonreducing terminus-specific exoglucanase, we have proposed the selection of the *A. cellulolyticus* EI, *T. reesei* CBH I, and *T. fusca* E_3 as components for our basic system.

Second, it should be possible to improve the kinetic efficiency of cellulases that work on pretreated biomass by using known principles of enzyme engineering (new strategies may also be required and are proposed for future work). We have proposed, for example, a two-phase approach to this problem by first modifying the native structure of EI by targeted amino acid replacement to ensure optimal enzyme-cellulose (biomass) surface interaction and by improving the catalytic efficiency of the active site, also by SDM. A similar approach will be used to improve the action of the exoglucanases on pretreated biomass, with the additional goal of improving thermal tolerance.

Third, the usefulness of "accessory" glycosyl hydrolases for the enhanced saccharification of pretreated biomass substrates should be investigated. For example, the new countercurrent pretreatment methodologies described earlier produce somewhat higher levels of cellulose hydrolysis; however, the material left insoluble is more resistant to conventional cellulase action. Also, the liquid waste streams from complete hydrolysis tend to harbor enhanced levels of xylooligodextrins, as well as other, as yet unidentified, oligosaccharides. An important example of the effectiveness of removing substituent chemical impediments to cellulases action was shown by Kong and co-workers (*91*) when they demonstrated that the cellulase digestion of aspen wood was substantially accelerated by prior chemical deactylation. This result could presumably also be achieved by the action of xylan acetyl esterases (*92*).

The overall strategy for developing engineered cellulase systems is depicted in Figure 2. The obvious benefit of routine reassessment of the specific enzyme candidates chosen for system membership is apparent, but resource intensive. A fundamental dilemma is encountered when limited resources dictate that choices be made between attempting to improve the operational characteristics of competitive enzymes (by SDM or mutation/selection) or simply returning to new rounds of screening from known cellulase producing organisms. A secondary benefit for the enzyme engineering approach is the acquisition of fundamental knowledge concerning cellulase action and structure relationships, which should improve chances of future success.

Relevant Cellulase Assays for Bioethanol Process Applications. The activities described above will be supported by the critical determination of the effectiveness of new enzymes using a novel assay method. Although cellulase enzymes are widely sold, and

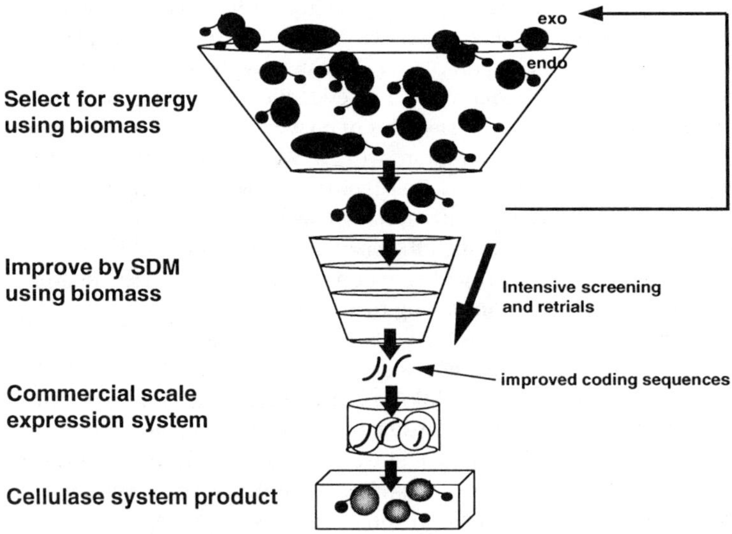

Figure 2. The scheme for selecting, mixing, and improving cellulases and related enzymes for biomass conversion processes.

their industrial utilization estimated, on the basis of the FPU of activity, the traditional "filter-paper assay" is severely limited as a predictor of cellulase performance in the extensive saccharification (80%-90%-plus) of actual industrial lignocellulosic substrates. These limitations are traceable both to the chemical and physical differences between filter paper and industrial substrates, and to the nonhomogeneous nature of most cellulosic substrates (filter paper included), which means that assays run to very limited extents of conversion (such as the 4% conversion target in the filter-paper assay) measure the digestibility of only the most easily digestible fraction of the substrate, and reveal little about the convertibility of the bulk of the substrate. Actual performance of cellulases is estimated better by assays that utilize the actual application substrate, *and* are run to the extents of conversion required in the process.

Because of the inhibitory nature of the products of cellulase action (primarily glucose and cellobiose), such high-conversion assays encounter the problem of significant product inhibition, if run in "closed" systems as simple saccharifications (*93*, *94*). A new saccharification assay has been devised at NREL in which a continuously buffer-swept membrane reactor is used to remove the solubilized saccharification products. The diafiltration saccharification assay (DSA) serves as a reliable predictor of the performance of combinations of cellulase and substrate under simulated SSF conditions but retains the analytically more direct and accurate nature of a saccharification reaction. This assay will be used to compare the effectiveness of commercial *T. reesei* and specially engineered (cloned) cellulase preparations in the saccharification of standard and novel dilute acid pretreated substrates.

For decades, cellulase biochemists have proposed that enzyme-secreting microorganisms should produce somewhat different hydrolytic enzyme systems when presented complex biomass substrates, than when grown on simple, soluble sugars. We have recently shown that a *T. reesei* mutant archived at NREL produces a cellulase system considerably more effective at hydrolyzing pretreated hardwood sawdust when grown in the presence of the same substrate (*95*).

Expression of Engineered Cellulase Genes. Our initial goal was to identify the cellulase genes that will be incorporated into the first-generation cellulase production system. This has been done (at least provisionally). The next goal is to select a host organism most likely to be able to produce a high specific activity enzyme preparation in a cost-effective manner. Initially, single-gene expression strains will be constructed to maximize expression of active gene product. Strains shown to express different target genes at desired levels will be crossed to combine two or more cellulase genes in the same strain. It is essential to maximize the specific activity (activity/g protein) of the genetically engineered cellulase preparation, which requires specific molar ratios of each particular combination of endoglucanases, exoglucanases, and β-glucosidases. Multi-gene expression strains will eventually be constructed to maximize expression of balanced cellulase expression systems.

Once a host system has been selected, the cellulase coding sequences must be incorporated into genetically engineered artificial gene constructs, which will be recognized and readily expressed in that organism. Because the cellulase genes targeted for expression in the desired host may have originated from either a bacterium or a fungus, the standard approach will be to use the coding sequence from the selected cellulase gene and place it downstream from a host-derived promoter known to express its native product at very high levels. Downstream from the coding sequence will be spliced appropriate transcription termination signals (and polyadenylation signals, if required) known to function effectively in the chosen host organism, possibly from the same gene as the promoter was derived. The exact DNA sequence context and proximity of the host promoter sequence and the foreign coding sequence is critical in any chimeric gene construction, because a single nucleotide change in sequence or distance can make a huge difference in the behavior of the gene. How changes in promoter sequence context will affect expression of chimeric genes cannot be accurately predicted for any organism at this time (except perhaps *E. coli*).

The cellulase coding sequences that have been selected for incorporation into the initial cellulase production system include:

- *A. cellulolyticus* EI endoglucanase (gene of bacterial origin; 61% G+C content)
- *T. reesei* CBH I (cDNA clone of fungal origin; ~50% G+C content)
- *T. fusca* E_3 (gene of bacterial origin; high G+C content)
- *A. cellulolyticus* β-glucosidase (gene not yet cloned)

Initial work in the area of gene cloning took place in the early 1970s using the gram-negative bacterium *E. coli* as the host organism. *E. coli* is still by far the most well-characterized organism in terms of its molecular genetics and the biochemistry of its genetic machinery. Within a short period it became possible to clone and express foreign DNA in other bacteria, such as *Bacillus subtilis*, *Streptomyces lividans,* and others, as well as brewer's yeast (*Saccharomyces cerevisiae*). During the past 10 years, cloning and expression of foreign DNA in fungi has become fairly routine (*Neurospora*, *Aspergillus*, and *Trichoderma,* for example). The ability to introduce foreign DNA and control its expression in a wide variety of higher organisms has also been achieved only recently (including many plant species, insects, and mammalian lactation system).

The most technically approachable systems for expressing foreign DNA at high levels include bacteria (*E. coli*, *S. lividans*, *B. subtilis*, *B. brevis*, *B. stearothermophilis*, and *B. licheniformis*), yeasts (*S. cerevisiae*, *Pichia pastoris*, and *P. stipidis*), filamentous fungi (*A. niger*, *A. oryzae*, *A. nidulans*, and *A. awamori*), higher plants (tobacco, alfalfa, and *Arabidopsis thaliana*), insect cells and larvae (Baculovirus), and mammalian milk. Two of the three bacterial systems listed have already been explored at NREL for their potential to express and secrete functional foreign cellulase gene products at high levels. Although perfectly adequate for producing reagent quantities of functional cellulases, both *E. coli* and *S. lividans* are inadequate as cellulase production systems on a scale required by the bioethanol process.

(1). *Escherichia coli.* *E. coli* is still the workhorse of modern molecular biology. As a genetic system, it is extremely well characterized. All cellulase genes obtained from various microorganisms under the sponsorship of the ethanol project during the past few years were first cloned and expressed in *E. coli*. Efforts to improve recombinant bacterial expression of cloned cellulases quickly gave way to other hosts, including *S. lividans* and *B. brevis* after it became clear that heterologous products were being degraded by proteases and incorporated into insoluble inclusion bodies in *E. coli*, and that *E. coli* has a limited capacity for secreting foreign proteins into the medium (*87*).

Although *E. coli* can synthesize a foreign protein to levels of 20%-30% of total cell protein, because of limitations of cell density in batch liquid culture, this amounts to only modest volumetric yields of protein (e.g., 1-2 g/L) (*96*). Fed-batch and continuous culture methods can achieve significantly higher specific volumetric yields (e.g., g/L/h). We employ *E. coli* only as a host for cloning and expressing new cellulase genes, for constructing expression plasmids which will be used in other host systems, and for producing reagent quantities of individual cellulases.

(2). *Streptomyces lividans.* Reputedly "strong" promoters isolated from various Streptomycetes were used at NREL to construct expression vectors for use in *S. lividans*, including tipA (a thiostrepton inducible promoter) and STI-II (*S. longisporus* soybean trypsin inhibitor II). Using these promoters, we have constructed various expression vectors in hopes of achieving g/L quantities of secreted, functional recombinant EI in *S. lividans*. We have successfully secreted fully active EI from most of these constructs, but none exceeds the production level of that produced by the native gene expressed in *S. lividans*. On the other hand, efforts to increase the expression level of a β-glucosidase cloned from *Microbispora bispora* successfully increased the expression level by a factor of 3-4, to more than 200 mg/L (Xiong, X., unpublished results, 1996). Additionally, Wilson reports that he has successfully increased the level of expression of one *T. fusca* exoglucanase gene, E_3, approximately five fold using the STI-II promoter (i.e., 150-200 mg/L) (Wilson, D., personal communication, 1996). Despite these successes, it is unlikely that the *S. lividans* system will be capable of much more than about 1 g/L production levels in batch culture.

(3). *Saccharomyces cerevisiae.* Despite being genetically well characterized, which provides the genetic engineer with numerous potential host strains, selectable markers, well-characterized promoters, and several vector alternatives, baker's yeast is not a good choice as a host for a cellulase expression system (*97*). Because of its notorious capacity for hyperglycosylation of foreign proteins and its limited capability for synthesizing and secreting foreign proteins, *S. cerevisiae* has not been considered a serious candidate for a cellulase production system at NREL.

(4). *Pichia pastoris.* *P. pastoris* can generate very high densities of protein-rich biomass in simple defined media, using methanol as a carbon source. For this reason, *P. pastoris* was first exploited commercially in a methanol-based process to produce single-cell protein (SCP) for use as a protein supplement in animal feeds. After characterization of the biochemical pathways for the metabolism of methanol in this and similar organisms, the two alcohol oxidase genes from *P. pastoris* were cloned and sequenced. Efficient transformation and expression systems were also developed for this organism, and are based on the methanol-inducible promoter from the alcohol oxidase 1 (*AOX1*) gene. The vectors used for *P. pastoris* transformation integrate into the yeast genome by homologous crossover at the *AOX1* locus. Because *P. pastoris* does not heavily glycosylate secreted proteins, as is common in *S. cerevisiae*, it is useful in the production of human pharmaceuticals that require glycosylation for biological activity. The state of the field of heterologous gene expression in *P. pastoris* has been very recently reviewed (*98*).

Several examples show that *Pichia* can routinely achieve percentage yields (5%-40% of total cell protein) much higher than baker's yeast, and often equivalent to *E. coli* or baculovirus (*99, 100*). Because *Pichia* is able to grow to much higher densities in liquid culture than any of the aforementioned systems, it can produce much higher volumetric yields (g/L). Scale-up of *Pichia* culture to extremely high cell density is simple and has resulted in enormous volumetric yields (e.g., 12 g/L for tetanus toxin fragment C [*101*] and >3 g/L secreted human serum albumin [*102*]).

We have tested a strain of *P. pastoris* designed to express and secrete of the *A. cellulolyticus* EI endoglucanase. In this construction, the mature EI coding sequence was joined in the same translational reading frame to the yeast alpha factor signal sequence present in pPIC9. Transformants have not yet been analyzed for gene copy number, but trial fermentations have already yielded 1.5 g/L of EI. Approximately 50% of the EI produced in *P. pastoris* is secreted into the medium and the remainder is found intracellularly (Thomas, S., unpublished results, 1995).

(5). Extracellular Production of Heterologous Fungal Proteins in Fungal Hosts. Several investigations concerned with the efficiency of secretion of heterologous proteins have focused on fungal enzymes. The results of published experiments to express foreign genes in filamentous fungi have recently been summarized by van den Hondel (*103*). Most of these enzymes are important for the industrial production of foodstuffs, animal feeds, and detergents. A large number of studies have been carried out using *Aspergillus awamori, A. niger* and *A. oryzae*, for which extensive experience in fermentation and downstream processing has been established.

The initial level of production of fungal proteins in heterologous hosts is usually in the range of 10 to 50 mg/L. Under nonoptimized conditions, similar production levels are observed for efficiently secreted homologous proteins. However, after optimization of the production process, levels of at least 3 g/L have be obtained. Clearly production yields can usually be improved considerably through the use of modified hosts, media optimization, use of appropriate large-scale fermentation conditions, and classical strain improvement procedures.

(6). Extracellular Production of Bacterial Proteins in Fungal Hosts. The literature reveals only a few reports of studies that deal with the expression and secretion of bacterial proteins in filamentous fungi are available to date. The *Cellulomonas fimi* endoglucanase, being a cellulase, is particularly relevant to the matter at hand. The 5' region of the inducible *A. nidulans alcA* gene was employed to direct expression of the *C. fimi* endoglucanase in an *A. nidulans* host strain previously modified to overproduce the *alcR* gene product, which positively regulates *alcA* transcription. This promoter is repressed in media that contains glucose, but can be induced when the carbon source is switched to ethanol. This work demonstrated production of approximately 20 mg/L of functional, secreted endoglucanase in shake flask cultures growing sub-optimally in minimal media at 37°C for 48 h (*104*).

Alternatively, Turnbull and co-workers (*105*) produced *E. coli* enterotoxin subunit B (LTB) at low levels (2 ng/μg soluble protein, or 24 μg/g wet weight of mycelia) in *A. nidulans* using an "up-regulated" inducible *amdS* promoter. This promoter can be induced by either acetate or acetamide. The fact that the pre-LTB product was properly processed to remove the bacterial signal peptide but not secreted into the medium indicates some sort of incompatibility with the fungal secretory apparatus, or that secreted protein was rapidly degraded by extracellular proteases.

(7). *Trichoderma reesei.* *T. reesei* can produce remarkable amounts of extracellular protein (20 to 50 g/L). Even so, until now, the use of filamentous fungi as general production hosts has been restricted mostly to research and industrial laboratories with special interest in these organisms. Compared with *E. coli* and *S. cerevisiae* , which serve as model organisms for basic research and are widely used in molecular biology, the efforts so far invested in the development of filamentous fungi as production hosts have been very limited. The major cellulase, cellobiohydrolase I (CBH I), which is produced from a single copy gene, represents ~50% of the total protein secreted. Thus, the *cbh1* promoter is extremely strong. The excellent synthesis and secretion capacity of the organism, together with established fermentation conditions, prompted development of

T. reesei as a host for production of heterologous proteins. *T. reesei* has the advantage of possessing a eukaryotic secretory machinery, and, most likely, similar protein modification properties (e.g., high mannose type N-glycosylation; *106*) to mammalian systems.

Recently, in a continued search for powerful promoters that are active in the presence of glucose-containing media, the group at VTT (Espoo, Finland) has screened a cDNA library for sequences that are highly abundant (*107*, *108*). It has then used those cDNAs to isolate the corresponding genomic clones and the promoters for those genes. This level of expression proved to be 20 to 50-fold higher than that of the *pgkl* gene. Promoters for two of these genes have been isolated and used to drive expression of the homologous EG I coding sequence in *T. reesei* strain QM9414 growing in glucose-containing medium.

(8). Higher Plants. There is sound reasoning behind the approach to produce bulk industrial proteins in crop plants. According to Pen and co-workers, "In terms of cost-effectiveness for producing biomass, the growing of crops in the field can generally compete with any other system. It is inexpensive, it can be done in bulk quantities, and it requires limited infrastructure. These observations suggest that the exploitation of arable crops for the production of food, feed, or processing materials would be very attractive" (*109*). Aside from various academic laboratories around the world, at least three plant genetic engineering companies are actively pursuing the expression of relatively low value bulk industrial proteins in crop plants of one sort or another: MOGEN International (Leiden, The Netherlands), human serum albumin (HSA) and α-amylase in tobacco and potato; Ciba-Geigy (Research Triangle Park, NC), cellulases in maize; Calgene, Inc. (Davis, CA), cellulases in tobacco.

The idea for production of industrial enzymes and other bulk proteins in plants is not unique to NREL. For example, others have already demonstrated expression of *Bacillus licheniformis* α-amylase in tobacco (*109*, *110*), *Clostridium thermocellum* xylanase (XynZ) in tobacco (*111*), and HSA in potato (*109*). These examples involve the stable transformation of plant nuclear DNA via *Agrobacterium*-mediated gene transfer. In all three cases, no effect on cell growth and development was observed, indicating the foreign DNA expression did not adversely affect the transformed plants. A patent suggests that a variety of proteins from many sources can be used as target proteins for expression in barley endosperm (*112*).

B. licheniformis α-amylase was expressed at 0.5% of total protein in tobacco. Alpha-amylase purified from tobacco exhibited a slightly higher molecular weight than the native enzyme (64 versus 55 kDa), and was entirely due to glycosylation in the plant system. Nevertheless, despite glycosylation, the α-amylase expressed in tobacco was active, secreted, stable at 95-100°C, and otherwise completely indistinguishable from the native enzyme. Perhaps just as important is that the starch content of transformed leaves (in chloroplasts) was unaffected by the cytoplasmically expressed α-amylase. This is an important point because it clearly demonstrates the ability to isolate a transgenic protein from its potential substrate by compartmentalization, thus avoiding potentially detrimental effects on the plant.

Herbers and co-workers (*111*) recently expressed a truncated version of a thermostable *C. thermocellum* xylanase in transgenic tobacco plants. The authors speculate that these plants might be useful for producing xylanase, which has numerous applications in the paper industry and agriculture. The xylanase was synthesized as a 37 kDa polypeptide and correctly targeted to the intercellular space by means of a proteinase inhibitor II signal peptide. The xylanase was one of the most abundant proteins in total extracts (4.1 +/- 1.6%) and represented more than 50% of protein present in the intercellular fluids. The transgenic plants, grown under greenhouse conditions, were not affected by the foreign enzyme, possibly because of the high temperature optimum of the xylanase and the low levels of xylan in tobacco cell walls.

In 1993, the highest published level of expression of a foreign protein by nuclear transformation in a plant system was 1.5% of soluble protein (*113*). Differences in the level of foreign gene expression seem largely gene dependent and may be due to efficiencies in transcription/translation or stability of the gene products. Further improvements in expression levels seem likely, given the modest effort expended in this area so far.

Another new approach to plant transgenic expression systems involves integrating the target gene into the tobacco mosaic virus RNA genome, followed by infecting tobacco and other solanaceous species (e.g., tomato). As the infection spreads systemically, the plant becomes a dedicated bioreactor for expressing the foreign gene (*114-116*). TMV-based vectors can produce heterologous proteins in tobacco at levels between 5% and 40% of total cell protein (della-Cioppa, G., personal communication, 1995).

Despite progress in the development of gene expression technology, significant problems remain in the manufacture of many complex proteins. Many post-translational modifications performed by animal cells can be performed by green plants which, like animals, are complex, eukaryotic organisms. Green plants are a promising, underexploited system for expressing new proteins, including cellulases. In addition, green plants are photoautotrophic, requiring only carbon dioxide, water, nitrogen, sulfur, phosphorus, and trace amounts of other elements for growth.

Advanced Fermentation Concepts at NREL

Virtually all fermentative microorganisms can ferment glucose to a mixture of fermentation products, but only a few can ferment glucose selectively to ethanol and thereby achieve high ethanol production yields. Even fewer microorganisms can ferment the pentose sugars in biomass (D-xylose and L-arabinose) to ethanol at high yields. A key challenge to developing advanced processes for converting biomass sugars to ethanol at high yield is to identify or construct microorganisms that exhibit the ability to convert all hexose and pentose biomass sugars to ethanol at high yield. Another is to develop bioprocesses using such microorganisms that achieve the high performance levels necessary to realize favorable overall process economics.

The objective of strain development and process development research–to maximize conversion of heterogeneous biomass sugars to ethanol– is discussed and recent advances in biomass sugar conversion are briefly reviewed. The major polysaccharides, which comprise hardwood and herbaceous biomass species and agricultural residues, are described to motivate the fact that glucose and xylose, and in some cases arabinose, must be efficiently fermented to achieve a high process yield and favorable process economics. The various processing options available for biomass saccharification and fermentation are then outlined and recent fermentation-related accomplishments achieved at NREL and other laboratories are reviewed. Finally, overviews of fermentation strain and fermentation process development research now being pursued at NREL are provided.

Introduction to Biomass Fermentation. Lignocellulosic materials that are potential feedstocks for bioethanol production include hardwoods, herbaceous crops, agricultural residues, and municipal solid waste (i.e., wastepaper and other fractions). These types of "biomass" are comprised primarily of cellulose, hemicellulose, and lignin (*76, 117*). Carbohydrates account for roughly 50% to 70% of the dry mass of such materials, with lignin accounting for approximately another 20%; ash and other minor components make up the balance.

Feedstock costs represent more than one-third of all processing costs in technoeconomic cost projections of large-scale bioethanol production using energy crop feedstocks costing $42 per dry ton (*43*). High feedstock costs mean that an integrated biomass-to-ethanol process must achieve high ethanol yields on the feedstock to be economical. Maximizing ethanol production requires hydrolyzing all cellulosic and hemicellulosic biomass sugars to fermentable forms (e.g., monomeric) and carrying out fermentation under conditions that produce high ethanol yields. Achieving high process is challenging because of the complex heterogeneous structure and relatively refractory nature of lignocellulosic biomass to complete saccharification and high yield fermentation.

Feedstock Composition. Table 2 lists the compositions of a variety of lignocellulosic biomass types. As this table shows, glucan (from cellulose) and xylan (from hemicellulose) are the two carbohydrates at the highest levels in these types of biomass. The contribution of pentosans (xylan and arabinan) to total carbohydrates varies with biomass type, ranging from about 25% in hardwoods to 33% in herbaceous crops and 40% in agricultural residues. The levels of the minor carbohydrates arabinan, galactan, and mannan also vary with biomass type. Hardwoods typically contain more mannan and are more highly acetylated, whereas herbaceous plants and agricultural residues generally contain higher levels of galactan and arabinan. In some herbaceous crops and agricultural residues arabinan levels are high enough that conversion of arabinan–in addition to glucan and xylan– is required to achieve high overall conversion yields (*118*).

Processing Options. A variety of processing schemes can be used to convert lignocellulose materials to ethanol. Most importantly, the process must achieve a high conversion yield on the hemicellulosic and cellulosic sugars to be economical (*43, 119*). Processing options differ in the method(s) chosen for hydrolyzing cellulose and hemicellulose to their component monomeric sugars (i.e., glucose, xylose, arabinose, galactose, and mannose). In addition, cellulose hydrolysis can either precede or be carried out simultaneously with fermentation. Processing schemes in which cellulose is hydrolyzed to glucose using either cellulase enzymes or concentrated acids at relatively low temperatures (lower than 100°C) typically degrade less carbohydrate than those based on higher temperature acid hydrolysis of cellulose. Although processes based on complete acid hydrolysis have a long history (*120-122*) and continue to be investigated (*123, 124*), enzyme hydrolysis-based processes are considered by many to offer the best potential to maximize overall process yields (*125*). Current efforts at NREL are directed at developing and demonstrating enzyme-based bioethanol process technology.

Economic analyses indicate that the most important factors influencing bioethanol production cost are overall process yield and final ethanol concentration; volumetric productivity is a significant but secondary factor (*119*). Economic analyses also show that substantial capital and operational savings are possible in advanced process designs in

Table 2
Composition of Representative Biomass Species Being Considered for Bioethanol Production

	Composition (wt%, dry basis)							Total
		Hexosan			Pentosan		Klason	Carbo-
Biomass type	Glucan	Galactan	Mannan	Xylan	Arabinan	Acetyl	Lignin	hydrate
Hardwoods								
silver maple	45.9	0	1.2	17.1	0.7	3.9	20.8	64.9
sycamore	44	0	0.9	16.3	0.6	3.6	22.8	61.8
black locust	49.4	0	1	16.2	0.4	3.8	21.5	67
poplar hybrid NE388	48.6	0.3	0.5	14.6	0.3	2.2	21.8	64.3
poplar hybrid N11	51.8	0.7	0.3	11.3	0.3	1.9	22.5	64.4
sweetgum	49.5	0.3	0.4	17.5	0.4	2.3	21.8	68.1
Herbaceous sp.								
switchgrass	36.6	1.2	0	16.1	2.2	1.1	21.9	56.1
weeping lovegrass	36.7	1.7	0	17.6	2.6	1.1	21.2	58.6
Sericea lespedeza	31.5	0.9	0	14.5	1.6	1.3	31.6	48.5
reed canary grass	26	0.1	0	9.8	2.4	0.9	15.6	38.3
flatpea hay	28.9	1.5	0.1	7.4	2	1.4	24.5	39.9
Ag. residues								
corn cobs	39.4	1.1	0	28.4	3.6	1.9	17.5	72.5
corn stover	40.9	1	0	21.5	1.8	1.9	16.7	65.2

Data from Torget and co-workers (*40- 42*).

which hexose and pentose sugars are fermented simultaneously (e.g., cofermentation) (Putsche, V., personal communication, 1996).

Figure 3 depicts a generic enzyme hydrolysis-based biomass-to-ethanol processing scheme. The dashed lines represent advanced processing options in which separate biologically mediated steps are combined. Processing begins with a "biomass handling" step wherein the biomass is milled to an appropriate size before being subjected to some type of thermochemical treatment, e.g., with steam and dilute sulfuric acid, to open up the lignocellulose pore structure and increase its susceptibility to enzymatic attack (*126*). This pretreatment step effectively hydrolyzes and solubilizes the relatively more labile hemicellulosic sugars, thereby producing a hydrolyzate liquor typically rich in pentose sugars and a cellulose-rich solid with greater porosity and dramatically increased enzymatic digestibility.

In the base-case process configuration shown in Figure 3, the insoluble pretreated lignocellulosic solids are converted to ethanol in an SSF bioreactor. Hydrolytic cellulase enzyme(s) and fermentative microorganism(s) are present in the SSF bioreactor. The cellulase enzymes hydrolyze the cellulose to glucose, which is then fermented to ethanol by a suitable fermentative microorganism (e.g., brewers' yeast). The SSF process is efficient because it reduces end-product inhibition of the cellulases by glucose through the continuous fermentative conversion of glucose to ethanol (*127*). It also reduces capital costs by reducing the number of vessels relative to those required to separately carry out enzymatic hydrolysis and fermentation. In the base-case scheme, the fermentative microorganism used in SSF does not ferment hemicellulosic sugars (e.g., xylose), so a separate pentose fermentation step is required. An enzyme production bioreactor is also required, because the fermentative microorganism used in SSF does not produce cellulolytic enzymes. Distillation is used to recover the ethanol produced in the separate SSF and pentose fermentation unit operations. The lignin fraction and other residual solids carried through the process are sent to a cogeneration plant to supply steam and electricity for the process (not shown).

The advanced processing option indicated in Figure 3 by the inner dashed rectangle (short dashes) incorporates a (novel) microorganism that can ferment hemicellulosic and cellulosic sugars to ethanol at high yield in an SSCF bioreactor. In this case, only a single step is required to enzymatically saccharify and ferment the process liquor and solid fractions of pretreated biomass. As in the base-case scheme, a separate enzyme production unit operation is required.

The scheme represented in Figure 3 by the outer dashed rectangle (long dashes) shows the most advanced processing option, one in which all biologically mediated steps–enzyme production, enzymatic cellulose hydrolysis, and biomass sugar fermentation–occur in a single bioreactor. This process, also known as direct microbial conversion (DMC), can be carried out to various extents by a number of microorganisms, including fungi such as *Fusarium oxysporum* (*128*) and bacteria such as thermophilic *Clostridia* (*129*, *130*). However, DMC strains exhibit relatively low ethanol selectivity; current strains produce at least one other fermentation product, typically acetic acid, at nearly equimolar levels with ethanol.

The designs shown in Figure 3 are considerably more complicated than those used for traditional starch-to-ethanol (or sugar-to-ethanol) processes already in commercial operation. This is because starch is much easier to hydrolyze to glucose than cellulose. And starch contains only glucose, whereas biomass substrates contain significant levels of other sugars, most notably xylose. Starch-based processes require neither cellulolytic enzyme production nor pentose fermentation to achieve high conversion yields. Another difference is that solids handling requirements are typically higher for biomass substrates, especially if cellulose hydrolysis is carried out in an SSF-mode rather than before fermentation (*131*). All these characteristics impose significant challenges for bioethanol process development beyond those encountered in starch-to-ethanol processing.

Trends in Fermentation Strain Development. Whereas a number of microorganisms can efficiently ferment glucose, the product of cellulose hydrolysis, to ethanol, only recently has conversion of the pentose sugars in the hemicellulose fraction become feasible. Until a few years ago, the few known organisms that could utilize either D-xylose or L-arabinose did not perform well, but typically grew slowly on pentoses and achieved relatively low ethanol yields and productivities. Therefore, identifying and developing microorganisms that can selectively convert D-glucose, D-xylose and L-arabinose to ethanol at high yield has been the focus of extensive research during the past 20 years. During the past decade, the sophistication of molecular biology has grown tremendously, and numerous attempts have been made to use recombinant DNA technologies to engineer superior microorganisms for bioethanol production. Only a few of these efforts have been highly successful; many are still in progress (*132*).

Metabolic pathway engineering is increasingly being recognized as a powerful approach for developing microorganisms that can efficiently convert biomass sugars to ethanol. In broad terms, superior ethanol-producing microorganisms can be developed by several metabolic engineering approaches.

- Increasing ethanol product selectivity in strains exhibiting broad substrate range but poor product selectivity
- Broadening the substrate range to include important biomass sugars besides glucose (e.g., arabinose, galactose, mannose or xylose) in strains that exhibit good product selectivity but that cannot ferment sugars other than glucose to ethanol

Of course, beyond these two basic approaches, significant metabolic engineering may also be required to stabilize "improved" strains or to enable such strains to achieve high ethanol yields and fermentation productivities.

Following the first approach, *E. coli* and *Klebsiella oxytoca* have been engineered to be highly effective ethanol producers by introducing the genes for ethanol production from *Z. mobilis* (*133-135*). Extensive evaluations of these "ethanologenic" strains have been carried out, both in media that contain pure sugars and in pretreatment hydrolysates derived from a variety of feedstocks (*136-141*).

The second approach–broadening the substrate utilization range of strains that are highly efficient ethanol producers–has been demonstrated at NREL by introducing the xylose assimilation and pentose phosphate pathway genes from *E. coli* into *Z. mobilis* to confer the ability to ferment xylose to ethanol at high yield (*142*). More recently, an arabinose-fermenting *Z. mobilis* strain has been similarly developed by introducing the arabinose assimilation and pentose phosphate pathway genes from *E. coli* into *Z. mobilis* (*143*). Both of these accomplishments are described in more detail below.

More recently, long term efforts to develop xylose-fermenting *Saccharomyces* sp. have also been successful, with xylose fermentation reported for strains transformed with the xylose reductase and xylitol dehydrogenase genes from *P. stipitis* and overexpressing xylulokinase (*144-146*). Other recent achievements in the metabolic engineering of superior ethanol producers include initial success at improving the performance of xylose-fermenting yeasts by optimizing the expression of genes that encode xylose assimilation and ethanol production pathways (*147-148*). Successful transformation of pentose-fermenting *Clostridium thermosaccharolyticum* has also been reported (*149*), which provides a key tool for further developing this microorganism by altering product selectivity to favor ethanol production.

The recombinant *E. coli* and *K. oxytoca* and one of NREL's recombinant *Z. mobilis* can ferment arabinose to ethanol, unlike recombinant xylose-fermenting *Saccharomyces* sp. and wild-type xylose-fermenting yeasts such as *P. stipitis*, which can aerobically grow on arabinose but can not anaerobically ferment arabinose.

Current Thrust of Fermentation Strain Development at NREL. Our near-term goal is to develop microbial catalysts that effectively convert sugar streams from hardwood sawdust in which glucose and xylose are the major sugar substrates. In the intermediate term, our goal is to develop strains for converting herbaceous energy crops such as switchgrass, which contain arabinose in addition to glucose and xylose. Our previous in-house research efforts have focused primarily on increasing the substrate range of *Z. mobilis*. This bacterium's high conversion yield, fermentation product selectivity, and ethanol tolerance levels are key attributes for a commercially viable ethanol-producing strain. To develop this organism for conversion of mixed hexose and pentose sugar streams derived from lignocellulosic feedstocks, we successfully engineered this bacterium to utilize both xylose and arabinose (*142*, *143*).

Xylose-fermenting *Z. mobilis* was developed through the coordinated expression of the *E. coli* xylose isomerase, xylulokinase, transketolase, and transaldolase genes (*142*). The recombinant *Z. mobilis* could grow on xylose as the sole carbon source and produced ethanol at 0.44 g/g of xylose consumed, corresponding to 86% of theoretical yield. The xylose assimilation and pentose phosphate metabolism genes were introduced on two operons under the control of strong constitutive promoters, which allowed the strain to rapidly ferment a mixture of glucose and xylose to ethanol at 95% of theoretical yield. In a medium that contained glucose and xylose, the xylose-fermenting strain utilized glucose at a faster rate than xylose, but both sugars were utilized simultaneously (*142*). This strain also performs well at converting cellulose and xylose to ethanol under SSCF conditions in which cellulase enzymes are present (*150*).

Arabinose-fermenting *Z. mobilis* was developed by introducing the L-arabinose isomerase, L-ribulokinase, L-ribulose-5-phosphate-4-epimerase, transaldolase, and transketolase genes from *E. coli* (*143*). The recombinant bacterium ferments both arabinose and a combination of glucose and arabinose to ethanol at more than 95% of the maximum theoretical ethanol yield on a consumed sugar basis. Arabinose-fermenting *Z. mobilis* does not yet perform as well on sugar mixtures as xylose-fermenting *Z. mobilis*. In media that contain glucose and arabinose, the arabinose-fermenting strain utilizes arabinose at a considerably slower rate than glucose, and only after glucose is nearly depleted. The arabinose-fermenting strain may be used in mixed culture with the xylose-fermenting *Z. mobilis* strain to efficiently ferment glucose, xylose, and arabinose, the predominant biomass sugars in agricultural residues and herbaceous energy crops (see Table 2).

Future directions for NREL's strain development research include continuing efforts to develop superior *Zymomonas* strains for bioethanol production. One focus is to use recombinant DNA technologies to combine xylose and arabinose fermentation capabilities into a single *Zymomonas* strain. Beyond this, we are pursuing development of metabolically engineered cofermenting *Zymomonas* strains that exhibit the high level of stability and robustness required for a commercial fermentation biocatalyst. We are currently developing methodologies for integrating genes into the *Zymomonas* chromosome. Ultimately, we plan to introduce the pentose-fermenting genes (without the current antibiotic resistance marker gene) directly into the *Zymomonas* chromosome to ensure that the strain will be stable and compatible with good large-scale manufacturing practices. We also plan to evaluate the potential to use *Zymomonas* strains that reportedly exhibit particularly high thermo- and ethanol tolerance as improved hosts for metabolic engineering of pentose fermentation capability. Some of these host strains will likely enable recombinant strains that are more tolerant of inhibitory pretreatment hydrolysates to be developed. The objective is to integrate genes that confer xylose and arabinose fermentation capabilities into the chromosomes of such strains. In addition to these activities, we plan to investigate other innovative molecular approaches to improving the performance of pentose-fermenting *Zymomonas* strains, including:

- Identifying and alleviating metabolic bottlenecks that limit fermentation performance
- Introducing improved pentose transport systems to increase sugar utilization rates
- Developing novel strategies to minimize by-product formation.

By implementing these efforts, we hope to eventually develop a "super" *Zymomonas* strain capable of robust high yield cofermentation of glucose, xylose, and arabinose mixtures to ethanol.

We are also attempting to develop other ethanologens. In a previous survey of prospective host microorganisms for developing superior fermentation strains, we identified *Lactobacillus* sp. as attractive candidates for metabolic engineering of pentose utilization and ethanol production. (*151*). Experiments carried out at NREL demonstrated that many *Lactobacillus* sp. exhibit higher tolerance to hardwood hemicellulose hydrolysates than *Zymomonas* sp. Although most of our work to date has focused on developing and improving pentose-fermenting *Zymomonas* strains, we have initiated research to develop *Lactobacillus* sp. that can ferment the major biomass sugars to ethanol. We have successfully introduced a xylose assimilation pathway into a homofermentative *Lactobacillus* strain that previously could not utilize xylose (Franden, M.A., personal communication, 1996). Our future efforts in this area will be to redirect carbon flow from pyruvate to ethanol.

Trends in Fermentation Process Development. Around the world, many efforts are under way to develop, scale up, and commercialize biomass conversion technologies. Active programs to develop cost-effective biomass production systems and biomass conversion technologies exist in many countries of the Americas, Europe, and Asia (*152-156*). In the United States, DOE/NREL has a particularly active ethanol project that has made significant accomplishments during the past several years. Several bioethanol development programs, including NREL's, are at the stage where integrated processes are beginning to be demonstrated at pilot scales (10-1000 kg/day dry biomass) (*124*, *157*). Most processes being developed are based on enzymatic hydrolysis of cellulose. Economic sensitivity analyses of enzyme hydrolysis-based processes show that improvements in biomass carbohydrate hydrolysis (i.e., saccharification by thermochemical pretreatment in combination with enzyme hydrolysis) offer the greatest potential to reduce bioethanol production costs (*158*).

The most notable breakthrough during the past several years is the development of improved fermentative microorganisms that can ferment pentose and hexose sugars to ethanol at high yield (*159*, *160*, *142*, *144-146*). Several of these novel microorganisms, including transformed strains of *E. coli* (*140*), *K. oxytoca* (*161*), and *Z. mobilis* (*150*) perform well under SSF-type conditions. Current efforts at NREL focus on developing processes in which pentose and hexose sugars are fermented together by xylose-fermenting *Z. mobilis* using an SSCF process configuration, as described above.

The enzyme hydrolysis-based processing option that arguably remains the least developed–but potentially the most promising–is the DMC process in which enzyme production, cellulose hydrolysis, and fermentation are carried out in a single vessel. NREL continues to fund subcontracts in the area of DMC processing. Although significant progress has been made during the past several years in demonstrating DMC-based conversion of lignocellulosic feedstocks (*162*), considerable progress in strain development remains to make DMC an economical processing option for bioethanol production.

Current Thrust of Fermentation Process Development at NREL. Current efforts at NREL focus on developing and demonstrating an integrated bioethanol process. Rigorous bench-scale experiments are being conducted to identify and characterize the performance of a hardwood sawdust-to-ethanol process based on an SSCF process configuration. In addition to SSCF, distinct processing elements under active development at NREL include pretreatment, conditioning of pretreated materials (detoxification), production of cellulase enzyme(s), and development of improved fermentation strains, including long-term adaptation of strains to develop strains that perform better under inhibitory processing conditions.

Previous process development efforts to optimize processing conditions for converting biomass to ethanol at NREL and other laboratories have generally focused on specific unit operations rather than on the whole process. As a result, previous studies do not provide sufficient information to predict how an integrated bioconversion process will perform. We believe, as do Rehn and co-workers (*163*), that integrated process testing is the most effective way to identify and optimize operating conditions for an overall bioethanol production process. Moreover, we believe that integrated process studies are most expedient for identifying important interactions that may exist between linked process components.

The major components of the integrated bioethanol process currently under development at NREL are summarized in Figure 4. These include: (1) dilute acid pretreatment of hardwood sawdust feedstock using a Sunds hydrolyzer pilot-scale pretreatment reactor (*124*, *157*); (2) slurry conditioning (detoxification) to improve the fermentability of pretreatment liquors and solids and thereby allow the integration of pretreatment with cellulase production and with enzymatic hydrolysis and fermentation; (3) cellulase production based on *T. reesei* utilizing pretreated (conditioned) hydrolysates and pretreated solids; and (4) cellulose saccharification and glucose and xylose cofermentation using cellulase enzyme(s) and NREL's xylose-fermenting *Z. mobilis* in an SSCF process configuration.

The SSCF process is at the heart of the overall process and achieving high SSCF

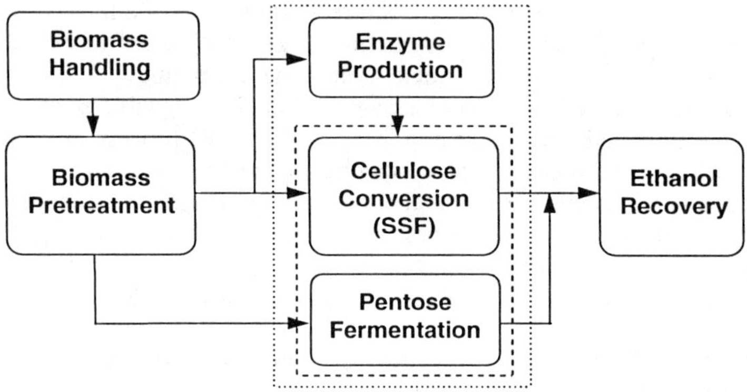

Figure 3. Processing options for bioethanol production. See text for discussion.

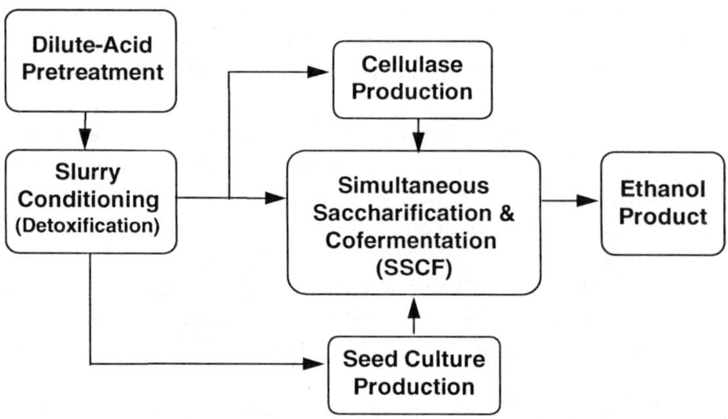

Figure 4. Major processing elements in NREL's integrated biomass-to-ethanol process.

conversion yields is the primary objective of fermentation process development efforts. Improved bioethanol process economics are achieved by maximizing SSCF performance and minimizing SSCF processing costs. Process economics benefit not only from reductions in direct SSCF operating costs, but also (and perhaps more dramatically) from improvements made in other processing elements that enable better SSCF performance to be achieved (e.g., improvements in pretreatment, enzyme production, detoxification).

We have made significant progress at NREL toward demonstrating cost-effective bioethanol process technology. Particular accomplishments related to fermentation process development include successfully combining separate xylose fermentation and cellulose conversion unit operations into a single SSCF process. During the past year, we characterized *Zymomonas*-based SSCF performance as a function of carbohydrate loading and pH, demonstrating that NREL's xylose-fermenting *Zymomonas* strain performs well in an SSCF process configuration (*150*). Our current focus is to integrate all major processing steps–pretreatment, hydrolyzate conditioning, cellulase enzyme production, and SSCF–at the bench scale. Most recently, we demonstrated a minimum level of performance for SSCF processing of conditioned (detoxified) pretreatment whole slurry. These experiments were carried using sterile-filtered cellulase enzyme broth produced using washed and pretreated hardwood sawdust solids and lactose rather than enzyme produced using only pretreated materials. Nonetheless, successfully integrating the main process technology elements is significant for two reasons: (1) it provides an excellent foundation for examining additional integrated process technology improvement; and (2) it greatly reduces uncertainty in technoeconomic projections of bioethanol manufacturing cost.

We plan to continue integrating the major processing steps at the bench scale. Integration at the mini-pilot scale will be pursued once the integrated process meets established performance criteria. Our immediate goal is to demonstrate an integrated process that achieves NREL's near-term objectives for bioethanol production cost. Integrated process testing is planned to examine the benefits of advanced processing concepts such as incorporating component recycle streams (e.g., recycle of process water, nutrients, substrates [solids], cells, or enzymes). Integrated process testing will also be carried out to evaluate and compare alternatives to SSCF processing such as sequential (enzymatic) hydrolysis and cofermentation (SHCF) or hybrid enzymatic saccharification/fermentation processes that fall between SSCF and SHCF. Such hybrid processes would only partially separate enzymatic saccharification and cofermentation processing steps. Maximizing performance in such hybrid processes will probably require that the sequentially linked saccharification and cofermentation unit operations be carried out at various temperatures and pHs to maximize cellulose hydrolysis in the first step and ethanol fermentation in the second step. Again, integrated process studies will be used to efficiently characterize interactions that are expected between the linked enzymatic cellulose hydrolysis and microbial ethanol production processing steps.

Technical Prospectus

We have reviewed critical features of the key unit operations necessary to economically operate bioethanol production plants; however, these technologies cannot be used to plan or model actual production plants until the process is integrated at reasonable scale. This view is commonly overlooked by researchers and planners involved in the important details of unit operations. Pilot plant verifications of biomass to ethanol processes have "discovered" critical problems that result from fully integrated operation that were not realized at bench or simulated integrated operation, especially contamination of cellulase production and SSF unit operations (*164*, *73*) and mass/O_2 transfer problems that result from lignocellulose utilization at large scale (*73*). Other unforeseeable scale-up issues may have equal impact on process economics; for example, the problems encountered in scaling pretreatment equipment are reflected (to some extent) only by experience in the pulp and paper industry.

For bioethanol production processes, in which yield is crucial to economic operation (*78*), all process-related diversions of fermentable sugars from ethanol are catastrophic. In the pretreatment section, for example, effects of wear and normal aging on equipment contacted by hot, acidic biomass are difficult to model. Char buildup on piping and reactor surfaces may pose a threat to biomass yield from unpredicted adsorption or condensation reactions, as well as the better understood effects of heat transfer reduction.

Another aspect of fully integrated plant operation not always considered is the consequence of allowing sugars, especially mono- and disaccharides, extended residence times following initial prehydrolysis or enzyme saccharification (*165*). Reducing sugars are reactive under virtually all conditions encountered in a biomass conversion facility! Mildly acidic conditions and moderate temperatures (i.e., conditions consistent with those following dilute acid pretreatment and storage) permit reversion and transglycosylation of glucose to generate populations of all possible configurations of alpha- and beta-linked disaccharides (i.e., 1-1, 1-2, 1-3, 1-4, and 1-6) and even higher-order oligomers in yields that approach 10% (*166*). Many of these oligosaccharides are not fermentable, and the nonreducing sugars, such as 1-1 beta- and alpha-linked glucose (α- and β-trehalose), are quite stable even after dilution and neutralization. When enzymes are added to the process, as in SHF or hybrid SSF schemes, reversion and transglycosylation reactions produce all possible configurations of disaccharides consistent with the anomeric requirement of the enzyme; i.e., α-glucosidase produces only alpha products and β-glucanases produce only beta products, but many linkage combinations are possible. Clear precedent exists for efficient production (20% to 37% yield) of transglycosylation products from 10% solutions of glycosyl donor (cellobiose) using a *Fusarium oxysporum* β-glucosidase (*167*) and from 10% solutions of maltose using an *A. niger* α-glucosidase (*168*). Enzymatic transglycosylation is also possible at glycosyl donor concentrations as low at 2% (*169*). Because most amylases and cellulases have active sites that can tolerate some diversity in the glycosyl acceptor group, it may be possible to find disaccharides or higher forms of glucose-xylose, glucose-galactose, and even glucose-mannose in biomass hydrolysis process streams. Of further concern to bioethanol plant efficiency are the observations that transglycosylation reactions that involve glycosyl transferases have been reported that utilize non-carbohydrate species as glycosyl acceptors, including alcohols (methanol, ethanol, and propanol) (*167*) and lignin model compounds (veratryl and vanillyl alcohols) (*170*). An unfortunate conclusion easily drawn from the literature is that the precise nature (chemical composition) of reversion and transglycosylation products formed during proposed bioethanol process operations is not currently known, nor is the resultant impact of this loss on fermentable sugars taken into account in process models. One critical reason for this dilemma lies in the advanced level of analytical capability required to properly detect and identify these products in biomass processing streams. A related and contributing problem is the inability of most laboratories to adequately assess fermentation process mass balances; thus, failure to achieve theoretical yields is often attributed to microbial or enzymatic performance problems, not to the presence of nonfermentable and unassayable sugars.

Executive Summary

At this writing, we are nearing the end of the seventeenth year of funding from DOE's Biofuels program to pursue and support the development of a biomass-to-ethanol industry in the United States. During this time, the benefits of a sustained level of funding to the program have become abundantly clear (although sometimes only in retrospect). We have taken a technology with roots in antiquity (i.e., fuel ethanol was produced by desperate governments embroiled in World Wars I and II) and brought it into the age of genetic engineering and advanced computer modeling. This was done for a very important reason–commercializable biomass-to-ethanol technology in the 1990s and beyond must be finely tuned and resembles the old fermentation processes in general form only. Herein lies the dilemma: at the surface bioethanol production appears technically simple; however, when considered in view of today's complex markets (e.g., costs and values for fuels and biomass) and business/environmental requirements, this initial impression fades and the need for technical acuity becomes obvious.

From our perspective, activities that occur from 1997 to the year 2000 will set the stage for the success or failure of the bioethanol industry. The primary driving factor for this opinion is the sudden and substantial increase in interest by the private sector in bioethanol production. To meet this challenge, biomass conversion researchers must address the issues urgently needed by first (alpha) plants. These plants will be sited in locations specifically chosen by investors for optimal access to infrastructure, including biomass and water resources, fermentation and electrical power facilities, by-product markets, and labor availability. The first processing plants will not be chosen for fit to feedstocks and technology. Even the impact of political mandates will overshadow technical preparedness. The biomass conversion research programs must, therefore, be keenly sensitive to the direction of industry and empowered to support it.

A sustained energy future for the United States also requires energetic support of technologies appropriate to longer-term energy feedstocks. Once the initial bioethanol industry has been established, technologies that provide incremental improvements in specific unit operations, as well as those that permit access to entirely new feedstocks, will ensure a robust and secure industry. A valuable lesson can be learned from the current corn-to-ethanol industry, in which the inability to process varied feedstocks (i.e., lignocellulose as well as corn and grain) has left processing plants precariously dependent on the volatile corn market. Future success of bioethanol plants lies in tolerance of feedstock diversity. Incremental technical improvements and new "leap-forward" technologies require active and vigorous advanced research programs. Apart from the obvious benefits to the bioethanol industry, and like all applied/basic research programs, these activities provide spinoff technical benefits that enhance the competitiveness of the general U.S. industry. Specific examples could include insights into cellulose and lignin chemistry from feedstock selection and analysis activities, insights into the kinetics of sugar hydrolysis, sugar destruction, and toxin formation from biomass pretreatment activities, insights into the biochemistry of enzymes that act on insoluble polymers from cellulase improvement activities, insights into the control and analysis of metabolic pathways in fermentative microorganisms from ethanologen improvement activities, insights into the microbiology of methanogenesis from waste treatment activities, and insights into animal feed/nutrient requirement methodologies from plant by-product recovery activities. Thus, the DOE-funded biomass-to-ethanol R&D program offers many levels of benefit to U.S. industry and represents a continuing success story for critical research managed by a national laboratory.

Acknowledgments

This work was funded by the Biochemical Conversion Element within the Biofuels Systems Program of the Office of Fuels Development of the U.S. Department of Energy.

Literature Cited

1. Steiner, G. *Strategic Planning: What Every Manager Must Know*; The Free Press: New York, NY, 1979.
2. Mintzberg, H. *The Rise and Fall of Strategic Planning*; The Free Press: New York, NY, 1994.
3. Sheehan, J. In *Enzymatic Conversion of Biomass for Fuels Production*; Himmel, M.E.; Baker, J.O.; Overend, R.P., Eds.; Symposium Series No. 566, The American Chemical Society: Washington, D.C., 1994; pp 1-52.
4. Carson, R.L. *Silent Spring*; Houghton Mifflin: New York, NY, 1994.
5. Colburn, T.; Dumanoski, D.; Myers, J.P. *Our Stolen Future: Are We Threatening Our Fertility, Intelligence & Survival?*; Dutton: New York, NY, 1996.
6. Fourier, J.-B. *Mem. Acad. R. Sci. Inst. France* **1827**, *7*, 569.
7. Arrhenius, S. *Philosophical Magazine and Journal of Science* **1896**, *41*, 237.
8. Revelle, R.; Suess. H. *Tellus* **1957**, *9/1*, 18-21.
9. Schneider, S. *Science* **1989**, *243*, 771-781.
10. Oreske, N.; Schrader-Frechette, K.; Belitz, K. *Science* **1994**, *263*, 641-646.
11. Houghton, R.; Woodwell, G. *Scientific American* **1989**, *260 (4)*, 36-44.
12. Michaels, P. *National Geographic Research and Exploration* **1993**, *9(2)*, 222-233.
13. Schneider, S. *Nature* **1992**, *356*, 11-12.
14. Schneider, S. *Science* **1994**, *263*, 341-347.
15. Kerr, R. *Science* **1993**, *250*, 890-892.
16. Overpeck, J. *Science* **1996**, *271*, 1820-1821.
17. Lents, J.; Kelly, W. *Scientific American* **1993**, *269 (4)*, 29-32.
18. Jenkins, S.; Hay, D. *Energy and Environment* **1996**, *7(1)*, 51-56.
19. Seinfeld, J. *Science* **1989**, *243*, 745-752.
20. Deluchi, M. *Emissions of Greenhouse Gases from the Use of Transportation Fuels and Electricity*; ANL/ESD/TM-22, Argonne National Laboratory: Chicago, IL, 1991.
21. *Office of Air and Radiation, National Air Quality and Emissions Trends Report, 1988*; PB-90-200114, U.S. Environmental Protection Agency: Research Triangle Park: NC, 1990.
22. Tyson, K. *Fuel Cycle Evaluations of Biomass-Ethanol and Reformulated Gasoline*; NREL/TP-231-3996; National Renewable Energy Laboratory: Golden, CO, 1993.

23. *Reinventing Energy: Making the Right Choices*; The American Petroleum Institute: Washington, D.C., 1995.
24. *Annual Energy Outlook: 1993*; Energy Information Administration, U.S. Department of Energy: Washington, D.C., 1993.
25. *America Speaks Out on Energy: A Survey of Public Attitudes on Sustainable Energy Issues*; Sustainable Energy Budget Coalition: Takoma Park, MD, 1996.
26. Farhar, B. *Trends in Public Perceptions and Preferences on Energy and Environmental Policy*; NREL/TP-461-485, National Renewable Energy Laboratory: Golden, CO, 1990.
27. Putsche, V; Sandor, D. In *Handbook on Bioethanol: Production and Utilization*; Wyman, C., Ed.; Taylor and Francis: Washington, DC, 1996; pp 19-31.
28. Gore, A. *Earth in the Balance: Ecology and the Human Spirit*; Houghton-Mifflin Company: New York, NY, 1992.
29. Drucker, P. *The New Realities: In Government and Politics/ In Economics and Business/In Society and World View*; Harper and Brothers: New York, NY, 1989.
30. Goodgame, D. *Time* **1993** February 15, 25-27.
31. Arrow, K.; Cropper, M.; Eads, G.; Hahn, R.; Lave, L.; Noll, R.; Portney, P.; Russell, M.; Schmalensee, R.; Smith, V.; Stavins, R. *Science* **1996**, *272*, 221-222.
32. Morgan, M. *Scientific American* **1993**, *269(1)*, 32-41.
33. Kempton, W.; Craig, P.; Kuennen, C. *Energy and Environment* **1995**, *6(2)*, 87-105.
34. Freundlich, N. *Business Week* **1996** August 19, 75.
35. *Annual Energy Outlook: 1995*; Energy Information Administration, U.S. Department of Energy: Washington, D.C., 1995.
36. Stevens, P. *Energy Policy* **1996,** *24(5)*, 391-402.
37. Bergeron, P. In *Handbook on Bioethanol: Production and Utilization*; Wyman, C., Ed.; Taylor and Francis: Washington, DC, 1996; pp 61-88.
38. *Biofuels: At the Crossroads*; Biofuels Systems Division, Office of Fuels Development, Office of Transportation Technologies, U.S. Department of Energy: Washington, D.C., 1994.
39. Grohmann, K.; Torget, R.; Himmel, M. *Biotechnol. Bioeng. Symp.* **1986**, *15*, 59-80.
40. Torget, R.; Werdene, P.; Himmel, M.; Grohmann, K. *Appl. Biochem. Biotechnol.* **1990**, *24/25*, 115-134.
41. Torget, R.; Walter, P.; Himmel, M.; Grohmann, K. *Appl. Biochem. Biotechnol.* **1991**, *28/29*, 75-86.
42. Torget, R.; Himmel, M.; Grohmann, K. *Appl. Biochem. Biotechnol.* **1992**, *34/35*, 115-124.
43. Hinman, N.; Schell, D.J.; Riley, C.J.; Bergeron, P.W.; Wyman, C.E. *Appl. Biochem. Biotechnol.* **1992**, *34/35*, 639-657.
44. Hsu, T.; Himmel, M.; Schell, D.; Farmer, J.; Berggren, M. *Appl. Biochem, Biotechnol.* **1996**, *57/58*, 3-18.
45. Connor, A.H. *Wood Fiber Sci.* **1984**, *16*, 268-277.
46. Maloney, M.T.; Chapman, T.W.; Baker, A.J. *Biotechnol. Prog.* **1986**, *2*, 192-202.
47. Kim, B.J.; Lee, Y.Y.; Torget, R. *Appl. Biochem, Biotechnol.* **1993**, *39/40*, 119-129.
48. Elander, R.; Torget, R.; Hatzis, C. *Kinetic Modeling and Process Simulations for Optimization of Dilute-Acid Pretreatment*; The AIChE 1995 Summer National Meeting, Boston, MA, July 30-August 2, 1995.

49. Torget, R.; Hatzis, C.; Hayward, T.K.; Hsu, T.; Philippidis, G.P. *Appl. Biochem. Biotechnol.* **1996**, *57/58*, 85-101.
50. Bobleter, O.; Bonn, G.; Concin, R. *Alt. Energy Sources* **1983**, *3*, 323-332.
51. Mok, W.S.; Antal, Jr., M.J. *Ind. Eng. Chem. Res.* **1992**, *31*, 1157-1161.
52. van Walsum, G.P.; Allen, S.G.; Spencer, M.J.; Laser, M.S.; Antal, Jr., M.J.; Lynd, L.R. *Appl. Biochem. Biotechnol.* **1996**, *57/58*, 157-170.
53. Saddler, J. In *Techno-Economic Modeling for Biomass to Ethanol*; International Energy Agency (IEA) Round Table Discussion, Eighteenth Symposium on Biotechnology for Fuels and Chemicals: Gatlinburg, TN, May 8, 1996.
54. Grethlein, H.E. *Process for Pretreating Cellulosic Substrates and for Producing Sugar Therefrom*; U.S. Patent # 4,237,226, Dec. 2, 1980.
55. Harris, J.F. *Two-Stage, Dilute Sulfuric Acid Hydrolysis of Wood: An Investigation of Fundamentals*; Gen. Tech. Rep. FPL-45, U.S. Department of Agriculture, Forest Service, Forest Products Laboratory: Madison, WI, October, 1985.
56. Bergeron, P.; Benham, C.; Werdene, P. *Appl. Biochem. Biotechnol.* **1989**, *20/21*, 119-134.
57. Wright, J.D.; Bergeron, P.W.; Werdene, P.J. *The Progressing Batch Hydrolysis Reactor*; SERI/TP-232-2803, Solar Energy Research Institute, Golden, CO, 1985.
58. Tatsumoto, K.; Baker, J.O.; Tucker, M.P.; Oh, K.K.; Mohagheghi, A.; Grohmann, K.; Himmel, M.E. *Appl. Biochem. Biotechnol.* **1988**, *18*, 159-174.
59. Ooshima, H.; Burns, D.S.; Converse, A.O. *Biotechnol. Bioeng.* **1990**, *36*, 446-452.
60. Czirnich, W.; Patt, R. *Holzforschung* **1976**, *30*, 124-132.
61. Hoffmann, P.; Patt, R. *Holzforschung* **1976**, *30*, 19-27.
62. Immergut, E.H. In *The Chemistry of Wood*; Browning, B.L., Ed.; Robert E. Krieger Publishing: Huntington, NY, 1975; pp 103-190.
63. Henrissat, B. *Cellulose* **1994**, *1(3)*, 169-196.
64. Davies, G.; Henrissat, B. *Structure* **1995**, *3*, 853-859.
65. Henrissat, B.; Bairoch, A. *Biochem. J.* **1993**, *293*, 781-788.
66. Henrissat, B. *Biochem. J.* **1991**, *280*, 309-316.
67. Henrissat, B.; Claeyssens, M.; Tomme, P.; Lemsle, L.; Mornon, J.-P. *Gene* **1989**, *81*, 83-95.
68. Sinnott, M.L. *Chem. Rev.* **1990**, *90*, 1171-1202.
69. Withers, S.G.; Tull, D.; Gebler, J.; Braun, C.; Aebersold, R.; Wang, Q.; Warren, T.; Kilburn, D.; Gilkes, N. In *Trichoderma reesei Cellulases and Other Cellulases*; Suominen, P.; Reinikainen, T., Eds.; Foundation for Biotechnical and Industrial Fermentation Research: Helsinki, Finland, 1993; pp 117-123.
70. Henrissat, B.; Callebaut, I.; Fabrega, S.; Lehn, P.; Mornon, J.-P.; Davies, G. *Proc. Natl. Acad. Sci. USA* **1995**, *92*, 7090-7094.
71. Dominguez, R.; Souchon, H.; Spinelli, S.; Dauter, Z.; Wilson, K.S.; Chauvaux, S.; Beguin, P.; Alzari, P.M. *Nature Struct. Biol.* **1995**, *2*, 569-576.
72. D'Amico, E. *Chemical Week* **1995**, *157*(14), 34.
73. Pokorny, M.; Zupancic, S.; Steiner, T.; Kreiner, W. In *Trichoderma reesei Cellulases: Biochemistry, Genetics, Physiology and Application*; Kubicek, C.P.; Eveleigh, D.E.; Esterbauer, H.; Steiner, W.; Kubicek-Pranz, E.M., Eds.; The Royal Society of Chemistry: London, UK, 1990; pp 168-184.

74. Douglas, L.J. *A Technical and Economic Evaluation of Wood Conversion Processes*; Entropy Associates Report of Contract File 051SZ.23283-8-6091, Efficiency and Alternative Energy Technology Branch, Energy, Mines and Resources Canada: Ottawa, Ontario, 1989.
75. Grethlein, H.E.; Dill, T. *The Cost of Ethanol Production from Lignocellulosic Biomass-A Comparison of Selected Alternative Processes*; Final Report for USDA, Michigan Biotechnology Institute: Lansing, MI, 1993; pp. 9-2.
76. Hayn, M.; Steiner, W.; Klinger, R.; Steinmuller, H.; Sinner, M.; Esterbauer, H. In *Bioconversion of Forest and Agricultural Plant Residues*; Saddler, J.N., Ed.; CAB International, 1993; pp 33-72.
77. Esterbauer, H.; Steiner, W.; Kreiner, W.; Sattler, W.; Hayn, M. *Biosource Technol.* **1992**, *39*, 117-123.
78. Lynd, L.; Cushman, J.; Nichols, R.; Wyman, C.E. *Science* **1991**, *253*, 1318-1323.
79. Baker, J.O.; Adney, W.S.; Thomas, S.R.; Nieves, R.A.; Chou, Y.-C.; Vinzant, T.B.; Tucker, M.P.; Laymon, R.A.; Himmel, M.E. In *Enzymatic Degradation of Insoluble Carbohydrates*; Saddler, J.N.; Penner, M.H., Eds.; Symposium Series No. 618, Amercian Chemical Society: Washington, DC, 1995; pp 113-141.
80. Irwin, D.C.; Spezio, M.; Walker, L.P.; Wilson, D.B. *Biotechnol. Bioeng.* **1993**, *42*, 1002-1013.
81. Nidetzky, B.; Hayn, M.; Macarron, R.; Steiner, W. *Biotechnology Lett.* **1993**, *15*, 71-76.
82. Converse, A.L.; Opekar, J.D. *Biotechnol. Bioeng.* **1993**, *42*, 145-148.
83. Nidetzky, B.; Steiner, W.; Claeyssens, M. In *Enzymatic Degradation of Insoluble Carbohydrates*; Saddler, J.N.; Penner, M.H., Eds.; Symposium Series No. 618, Amercian Chemical Society: Washington, DC, 1995; pp 90-112.
84. Mohagheghi, A.; Grohmann, K.; Himmel, M.E.; Leighton, L.; Updegraff, D. *Int. J. Syst. Bacteriol.* **1986**, *36*, 435-443.
85. Sakon, J.; Adney, W.S.; Himmel, M.E.; Thomas, S.R.; Karplus, P.A. *Biochemistry* **1996**, *35*, 10648-10660.
86. Himmel, M.E.; Adney, W.S.; Grohmann, K.; Tucker, M.P. *Thermostable Purified Endoglucanase from Acidothermus cellulolyticus*; United States patent No. 5,275,944, January 4, 1994.
87. Thomas, S.R.; Laymon, R.A.; Chou, Y.-C.; Tucker, M.P.; Vinzant, T.B.; Adney, W.S.; Baker, J.O.; Nieves, R.A.; Mielenz, J.R.; Himmel, M.E. In *Enzymatic Degradation of Insoluble Polysaccharides*; Saddler, J.N.; Penner, M.H., Eds.; Amercian Chemical Society: Washington, DC, 1995; pp 208-236.
88. Wang, Q.; Tull, D.; Meinke, A.; Gilkes, N.R.; Antony, R.; Warren, J.; Aebersold, R.; Withers, S.G. *J. Biol. Chem.* **1993**, *268*, 14096-14102.
89. Enari, T.-M. In *Microbial Enzymes and Biotechnology*; Fogarty, W.M., Ed.; Applied Science Publishers: London, 1983; pp 183-223.
90. Bairoch, A. SWISS-PROT Protein Sequence Data Bank, (http://expasy.hcuge.ch/cgi-bin/lists?glycosid.txt), Release 34.0, June 1996.
91. Kong, F.; Engler, C.R.; Soltes, E.J. *Appl. Biochem. Biotechnol.* **1992**, *34/35*, 23-35.
92. Linden, J.; Samara, M.; Decker, S.; Johnson, E.; Boyer, M.; Pecs, M.; Adney, W.; Himmel, M.E. *Appl. Biochem. Biotechnol.* **1994**, *45/46*, 383-393.
93. Mandels, M. *Biochem. Soc. Trans.* **1985**, *13*, 414-416.

94. Klyosov, A.A. In *Biochemistry and Genetics of Cellulose Degradation*; Aubert, J.P.; Beguin, P.; Millet, J., Eds.; FEMS Symposium Series 43, Academic Press: London, UK, 1988; pp 87-99.
95. Baker, J.O.; Vinzant, T.B.; Ehrman, C.I.; Adney, W.S.; Himmel, M.E. *Biotechnol. Appl. Biochem.* **1997**, In press.
96. Gold, L. *Methods Enzymol.* **1990**, *185*, 11-14.
97. Bailey, M.J.; Siika-aho, M.; Valkeajarvi, A.; Penttila, M.E. *Biotechnol. Appl. Biochem.* **1993**, *17*, 65-76.
98. Ramanos, M. *Current Opinion Biotechnol.* **1995**, *6*, 527-533.
99. Cregg, J.M.; Vedvick, T.S.; Raschke, W.C. *Bio/Technology* **1993**, *11*, 905-910.
100. Ramanos, M.A.; Scorer, C.A.; Clare, J.J. *Yeast* **1992**, *8*, 423-488.
101. Clare, J.J.; Rayment, F.B.; Ballantine, S.P.; Sreekrishna, K.; Romanos, M.A. *Bio/Technology* **1991**, *9*, 455-460.
102. Barr, K.A.; Hopkins, S.A.; Sreekrishna, K. *Pharm. Eng.* **1992**, *12*, 48-51.
103. van den Hondel, C.A.M.J.J. In *More Gene Manipulations in Fungi*; Bennett, J.W.; Lasure, L.A., Eds.; Academic Press: San Diego, CA, 1991; pp 396-428.
104. Gwynne, D.J.; Buxton, F.P.; Williams, S.A.; Garven, S.; Davies, R.W. *Bio/Technology* **1987**, *5*, 713-719.
105. Turnbull, I.F.; Rand, K.; Willets, N.S.; Hynes, M.J. *Bio/Technology* **1989**, *7*, 169-174.
106. Salovuori, I.; Makarow, M.; Rauvala, H.; Knowles, J.; Kaariainen, L. *Bio/Technology* **1987**, *5*, 152-156.
107. Nakari, T.; Onnela, M.-L.; Ilmen, M.; Penttila, M. In *Proceedings of the 2nd Tricel Meeting, Majvik, Finland*; Reinikainen, T.; Suominen, P., Eds.; Foundation for Biotechnical and Industrial Fermentation Research: Helsinki, Finland, 1993a; pp 239-246.
108. Nakari, T.; Alatalo, E.; Penttila, M. *Gene* **1993b**, *136*, 313-318.
109. Pen, J.; Sijmons, P.C.; van Ooijen, A.J.J.; Hoekema, A. In *Transgenic Plants - Fundamentals and Applications*; Hiatt, A., Ed.; Marcel Dekker, New York, NY, 1993; pp 239-251.
110. Pen, J.; Molendijk, L.; Quax, W.J.; Sijmons, P.C.; van Ooijen, A.J.J.; van den Elsen, P.J.M.; Rietveld, K.; Hoekema, A. *Bio/Technology* **1992**, *10*, 292-296.
111. Herbers, K.I.; Wilke, I.; Sonnenwald, U. *Bio/Technology* **1995**, *13*, 63-66.
112. Rogers, J.C. *Producing Commercially Valuable Polypeptides with Genetically Transformed Endosperm Tissue*, Australian Patent Application #AU-B-40372/89, 1993.
113. Hiatt, A.; Cafferkey, R.; Bowdish, K. *Nature* **1989**, *342*, 76-78.
114. Donson, J.; Kearney, C.M.; Hilf, M.E.; Dawson, W.O. *Proc. Natl. Acad. Sci. USA* **1991**, *88*, 7204-7208.
115. Kumagai, M.H.; Turpen, T.H.; Weinzettl, N.; della-Cioppa, G.; Turpen, A.M.; Donson, J.; Hilf, M.E.; Grantham, G.L.; Dawson, W.O.; Chow, T.P.; Piatak, M.; Grill, L.K. *Proc. Natl. Acad. Sci. USA* **1993**, *90*, 427-430.
116. Kumagai, M.H.; Donson, J.; della-Cioppa, G.; Harvey, D.; Hanley, K.; Grill, L.K. *Proc. Natl. Acad. Sci. USA* **1995**, *92*, 1679-1683.
117. Wenzl, H.F.J. *The Chemical Technology of Wood*; Academic Press: New York, NY, 1970.

118. McMillan, J.D.; Boynton, B.L. *Appl. Biochem. Biotechnol.* **1994**, *45/46*, 569-584.
119. Hinman, N.D.; Wright, J.D.; Hoagland, W.; Wyman, C.E. *Appl. Biochem. Biotechnol.* **1989,** *20/21*, 391-401.
120. Dunning, J.W.; Lathrop, E.C. *Ind. & Eng. Chem.* **1945**, *37*, 24-29.
121. Harris, E.E.; Beglinger, E.; Hajny, G.J.; Sherrard, E.C. *Ind. & Eng. Chem.* **1945**, *37*, 12-23.
122. Sherrard, E.C.; Kressman, F.W. *Ind. & Eng. Chem.* **1945**, *37*, 4-8.
123. Worthy, W. *Chem. & Eng. News* **1981**(Dec. 7), 35-42.
124. Schell, D.J.; Duff, B. In *Handbook on Bioethanol: Production and Utilization*; Wyman, C.E., Ed.; Taylor & Francis: Washington, DC, 1996; pp 381-394.
125. Galbe, M.; Zacchi, G. In *Bioconversion of Forest and Agricultural Plant Residues*; Saddler, J.N., Ed.; CAB International: Wallingford, UK, 1993; pp 291-319.
126. McMillan, J.D. In *Enzymatic Conversion of Biomass for Fuels Production*; Himmel, M.E.; Baker, J.O.; Overend, R.P., Eds.; Symposium Series 566, American Chemical Society: Washington, DC, 1994; pp 292-324.
127. Grohmann, K. In *Bioconversion of Forest and Agricultural Plant Residues*; Saddler, J.N., Ed.; CAB International: Wallingford, UK, 1993; pp 183-209.
128. Singh, A.; Kumar, P.K.R. *Crit. Rev. Biotechnol.* **1991**, *11*, 129-147.
129. Avgerinos, G.C.; Wang, D.I.C. *Ann. Rep. Ferm. Processes* **1980**, *4*, 165-191.
130. Lynd, L.R. *Adv. Biochem. Eng/Biotechnol.* **1989**, *38*, 1-52.
131. Mohagheghi, A.; Tucker, M.; Grohmann, K.; Wyman, C. *Appl. Biochem. Biotech.* **1992**, *33*, 67-81.
132. Picataggio, S.K.; Zhang, M. In *Handbook on Bioethanol: Production and Utilization*; Wyman, C.E., Ed.; Taylor & Francis: Washington, DC, 1996; pp163-178.
133. Ingram, L.O.; Conway, T.; Clark, D.P.; Sewell, G.W.; Preston, J.F. *Appl. Environ. Microbiol.* **1987**, *53*, 2420-2425.
134. Ohta, K.; Beall, D.S.; Mejia, J.P.; Shanugham, K.T.; Ingram, L.O. *Appl. Environ. Microbiol.* **1991**, *57*, 893-900.
135. Ohta, K.; Beall, D.S.; Mejia, J.P.; Shanugham, K.T.; Ingram, L.O. *Appl. Environ. Microbiol.* **1991**, *57*, 2810-2815.
136. Beall, D.; Ohta, K.; Ingram, L.O. *Biotech. Bioeng.* **1991**, *38*, 296-303.
137. Beall, D.; Ohta, K.; Ingram, L.O. *Biotech. Lett.* **1992**, *14*, 857-862.
138. Lawford, G. H.; Rousseau, J. *Biotech. Lett.* **1992**, *14*, 421-426.
139. Hahn-Hägerdal, B.; Jeppsson, H.; Olsson, L.; Mohagheghi, A. *Appl. Microbiol. Biotechnol.* **1994**, *41*, 62-72.
140. Padukone, N.; Evans, K.W.; McMillan, J.D.; Wyman, C.E. *Appl. Microbiol. Biotechnol.* **1995,** *43*, 850-855.
141. Asghari, A.; Bothast, R. J.; Doran, J. B.; Ingram, L.O. *J. Ind. Microbiol.* **1996**, *16*, 42-47.
142. Zhang, M.; Eddy, C.; Deanda, K.; Finkelstein, M.; Picataggio, S. *Science* **1995,** *267*, 240-243.
143. Deanda, K.A.; Eddy, C.; Zhang, M.; Picataggio, S. *Metabolic Engineering of* Zymomonas mobilis *for Ethanol Production from Arabinose Containing Feedstocks;* Poster P3, Presented at the Society for Industrial Microbiology Annual Meeting, San Jose, CA, Aug. 6-11, 1995.

144. Ho, N.W.Y.; Brainard, A.; Chen, Z. *Recombinant Xylose-Fermenting Saccharomyces Capable of Effective Fermentation of Sugars from Lignocellulosic Hydrolyzates*; Poster 73, Presented at the Seventeenth Symposium on Biotechnology for Fuels and Chemicals, Vail, CO, May 7-11, 1995.
145. Ho, N.W.Y.; Chen, Z.; Brainard, A. *The Key to Metabolic Engineering of Saccharomyces Yeasts for Effective Fermentation of Xylose*; Paper 25b, Presented at the AIChE 1995 Summer National Meeting, Boston, MA, Jul.30-Aug.2, 1995.
146. Toon, S.; Philippidis, G.P.; Ho, N.W.Y.; Chen; Z.D.; Brainard; A.; Lumpkin, R.E.; Riley, C.J. *Enhanced Cofermentation of Glucose and Xylose by Recombinant Saccharomyces Yeast Strains in Batch and Continuous Operating Modes*; Paper 13, Presented at the Eighteenth Symposium on Biotechnology for Fuels and Chemicals, Gatlinburg, TN, May 5-9, 1996.
147. Jeffries, T.W.; Dahn, K.; Davis, B.P.; Cho, J.Y.; Lu, P.; Shi, N.Q.; Sreenath, H.K. *Development of Improved Xylose Fermenting Yeasts*; Presented at the Second Biomass Congress of the North American Continent, Portland, Oregon, Aug. 21-24, 1995.
148. Jeffries, T.W.; Dahn, K.; Sreenath, H.K.; Cho, J.Y.; Davis, B.P. *Metabolic Engineering of Pichia stipitis for Improved Fermentation of Pentose Sugars*; Paper S112, Presented at the Society for Industrial Microbiology Annual Meeting, San Jose, CA, Aug. 6-11, 1995.
149. Klapatch, T.R.; Guerino, M.L.; Lynd, L.R. *J. Ind. Microbiol.* **1996,** *16*, 342-347.
150. McMillan, J.D.; Mohagheghi, A.; Newman, M.M.; Picataggio, S. *Process Development for Ethanol Production from Biomass Sugars Using Metabolically Engineered Zymomonas mobilis*; Paper 216c, Presented at the Annual Meeting of the American Institute of Chemical Engineers, Miami, FL November 12-17, 1995.
151. Zhang, M.; Franden, M.A.; Newman, M.N.; McMillan, J.D.; Finkelstein, M.; Picataggio, S. *Appl. Biochem. Biotechnol.* **1995**, *51/52*, 527-536.
152. *Biofuels: Project Summaries*; U.S. Department of Energy, Publication DOE/CH10093-297, National Renewable Energy Laboratory: Golden, CO, 1993. (Extensive information on research activities is also available from NREL's Alternative Fuels Data Center homepage at URL http://www.afdc.nrel.gov).
153. Wright, L.L.; Hohenstein, W.G. *Biomass Bioenergy* **1994**, *6*, 161-241.
154. *Technical Program Summary and Books of Abstracts*; 1995 International Chemical Congress of Pacific Basin Societies (Pacifichem '95), Honolulu, Hawaii, December 17-22, 1995.
155. Mitchell, C.P. *Biomass Bioenergy* **1995**, *9*, 1-440.
156. *International Energy Agency Bioenergy Annual Report 1995*; Publication No. B1996:1, Swedish National Board for Industrial and Technical Development: Sweden, 1996; pp 5-6, 32-37.
157. Nguyen, Q.A.; Dickow, J.H.; Duff, B.W.; Farmer, J.D.; Glassner, D.A.; Ibsen, K.N.; Ruth, M.F.; Schell, D.J.; Thompson, I.B.; Tucker, M.P. *Bioresource Technology*, 1996, In press.
158. Philippidis, G.P. In *Enzymatic Conversion of Biomass for Fuels Production*; Himmel, M.E.; Baker, J.O.; Overend, R.P., Eds.; Symposium Series 566, American Chemical Society: Washington, DC, 1994; pp 188-217.
159. Ingram, L.O.; Conway,T. *Appl. Environ. Microbiol.* **1988**, *55*, 1943-1948.

160. Ingram, L.O.; Alterthum, F.; Ohta, K.; Beall, D.S. *Develop. Industrial Microbiol.* **1990**, *31*, 21-30.
161. Doran, J.B.; Ingram, L.O. *Biotechnol. Prog.* **1993,** *9*, 533-538.
162. Baskaran, S.; Ahn, H.-J.; Lynd, L.R. *Biotechnol. Prog.* **1995**, *11*, 276-281.\
163. Rehn, C.; Stenberg, K.; Palmquist, E.; Larsson, M.; Zacchi, G. *Design and Operation of a Bench-Scale Unit for Enzymatic Production of Ethanol from Lignocellulosic Materials*; Paper 215e, Presented at the Annual Meeting of the American Institute of Chemical Engineers, Miami, FL, November 12-17, 1995.
164. Emert, G.H. *Develop and Demonstrate the Cellulose Process*; Final Technical Report for DOE/SERI subcontract No. XR-0-9248-1; Vols. I and III, Biomass Research Center, University of Arkansas: Fayetteville, AR, 1982.
165. Lewis, S.M. In *Industrial Enzymology*; Godfrey, T.; West, S. Eds.; Stockton Press: New York, NY, 1996; pp 13-48.
166. Thompson, A.; Anno, K.; Wolfrom, M.L.; Inatome, M. *J. Amer. Chem. Soc.* **1954**, *76*, 1309.
167. Christakopoulos, P.; Bhat, M.K.; Kekos, D.; Macris, B.J. *Int. J. Biol. Macromol.* **1994**, *16*, 331-334.
168. Christakopoulos, P.; Kekos, D.; Macris, B.J.; Goodenough, P.W.; Bhat, M.K. *Biotechnol. Lett.* **1994**, *16*, 587-592.
169. Duan, K.-J.; Sheu, D.-C.; Lin, C.-T. *Ann. New York Acad. Sci.* **1995**, *750*, 325-328.
170. Bekker, E.G.; Gusakov, A.V.; Sinitsyn, A.P. *Prikladnaya Biokhimiya i Mikrobiologiya* **1991**, *27*, 482-485.

Chapter 2

Enzymes in Lignocellulosic Biomass Conversion

Badal C. Saha and Rodney J. Bothast

Fermentation Biochemistry Research Unit, National Center for Agricultural Utilization Research, Agricultural Research Service, U.S. Department of Agriculture, 1815 North University Street, Peoria, IL 61604

Advances in enzymes and lignocellulosic biomass processing are necessary to lower the cost of fuels and chemicals production from biomass. Recent developments in lignocellulosic biomass conversion enzymology and process technology are reviewed. Current problems of these multi-enzymes based complex processes, economic assessment, regulatory issues, strategies for development of improved enzymes and processes, and directions of future research are discussed. Results of our endeavor to develop novel enzymes for biomass conversion are presented.

Currently, more than one billion gallons of ethanol are produced annually in the United States, with approximately 95% derived from fermentation of corn starch (*1*). Enzymes play an important part in the conversion of corn starch to glucose that is then fermented to ethanol by yeast. In fact, application of amylases in starch conversion is a great example of the successful use of enzymes in biotechnology. With increased attention to clean air and oxygenates for fuels, opportunities exist for rapid expansion of the fuel ethanol industry. Various lignocellulosic biomass such as agricultural residues, wood, municipal solid wastes and wastes from pulp and paper industry have potential to serve as low cost and abundant feedstocks for production of fuel ethanol or chemicals. Right now, the use of lignocellulosic biomass to produce fuel ethanol represents significant technical and economic challenges, and its success depends largely on the development of highly efficient and cost-effective biocatalysts for conversion of pretreated biomass to fermentable sugars. In this article, we describe briefly current knowledge on the application of enzymes in various lignocellulosic biomass conversion.

Lignocellulosic Biomass

Lignocellulosic biomass includes various agricultural residues (straws, hulls, stems, stalks), deciduous and coniferous woods, municipal solid wastes (MSW, paper, cardboard, yard trash, wood products), waste from pulp and paper industry and herbaceous energy crops. The compositions of these materials vary. The major component is cellulose (35-50%), followed by hemicellulose (20-35%) and lignin (10-25%). Proteins, oils and ash make up the remaining fraction of lignocellulosic biomass (*2*). The structure of these materials is very complex and native biomass is resistant to an enzymatic hydrolysis. In the current model of the structure of lignocellulose, cellulose fibers are embedded in a lignin-polysaccharide matrix. Xylan may play a significant role in the structural integrity of cell walls by both covalent and non-covalent associations (*3*). The pretreatment of lignocellulosic biomass is crucial before enzymatic hydrolysis. Various pre-treatment options are available now to fractionate, solubilize, hydrolyze and separate cellulose, hemicellulose and lignin components (*4-7*). These include steam explosion, dilute acid treatment, concentrated acid treatment, alkaline treatment, treatment with SO_2, treatment with hydrogen peroxide, ammonia fiber explosion, and organic solvent treatments. In each option, the biomass is treated to reduce its size and open its structure. Pretreatment usually hydrolyzes hemicellulose to its sugars (xylose, L- arabinose, and other sugars) that are water soluble (*4*). The residue contains cellulose and lignin. The lignin can be extracted with solvents such as ethanol, butanol or formic acid. Alternatively, hydrolysis of cellulose with lignin present makes water-soluble sugars and the residues are lignin plus unreacted materials. The use of SO_2 as a catalyst during steam pretreatment resulted in the enzymatic accessibility of cellulose and enhanced recovery of the hemicellulose derived sugars (*8*). Steam pretreatment at 200-210°C with the addition of 1% SO_2 (w/w) was superior to other forms of pretreatment of willow (*9*). A glucose yield of 95%, based on the glycan available in the raw material, was achieved. A summary of various pretreatment options is given in Table I. Recently, supercritical carbon dioxide explosion was found to be very effective for pretreatment of cellulosic materials before enzymatic hydrolysis (*10*). The sequential steps for production of fuels and chemicals from lignocellulosic biomass involve feedstock preparation, pretreatment, fractionation, enzyme production, hydrolysis, fermentation, product recovery, and waste treatment. The pretreatment of lignocellulosic biomass is an expensive procedure with respect to cost and energy.

Cellulose conversion

Cellulose is a linear polymer of 8,000-12,000 D-glucose units linked by 1,4-ß-D-glucosidic bonds. The enzyme system for the conversion of cellulose to glucose comprises endo-1, 4-ß-glucanase (EC 3.2.1.4), exo-1, 4-ß-glucanase (EC 3.2.1.91) and ß-glucosidase (EC 3.2.1.21). Cellulolytic enzymes with ß-glucosidase act sequentially and cooperatively to degrade crystalline cellulose to glucose. Endoglucanase acts in a random fashion on the regions of low crystallinity of the cellulosic fiber whereas exoglucanase removes cellobiose (ß-1, 4 glucose dimer) units from the non-reducing ends of cellulose chains. Synergism between these two enzymes is attributed to the *endo-exo* form of cooperativity and has been studied extensively between cellulases in the degradation of cellulose in *Trichoderma reesei* (*11*). ß-Glucosidase hydrolyzes cellobiose and in some cases cellooligosaccharides to glucose. The enzyme is generally responsible for the regulation of the whole cellulolytic process and is a rate limiting factor during enzymatic hydrolysis of cellulose as both endoglucanase and

Table I. Methods for pretreatment of lignocellulosic biomass

Method	Example
Thermo-mechanical	Grinding, milling, shearing, extruder.
Autohydrolysis	Steam pressure, steam explosion, supercritical carbon dioxide explosion.
Acid Treatment	Dilute acid (H_2SO_4, HCl), concentrated acid (H_2SO_4, HCl), acetic acid.
Alkali treatment	Sodium hydroxide, ammonia, alkaline hydrogen peroxide.
Organic solvents treatment	Methanol, ethanol, butanol, phenol.

cellobiohydrolase activities are often inhibited by cellobiose (*12-14*). Thus, β-glucosidase not only produces glucose from cellobiose but also reduces cellobiose inhibition, allowing the cellulolytic enzymes to function more efficiently. However, like ß-glucanases, most ß-glucosidases are subject to end-product (glucose) inhibition. The kinetics of the enzymatic hydrolysis of cellulose including adsorption, inactivation and inhibition of enzymes have been studied extensively (*15*). For a complete hydrolysis of cellulose to glucose, the enzyme system must contain the three enzymes in right proportions.

Product inhibition, thermal inactivation, substrate inhibition, low product yield and high cost of cellulase are some barriers to commercial development of the enzymatic hydrolysis of cellulose. Many microorganisms are cellulolytic. However, only two microorganisms (*Trichoderma* and *Aspergillus*) have been studied extensively for cellulase. There is an increasing demand for the development of thermostable, environmentally compatible, product and substrate tolerant cellulase with increased specificity and activity for application in the conversion of cellulose to glucose in the fuel ethanol industry. Thermostable cellulases offer certain advantages such as higher reaction rate, increased product formation, less microbial contamination, longer shelf-life, easier purification and better yield.

In our work, forty-eight yeast strains belonging to the genera *Candida*, *Debaryomyces*, *Kluyveromyces* and *Pichia* (obtained from the ARS Culture Collection, Peoria, IL) were screened for production of extracellular glucose tolerant and thermophilic ß-glucosidase activity using p-nitrophenyl-ß-D-glucoside as substrate (*16*). Enzymes from 15 yeast strains showed very high glucose tolerance (< 50% inhibition at 30%, w/v glucose). The optimal temperatures and pH for these ß-glucosidase activities varied from 30 to 65°C and pH 4.5 to 6.5. The ß-glucosidase from *D. yamadae* Y-11714 showed highest optimal temperature at 65°C followed by *C. chilensis* Y-17141(60°C) and *K. marxianus* Y-1195 (60°C). The optimal pH of these three enzyme preparations were 6.5, 6.0 and 6.5, respectively. The temperature and pH profiles of β-glucosidases from *C. chilensis* Y-17141, *D. yamadae* Y-11714 and *K. marxianus* Y-1195 are shown in Figure 1. The β-glucosidases from all these yeast strains hydrolyzed cellobiose. Novel glucose tolerance and thermoactivity found in the enzyme preparations from *D. yamadae*, *K. marxianus* and *C. chilensis* are desired

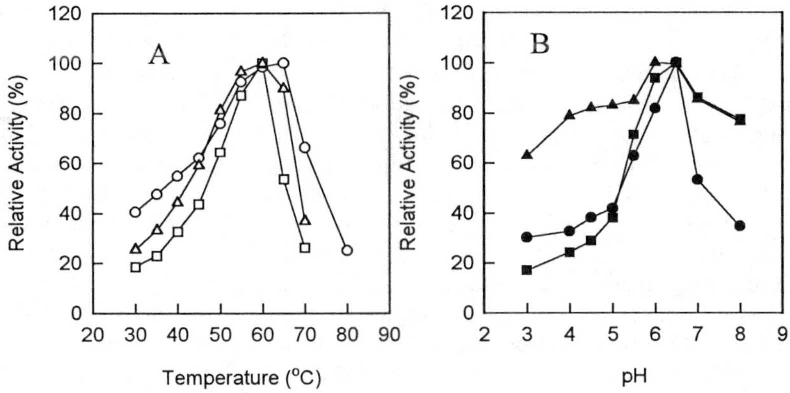

Figure 1. Effect of temperature (A) and pH (B) on the activity of extracellular β-glucosidases from *Candida chilensis* Y-17141, *Debaryomyces yamadae* Y-11714 and *Kluyveromyces marxianus* Y-1195. Each value is average of duplicate experiments for each enzyme preparation. Buffers (0.1 M) used: acetate (pH 3.0-5.5) and phosphate (6.0-8.0). Symbols: (□,■), *C. chilensis* Y-17141; (○,●), *D. yamadae* Y-11714; (△,▲), *K. marxianus* Y-1195.

Table II. Biochemical characteristics of thermostable ß-glucosidase from *Aureobasidium pullulans* NRRL Y-12974 (*17*)

Specific activity	315 U/mg protein
Molecular weight	340,000 (2 subunits), glycoprotein
Optimum temperature	75-80°C
Optimum pH	4.5
Specificity	Hydrolyzes cellobiose and cello-oligosaccharide
Half-life (crude enzyme)	24 h at 80°C; 72 h at 75°C
K_m value (mM)	
pNPßG (at pH 4.5, 75°C)	1.17
Cellobiose (at pH 4.5, 75°C)	1.00
Metal ion requirement	None
Substrate inhibition	
pNPßG (20 mM)	No inhibition
Cellobiose (6 %, w/v)	No inhibition
Inhibition by glucose	Competitive (K_i=5.65 mM)

attributes of a β-glucosidase suitable for industrial application for enzymatic hydrolysis of cellulose to glucose. We have purified and characterized a highly thermophilic β-glucosidase from a color variant strain of *Aureobasidium pullulans* (*17*). Some properties of this enzyme are summarized in Table II.

The cellulose hydrolysis step is a significant component of the total production cost of ethanol from wood (*18*). Achieving a high glucose yield is necessary (>85% theoretical) at high substrate loading (>10% w/v) over short residence times (< 4 days). It was shown that simultaneous saccharification (hydrolysis) of cellulose to glucose and fermentation of glucose to ethanol (SSF) improve the kinetics and economics of biomass conversion by reducing accumulation of hydrolysis products that are inhibitory to cellulase and β-glucosidase, reducing the contamination risk because of the presence of ethanol, and reducing the capital equipment requirements (*19*). An important drawback of SSF is that the reaction has to operate at a compromised temperature of around 30°C instead of enzyme optimum temperature of 45-50°C. Enzyme recycling, by ultrafiltration of the hydrolyzate, can reduce the net enzyme requirement and thus lower costs (*20*). Hinman et al. (*21*) reported that a preliminary estimate of the cost of ethanol production for SSF technology based on wood-to-ethanol process is $ 1.22/gal of which the wood cost is $ 0. 459/gal. Wright et al. (*22*) evaluated a separate fungal enzyme hydrolysis and fermentation process for converting lignocellulose to ethanol. The cellulase enzyme was produced by the fungal mutant *Trichoderma* Rut C-30 (the first mutant with greatly increased β-glucosidase activity) in a fed batch production system that is the single most expensive operation in the process. The conversion of lignocellulosic biomass to fermentable sugars requires the addition of complex enzyme mixtures tailored for the process and parallel reuse and recycle the enzymes until the cost of enzymes comes down. Enzyme recycling can increase the rates and yields of hydrolysis, reduce the net enzyme requirements and thus lower costs (*23*). The first step in cellulose hydrolysis is considered as the adsorption of cellulase onto cellulosic substrate. As the cellulose hydrolysis proceeds, the adsorbed enzymes (endo- and exo-glucanase components) are gradually released in the reaction mixture. The β-glucosidase does not adsorb onto the substrate. These enzymes can be recovered and reused by contacting the hydrolyzate with the fresh substrate. However, the amount of enzyme recovered is limited because some enzymes remain attached to the residual substrate and some enzymes are thermally inactivated during hydrolysis. It has been shown that several substrates containing a high proportion of lignin result in the poor recovery of cellulase (*24*).

Gusakov et al. (*25*) found that cellolignin was completely converted to glucose by cellulase from *T. viride* and *A. foetidus*. Cellolignin was an industrial residue obtained during the production of furfural from wood and corn cobs when pretreated by dilute H_2SO_4 at elevated temperature. The concentration of glucose in the hydrolyzate reached 4-5.5%, cellulose conversion being not less than 80%. Kinetic analysis of cellolignin hydrolysis, using a mathematical model of the process, has shown that, with product inhibition, nonspecific adsorption of cellulase onto lignin and substrate induced inactivation seem to affect negatively the hydrolysis efficiency. Borchert and Buchholz (*26*) investigated the enzymatic hydrolysis of different cellulosic materials (straw, potato pulp, sugar beet pulp) with respect to reactor design. The kinetics was studied including enzyme adsorption, inhibition, and inactivation. The results suggest the use of reactors with plug flow characteristics to achieve high substrate and product concentrations and to avoid back-mixing to limit the effect of product inhibition. For efficient use of cellulases, a reactor with semipermeable hollow fiber or an ultrafilter membrane was used and this allowed cellulases to escape end-product inhibition (*27-30*). A totally integrated biotechnology of rice straw conversion into ethanol was reported (*31*). It dealt with (i) ethanol refining of rice straw to segregate cellulose from pentose sugars and lignin, (ii) preparation of highly active mixed cellulase enzymes, (iii) a novel reactor system allowing rapid product formation involving enzymatic hydrolysis of cellulose to sugars followed by microbial conversion of the later into ethanol and its simultaneous flash separation employing a programmed recompression of ethanol vapors and condensation, and (iv) concentration of ethanol via alternative approaches.

In direct microbial conversion of lignocellulosic biomass into ethanol that could simplify the ethanol production process from these materials and reduce ethanol production costs, *Clostridium thermocellum,* a thermoanaerobe was used for enzyme production, hydrolysis and glucose fermentation (*32*). Cofermentation with *C. thermosaccharolyticum* simultaneously converted the hemicellulosic sugars to ethanol. However, the formations of by-products such as acetic acid and low ethanol tolerance are some drawbacks of the system. Several recent reviews have dealt with the molecular biology of cellulose degradation, cellulolytic enzyme systems, and the structure and function of various domains found in the enzymes involved (*33-36*).

Hemicellulose conversion

Hemicelluloses are heterogeneous polymers of pentoses (xylose and L-arabinose), hexoses (mannose) and sugar acids. Xylans, major hemicelluloses of many plant materials, are heteropolysaccharides with a homopolymeric backbone chain of 1,4-linked β-D-xylopyranose units. Besides xylose, xylans may contain L-arabinose, D-glucuronic acid or its 4-o-methyl ether, and acetic, p-coumaric, and ferulic acids.

The total hydrolysis of xylan requires endo β-1,4-xylanase (EC 3.2.1.8), β-xylosidase (EC 3.2.1.37), and several accessory enzyme activities such as α-L-arabinosidase (EC 3.2.1.55), α-glucuronidase (EC 3.2.1.1), acetyl xylan esterase (EC 3.2.1.6), feruloyl esterase and p-coumaroyl esterase which are necessary for hydrolyzing various substituted xylans. The endo-xylanase randomly attacks the main chains of xylans and β-xylosidase hydrolyzes xylooligosaccharides to xylose. The α-L-arabinosidase and α-glucuronidase remove the arabinose and 4-O-methyl glucuronic acid substituents, respectively from the xylan backbone. The esterases hydrolyze the ester linkages between xylose units of the xylan and acetic acid (acetyl xylan esterase) or between arabinose side chain residues and phenolic acids such as ferulic acid (feruloyl esterase) and p-coumaric acid (p-coumaroyl esterase). Synergistic action of depolymerizing and side-group cleaving enzymes has been proved using acetylated xylan as substrate (*37*). Bachmann and McCarthy (*38*) reported significant synergistic interaction between endo-xylanase, β-xylosidase, α-L-arabinofuranosidase, and acetyl xylan esterase enzymes of the thermophilic actinomycete *Thermomonospora fusca.* Many xylanases do not cleave glycosidic bonds between xylose units which are substituted. The side chains must be cleaved before the xylan backbone can be completely hydrolyzed (*39*). On the other hand, several accessory enzymes only remove side-chains from xylooligosaccharides. These enzymes require xylanases to hydrolyze hemicellulose partially before side-chains can be cleaved (*40*). Although the structure of xylan is more complex than cellulose and requires several different enzymes with different specificities for a complete hydrolysis, the polysaccharide does not form tightly packed crystalline structures and is thus more accessible to enzymatic hydrolysis (*41*). The yeast-like fungus *Aureobasidium* is a promising source of xylanase (MW 20 kDa) with an exceptionally high specific activity (2100 U/mg protein) (*42*). Xylanase represented nearly half the total extracellular protein, with a yield of up to 0.3 g of xylanase per liter (*43*). A few recent reviews have dealt with the multiplicity, structure and function of microbial xylanases, and molecular biology of xylan degradation (*3, 44, 45*).

The utilization of hemicellulosic sugars is essential for efficient conversion of lignocellulose to ethanol. The commercial exploitation of the pentose fermenting yeasts for ethanol production from xylose is restricted mainly by their low ethanol tolerance, slow rates of fermentation, difficulty to control the rate of oxygen supply at the optimal level plus sensitivity to microbial inhibitors, particularly those liberated during pretreatment and hydrolysis of lignocellulosic substrates (*46, 47*). Xylose can also be converted to xylulose using the enzyme xylose isomerase and traditional yeasts can ferment xylulose to ethanol (*48, 49*). Xylose can be easily converted into xylitol, a potentially attractive sweetening agent by a variety of microorganisms (yeasts, fungi and bacteria) (*50*).

Lignin conversion

Lignin is a long chain heterogeneous polymer composed largely of phenylpropane units most commonly linked by ether bonds. It effectively protects the woody plants against microbial attack and only a few organisms including rot-fungi and some bacteria can degrade it (*51*). The conversion of cellulose and hemicellulose to fuels and chemicals will generate lignin as a by-product that can be burned to provide heat and electricity, converted to low-molecular weight chemicals, and used in the manufacture of various polymeric materials. As lignin makes up 15-25% in some lignocellulosic biomass, the selling price of lignin has a very large impact on ethanol price (*18*).

In recent years, removal of lignin from lignin-carbohydrate complex (LCC) has received much attention because of potential application in pulp and paper industry. The lignin barrier can be disrupted by a variety of pretreatments rendering the cellulose and hemicellulose more susceptible to enzymatic attack (*52*). There are many papers about microbial breakdowns of lignin, the enzymes and the pathways (*53-56*). The degradation of lignin by the basidiomycete *Phanerochaete chrysosporium* is catalyzed by extracellular peroxidases (lignin peroxidase, Lip and manganese peroxidase, MnP) in a H_2O_2-dependent process (*57, 58*). However, due to extreme complexity of the problem, a vast amount of research needs to be done to understand all the factors involved in lignin biodegradation process (*59*).

Screening for enzymes with targeted properties

The cost of the enzymes for enzymatic hydrolysis of cellulosic biomass is clearly the critical parameter from an economic point of view. Most of the industrial enzymes are produced by organisms isolated from natural sources by a labor intensive, unpredictable classical screening, strain selection, medium optimization for over production, fermentation and recovery process development. Screening of naturally occurring microorganisms still may be the best way to obtain new strains and/or enzymes for commercial applications (*60*). Fundamental tasks and strategies for commercial development of an enzyme from natural sources are shown in Figure 2. Recombinant DNA technology and protein engineering have also proven as effective means of production of industrial enzymes (*61*). The marketing of all enzymes is subject to a

Screening for microorganisms

⇓

Culture selection

⇓

Fermentation studies (preliminary)

⇓

Isolation, purification and final characterization

⇓

Evaluation

⇓

Toxicology

⇓

Regulatory agency

⇓

Improvement of fermentation and recovery process development

⇓

Product formulation

⇓

Marketing

Figure 2. Strategies for commercial development of an enzyme

variety of Federal laws and regulations. The generally recognized as safe (GRAS) status of an industrial enzyme depends on the source of its origin. Federal laws, regulations and policies that have an impact on industrial enzymes have been reviewed by Fordham and Block (*62*).

Concluding remarks

The industrial enzyme market approaches approximately one billion US dollars annually. Several enzymes have already become commodity chemicals for many industrial application purposes such as in the production of various corn syrups and sweeteners and fuel ethanol from starch. Right now, the market for the enzymes involved in various lignocellulosic biomass conversion is limited and depends entirely on their use in the conversion of various lignocellulosic feedstocks to fermentable sugars for the subsequent production of fuel alcohols and value-added chemicals. Currently, cellulolytic enzymes are expensive and their hydrolysis rates are very slow. The development of an environmentally compatible highly efficient enzyme system free from product and substrate inhibitions for conversion of various pretreated agricultural residues to glucose is very important for use of these materials for production of fuel alcohol. The market for these enzymes will expand rapidly if certain properties of them can be improved and if these enzymes are made available for biomass conversion at a competitive price like starch degrading enzymes. On the other hand, the development of a very efficient substrate pretreatment that increases the susceptibility of crystalline cellulose to enzymatic hydrolysis significantly will lower the cost of producing ethanol from lignocellulosic biomass.

Literature cited

1. Bothast, R. J. *Proc. Corn Utilization Conf.*, **1994**, National Corn Growers Assoc., St. Louis, MO.
2. Wyman, C. E. *Bioresource Technology* **1994**, 50, 3-16.
3. Thomson, J. A. *FEMS Microbiology Reviews* **1993**, 104, 65-82.
4. Bungay, H. *Enzyme Microb. Technol.* **1992**, 14, 501-507.
5. Wyman, C. E. *Appl. Biochem. Biotechnol.* **1994**, 45/46, 897-915.
6. Dale, B. E.; Moreira, M. J. *Biotechnol. Bioeng. Symp.* **1982**, 12, 31-43.
7. Weil, J.; Westgate, P.; Kohlmann, K.; Ladisch, M. R. *Enzyme Microb. Technol.* **1994**, 16, 1002-1004.
8. Brownell, H. H.; Saddler, J. N. *Biotechnol. Bioeng. Symp.* **1984**, 14, 55-68.
9. Eklund, R.; Zacchi, G. *Enzyme Microb. Technol.* **1995**, 17, 255-259.
10. Zheng, Y.; Lin, H.-M.; Wen, J.; Cao, N.; Yu, X.; Tsao, G. T. *Biotech. Letts.* **1995**, 17, 845-850.
11. Henrissat, B.; Driguez, H.; Viet, C.; Schulein, M. *Bio/Technology* **1985**, 3, 722-726.
12. Woodward, J.; Wiseman. A. *Enzyme Microb. Technol.* **1982**, 2, 73-79.
13. Coughlan, M. P. *Biotechnol. Genet. Eng. Rev.* **1985**, 3, 39-109.
14. Kadam, S. K.; Demain, A. L. *Biochem. Biophys. Res. Commun.* **1989**, 161, 706-711.
15. Ladisch, M. R.; Lin, K. W.; Voloch, M.; Tsao, G. T. *Enzyme Microb. Technol.* **1983**, 5, 82-102.
16. Saha, B. C.; Bothast, R. J. *Biotech. Letts.* **1996**, 18, 155-158.
17. Saha, B. C.; Freer, S. N.; Bothast, R. J. *Appl. Environ. Microbiol.* **1994**, 60, 374-3780.
18. Nguyen, Q. A.: Saddler, J. N. *Bioresource Technology* **1991**, 35, 275-282.
19. Philippidis, G. P.; Smith, T. K.; Wyman, C. E. *Biotechnol. Bioeng.* **1993**, 41, 846-853.
20. Tan, L. U. L.; Yu, E. K. C.; Mayers, P.; Saddler, J. N. *Appl. Microbiol. Biotechnol.* **1987**, 26, 21-27.
21. Hinman, N. D.; Schell, D. J.; Rieley, C. J.; Bergeron, P. W.; Walter, P. J. *Appl. Biochem. Biotechnol.* **1992**, 34/35, 639-649.
22. Wright, J. D.; Power, A. J.; Douglas, L. J. *Biotechnol. Bioeng. Symp.* **1986**, 17, 285-302.
23. Lee, D.; Yu, A. H. C.; Saddler, J. N. *Biotechnol. Bioeng.* **1995**, 45: 328-336.
24. Tanaka, M.; Fukui, M.; Matsuno, R. *Biotechnol. Bioeng.* **1988**, 32, 897-902.
25. Gusakov, A. V.; Sinitsyn, A. P.; Manenkova, J. A.; Protas, O. V. *Appl. Biochem. Biotech.* **1992**, 34/35, 625-637.

26. Borchert, A.; Buchholz, K. *Process Biochem.* **1987**, 22, 173-180.
27. Henley, R. G.; Young, R. Y. K.; Greenfield, P. F. *Enzyme Microb. Technol.* **1980**, 2, 206-208.
28. Klei, H. E.; Sundstrom, D. W.; Coughlin, K. W.; Ziolkowski, K. *Biotechnol. Bioeng. Symp.* **1981**, 11, 593-601.
29. Ohlson, I.; Trayardh, G.; Hahn-Hagerdal, B. *Biotechnol. Bioeng.* **1984**, 26, 647-653.
30. Kinoshita, S.; Chua, J. W.; Kato, N.; Yoshida, T.; Taguchi, T. *Enzyme Microb. Technol.* **1986**, 8, 691-695.
31. Ghose, T. K.; Ghosh, P. In: *Harnessing Biotechnology for the 21st Century*, Eds. Ladisch, M. R. and Bose, A., American Chemical Society, Washington, D. C., **1992**, pp. 505-506.
32. Lynd, L. In: *Adv. Biochem. Eng. Biotechnol.* Fiechter, A., Ed. **1989**, 38, pp. 1-52.
33. Beguin, P. *Annu. Rev. Microbiol.* **1990**, 44, 219-248.
34. Knowles, J.; Lehtovaara, P.; Teeri, T. *Trends Biotechnol.* **1987**, 5, 255-261.
35. Robson, L. M.; Chambliss, G. H. *Enzyme Microb. Technol.* **1989**, 11, 626-644.
36. Gilkes, N. R.; Henrisatt, B.; Kilburn, D. G.; Miller, R. C. Jr.; Warren, R. A. J. *Microbiol. Rev.* **1988**, 55, 303-315.
37. Poutanen, K.; Puls, J. In *Biogenesis and biodegradation of Plant Cell Wall Polymers*, Lewis, G.; Paice, M. Eds. American Chemical Society, Washington, D.C., **1989**, pp. 630-640.
38. Bachmann, S. L.; McCarthy, A. J. *Appl. Environ. Microbiol.* **1991**, 57, 2121-2130.
39. Lee, S. F.; Forsburg, C. W. *Can. J. Microbiol.* **1987**, 33, 1011-1016.
40. Poutanen, K.; Tenkanen, M.; Korte, H.; Puls, J. In: *Enzymes in Biomass Conversion*, Leatham, G. F. and Himmel, M. E., Eds. American Chemical Society, Washington, D. C., **1991**, pp. 426-436.
41. Gilbert, H. J.: Hazlewood, G. P. *J. Gen. Microbiol.* **1993**, 139, 187-194.
42. Leathers, T. D. *J. Ind. Microbiol.* **1989**, 4, 341-348.
43. Leathers, T. D. *Appl. Environ. Microbiol.* **1986**, 52, 1026-1030.
44. Wong, K. K. Y.; Tan, L. U. L.; Saddler, J. N. *Microbiol. Rev.* **1988**, 52, 305-317.
45. Christov, L. P.; Prior, B. A. *Enzyme Microb. Technol.* **1993**, 15, 460-475.
46. du Preez, J. C. *Enzyme Microb. Technol.* **1994**, 16, 944-956.
47. Hahn-hagerdal, B.; Jeppsson, H.; Skoog, K.; Prior, B. A. *Enzyme Microb. Technol.* **1994**, 16, 933-943.
48. Gong, C-S. ; Chen, L.-F., Flickinger, M. C.; Chiang, L.-C.; Tsao, G. T. *Appl. Environ. Microbiol.* **1981**, 41, 430-436.
49. Hahn-hagerdal, B.; Berner, S.; Skoog, K. *Appl. Microbiol. Biotechnol.* **1986**, 24, 287-293.
50. Nigam, P.; Singh, D. *Process Biochem.* **1995**, 30, 117-124.
51. Polvinen, K.; Lehtonen, P.; Leisola, M.; Visuri, K. In: *Enzymes in Biomass Conversion*, Eds. Leatham, G. F. and Himmel, M. E., American Chemical Society, Washington, D. C., **1991**, pp. 225-235.
52. Chang, M. M.; Chou, T. Y. C.; Tsao, G. T. *Adv. Biochem. Eng.* **1981**, 20, 15-42.
53. Tien, M.; Kirk, T. K. *Science* **1983**, 221, 661-663.
54. Tien, M. *CRC Crit. Rev. Microbiol.* **1987**, 15, 141-168.
55. Umezawa, T.; Higuchi, T. In: *Enzymes in Biomass Conversion*, Leatham, G. F. and

Himmel, M. E. Eds. American Chemical Society, Washington, D. C., **1991**, pp. 236-246.
56. Fiechter, A. *J. Biotechnol.* **1993**, 30, 49-55.
57. Shoemaker, H. E.; Leisola, M. S. A. *J. Biotechnol.* **1990**, 13, 101-109.
58. Zimmermann, W. J. Biotechnol. 1990, 13, 119-130.
59. Vicuna, R. *Enzyme Microb. Technol.* **1988**, 10, 646-655.
60. Cheetham, P. J. *Enzyme Microb. Technol.* **1987**, 9, 194-212.
61. Arbige, M. V.; Pitcher, W. H. *Trends Biotechnol.* **1989**, 7, 330-335.
62. Fordham, J. R.; Block, N. H. *Developments in Industrial Microbiology* **1987**, 28, 25-31.

Chapter 3

Fuel Ethanol Production from Lignocellulose Using Genetically Engineered Bacteria

L. O. Ingram, X. Lai, M. Moniruzzaman, B. E. Wood, and S. W. York

Department of Microbiology and Cell Science, Institute of Food and Agricultural Sciences, University of Florida, Gainesville, FL 32611

The technologies to convert carbohydrate components of lignocellulose into fuel ethanol are currently available. Today's challenge is to assemble these into a cost-effective process for commercial demonstration while mounting a vigorous research and development program to achieve incremental improvements. Bacteria such as *Escherichia coli* KO11 have been engineered to produce ethanol from all of the sugars present in lignocellulose by adding the genes (*pdc, adhB*) encoding the ethanol pathway from *Zymomonas mobilis*. These *Z. mobilis* genes have been integrated into the *E. coli* chromosome to produce a stable organism for industrial applications. The effectiveness of KO11 has been demonstrated at 150-L scale with hemicellulose syrups and at 10,000-L scale with laboratory sugars, producing over 40 g ethanol/L within 48 h (greater than 90% of theoretical yield). Additional organisms have been engineered for the fermentation of cellulose using an analogous approach. One of these, *Klebsiella oxytoca* P2, has been investigated in some detail. Cellulase enzymes represent a major cost associated with ethanol production from lignocellulose. *K. oxytoca* P2 contains native enzymes for the uptake and metabolism of cellobiose and cellotriose, eliminating the need for one of the major cellulase components, ß-glucosidase. Additional studies have shown that the cellulase requirement for ethanol production can be reduced by enzyme recycling during simultaneous saccharification and fermentation. With mixed waste office paper, projected cellulase costs (production on site) can be as low as $0.085 cents (U.S.)/L with a yield of 539 L/metric ton. Recent studies have identified inexpensive nutrients which can be used for industrial processes and opportunities for synergy between lignocellulose to ethanol plants and grain-based or cane-based ethanol plants.

The United States is dependent upon a secure supply of inexpensive oil. Today, over half of this oil is imported at a direct cost equivalent to half of the national trade imbalance and a military cost exceeding the annual trade imbalance. Although much of the U.S. money used to purchase foreign oil returns through the sale of arms, this proliferation of arms has created further problems and expenses. The strategic and societal benefits from replacing imported oil in the U.S. with renewable domestic fuels such as ethanol would be tremendous (*1-4*). Lignocellulosic residues from landfills, agriculture, and wood-processing represent both a disposal problem and a potential resource for bioconversion processes. Many excellent books and reviews have recently been published concerning progress in this area. The reader is referred to these for information concerning progress with other systems (*5-13*). The current paper will be restricted to our studies with genetically engineered bacteria and comparison to cellulose fermentation by yeasts.

Lignocellulose is a complex material composed of cellulose fibers embedded in a molecular matrix of lignin and hemicellulose. In native lignocellulose, cellulose typically represents 45% and hemicellulose approximately 30% of the dry weight. Both of these carbohydrate polymers must be converted to soluble sugars prior to fermentation. Lignin is a phenolic polymer which can be used as boiler fuel for product recovery. In wood-based ethanol plants dating back to the 1900s, high concentrations of mineral acids were used for depolymerization. However, environmental problems associated with the use and disposal of large amounts of mineral acids prevent a similar process today. Variations of this process could be used, however, provided cost-effective methods of acid recovery are employed. Total enzymatic solubilization of lignocellulose represents the other extreme. Both cellulose and hemicellulose can be depolymerized using microbial enzymes but may include high costs for enzymes. Our focus has been to develop a hybrid approach in which dilute sulfuric acid is used to hydrolyze hemicellulose to monomer sugars and to expose cellulose for enzymatic hydrolysis (Figure 1).

Fermentation of Hemicellulose Sugars after Dilute Acid Pretreatment of Lignocellulose

Effect of Dilute Acid Pretreatment. Dilute acid pretreatment (0.2-2.0% sulfuric acid, 120-220°C) of native lignocellulose serves three important functions in a lignocellulose conversion process: 1) hydrolysis of the hemicellulose components to produce a syrup of monomeric sugars; 2) exposure of cellulose for enzymatic digestion (removal of hemicellulose and part of the lignin); and 3) solubilization of heavy metals which may be contaminating feedstocks. This treatment also causes potential problems such as the production of an acid stream which must either be recovered for reuse or neutralized to produce inert solids such as gypsum (sold or disposed). However, with feedstocks which are contaminated with heavy metals (ie., lignocellulosics from municipal solid waste), these metals are concentrated and recovered in the gypsum fraction providing an environmental advantage. A second problem associated with dilute acid pretreatments is the formation of compounds in the syrups which are toxic to biocatalysts such as bacteria and yeasts. Indeed, considerable art exists in the area of toxin amelioration and optimization of hydrolysis conditions to minimize the generation of these toxins.

The production of a hemicellulose syrup which contains high concentrations of sugars is particularly challenging (Figure 2). Since the sugar yield (hemicellulose) from lignocellulosic feedstocks is typically 0.2-0.3 g sugar/g feedstock (dry weight), hydrolysis must be carried out with 25%-35% solids and little free liquid (Table I). One exception is corn hulls and fibers which have an unusually high content of hemicellulose plus starch. Hemicellulose hydrolysis of woody substrates under more dilute conditions is potentially more expensive since it would increase the amount of required acid (neutralization and disposal as well) and necessitate the addition of process steps for the concentration of sugars prior to fermentation.

Table I. Hemicellulose Syrups from Dilute Acid Hydrolysis

Sugar Composition (%)				Total Sugar	Yield
xylose	arabinose	glucose	galactose	(g/L)	(g sugar/g biomass)
1. Corn hulls plus fiber hemicellulose syrup					
39	23	27	11	100-140	0.50-0.70
2. Corn stover hemicellulose syrup					
61	12	19	7	80-130	0.22-0.27
3. Bagasse hemicellulose syrup					
89	14	6	0	70-110	0.20-0.25

Metabolic Engineering of Recombinant *Escherichia coli* for Pentose Fermentations. The complex mixtures of sugars which comprise hemicellulose hydrolysates pose a special problem for bioconversion into useful products. Typically, these contain predominantly pentose sugars although in some cases equal amounts of hexose and pentose sugars are present. In 1987, we developed *E. coli* strains which efficiently ferment all of the hexose and pentose sugars present in hemicellulose syrups at near theoretical yields (*14,15*). This was done by expressing the two *Zymomonas mobilis* genes encoding the ethanol pathway (*pdc, adhB*) at high levels in *E. coli* using plasmids (*16-18*). In the initial constructs, both the native transcriptional promoter and terminator were removed from the *pdc* gene. The remaining *pdc* coding region was ligated upstream from a promoterless *adhB* gene (including the *adhB* transcriptional terminator). This artificial operon was completed by its insertion behind the *lac* promoter in pUC18. The resulting construct contains a portable ethanol production pathway which, with minor modifications, can be engineered for expression in many different host organisms. The rationale behind this approach lies in the relative simplicity in genetically engineering the production of soluble, cytoplasmic enzymes which can effectively redirect central metabolism in microorganisms that contain other useful properties such as the utilization of many different types of sugars, secretion of hydrolases, tolerance to environmental extremes or toxins, etc.

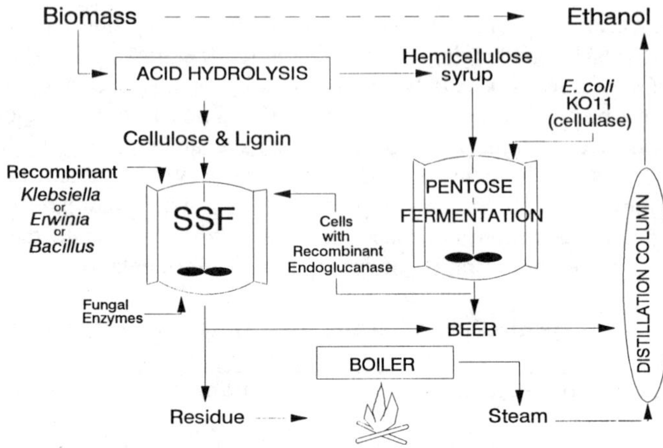

Figure 1. Generalized process for the conversion of lignocellulose to ethanol. Hydrolysis of hemicellulose produces a syrup containing hexose and pentose sugars. This syrup is then fermented to ethanol by *E. coli* strain KO11 which optionally contains genes for the production of intracellular cellulases as co-products. The acid-treated fibrous residue composed of lignin and cellulose is then converted to ethanol by simultaneous saccharification and fermentation using *K. oxytoca* strain P2 (or other recombinant bacterium) and cellulases from fungi and bacteria. The lignin-rich residues can be burned to provide energy.

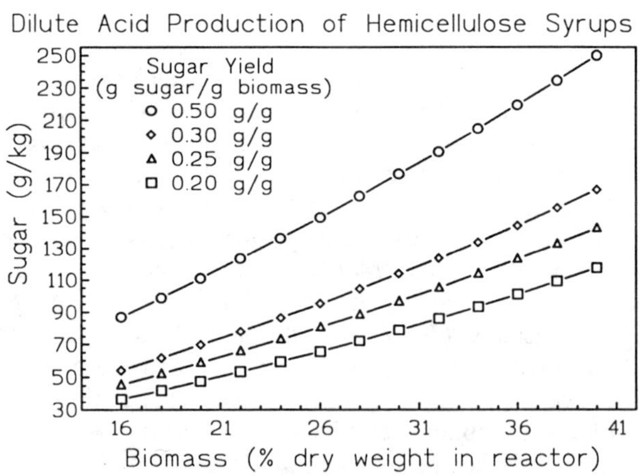

Figure 2. Effect of lignocellulose concentration (% dry weight) of the charge which is processed in a biomass hydrolyzer on the predicted sugar concentration. Lines represent substrates with differing sugar yields (g sugar/g biomass). Sugar concentration is expressed on a weight/weight basis. Sugar yields of 0.50 g/g and higher can be obtained with corn hulls and fibers, 0.25-0.30 g/g with corn stover, and 0.20-0.25 with bagasse.

A variety of *E. coli* hosts strains were investigated for ethanol and environmental tolerance prior to selecting ATCC11303 as the host for further investigation (*19,20*). This host, strain B, is quite safe and has been the subject of research investigations for over 50 years. An improved biocatalyst was developed by integration of the *Z. mobilis* ethanol genes into the chromosome of *E. coli* B to produce strain KO3 (*21*). These genes were integrated behind the *pfl* promoter, a strong promoter which is always active in *E. coli*. Prior to integration, a chloramphenicol gene was added downstream the adhB gene to facilitate selection which proved very fortuitous. The original integrated strain did not express ethanol pathway genes at sufficiently high levels to effectively divert metabolism. Single-step mutants were subsequently selected by plating on solid medium containing 600 mg chloramphenicol/L. Resulting mutants expressed the ethanol production genes at a 10-fold higher level and effectively diverted pyruvate metabolism to ethanol. One of these, strain KO4, was further improved by inserting a *frd* mutation (fumarate reductase) to eliminate the production of succinate, creating strain KO11. Studies were conducted to optimize fermentation conditions and investigate nutrient requirements (*22,23*). A variety of subsequent mutations and selections have yielded strains incapable of fermenting glucose while retaining the ability to ferment galactose and all pentose sugars, and other mutants which ferment sugar mixtures with minimal diauxie (*24*). Additional mutants have been selected with improved ethanol tolerance.

Fermentation of hemicellulose syrups with *E. coli* KO11. A pilot plant for hemicellulose fermentation was constructed in Gainesville, Florida by BioEnergy International (University of Florida licensee purchased by B.C. International, Hingham, MA in 1995). Both a rotating batch reactor similar to a small cement mixer (6 kg charge) and a continuous, dual horizontal screw reactor (1 ton per day) were developed and used to define process conditions needed to yield fermentable hemicellulose syrups containing high sugar concentrations (Table I). Many substrates were tested including bagasse, corn stover, corn cobs, corn fiber from wet milling, sawdust, etc. (*25-30*). Sugar content was maximized and acid utilization minimized by working at very high solids (Figure 2). After proprietary treatments to ameliorate toxins, hemicellulose syrups were fermented to ethanol at a 150-L scale with high yields (Figure 3; Table II) (*26*). Other studies confirmed effective fermentations with laboratory sugars at a 10,000-L scale. Corn steep liquor (*26*), crude yeast hydrolysates (*31*) or crude soy hydrolysates which can be produced on site (*32*), and other industrial nutrients have been investigated and optimized to minimize costs of ethanol production. In many cases, ethanol concentration was limited by sugar content rather than ethanol tolerance. Beer concentrations above 40 g ethanol/L were readily produced when sufficient sugars were present. Materials costs for acid, base, and nutrients are estimated to be from $0.05-0.10/L ethanol and represent opportunities for improvements.

The Modified Gulf Process: Simultaneous Saccharification and Fermentation (SSF) of Cellulose using Genetically Engineered Bacteria as the Biocatalyst.

The cost of fungal cellulase (purchase or production on site) represents a major barrier for the commercial production of ethanol from cellulose. Due to the crystalline

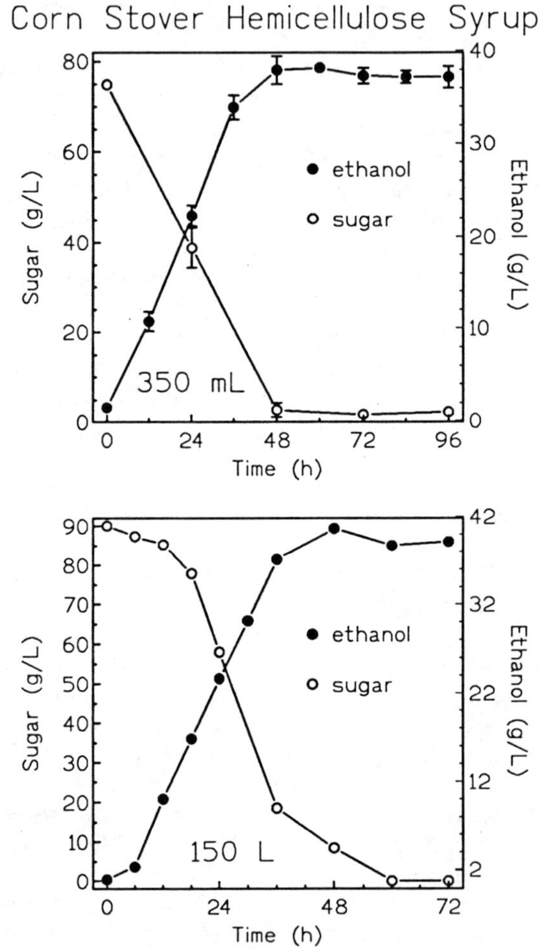

Figure 3. Fermentation of pentose hydrolysates (350 ml and 150 L) from corn stover by *E. coli* strain KO11. Corn steep liquor (CSL) and crude yeast autolysate (YA) were added as nutrients.

Table II. Fermentation of Hemicellulose Syrups by *E. coli* Strain KO11

Nutrients	Volume (L)	Sugar (g/L)	Base[b] (ml/L)	Max. EtOH (g/L)	Time (h)	Total EtOH[c] (g/L)	Q_p (gE/L/h)	Yield (gTE/gS)	Efficiency of[d] Conversion (%)
1. Corn hulls plus fiber hemicellulose syrup									
CSL, YA	0.35	94	27	44.0	72	45.0	0.63	0.48	94
CSL, YA	0.35	94	27	42.1	72	43.2	0.60	0.46	90
2. Corn stover hemicellulose syrup									
Difco	0.35	77	10	37.9	36	39.0	1.08	0.51	100
CSL, YA	0.35	75	8	38.1	48	39.0	0.81	0.52	103
CSL, YA	1.6	80	0	42.4	48	42.4	0.88	0.53	104
CSL, YA	25.0	69	nm	35.0	48	35.0	0.73	0.51	100
Difco	150.0	90	nm	40.8	48	40.8	0.85	0.45	88
3. Bagasse hemicellulose syrup									
Difco	0.35	80	nm	35.0	72	35.0	0.49	0.44	86
CLS, YA	0.35	80	nm	36.9	60	36.9	0.62	0.46	90
CSL, YA	25.0	71	nm	36.5	48	36.5	0.61	0.51	100
Difco	1.6	75	nm	36.2	48	36.2	0.75	0.48	94

[a] Abbreviations: Difco, 5 g Difco Yeast Extract and 10 g Difco Tryptone/L; CSL, corn steep liquor; YA, crude yeast autolysate; Max. EtOH, highest concentration of ethanol achieved during fermentation; nm, not measured; Q_p, average volumetric productivity calculated by dividing total ethanol by fermentation time; gTE/gS, g total ethanol divided by the grams of sugar.

[b] 2N KOH consumed to maintain pH 6.0-6.5 during fermentation.

[c] Corrected for dilution by base.

[d] Ethanol Yield (g/g) divided by 0.51 X 100.

nature of the substrate, extensive hydrogen bonding of chains, etc., large amounts of cellulase protein are required (*6*). Enzymatic hydrolysis is further complicated by the accumulation of soluble products (glucose, cellobiose, cellotriose) which act as competitive inhibitors of hydrolysis. The problem of glucose inhibition was substantially reduced by development of the Gulf Simultaneous Saccharification and Fermentation (SSF) process in 1976 (*33,34*). In this process, fermentation with yeast and saccharification occur together within the same vessel. Cellobiose and cellotriose accumulation were prevented by supplying high levels of supplemental ß-glucosidase, but at an additional cost. With minor modification such as the development of cellobiose-fermenting biocatalysts which do not require supplemental ß-glucosidase (*35,36*), process optimization (*37*), and enzyme recycle (*38*), a Modified Gulf SSF process remains the best available technology for cellulose conversion to ethanol.

Development and Use of Bacterial Biocatalysts for Cellulose Fermentation

Genetic Engineering of *Klebsiella oxytoca* and other Bacteria for the SSF Process. Many different bacteria have the native ability for cellobiose uptake and metabolism. However, none of these organisms produce ethanol efficiently without genetic modification. Research at the University of Florida has focussed on the genetic engineering of improved organisms for the SSF process using the portable ethanol pathway derived from *Z. mobilis*. Three organisms have been initially targeted: 1) *Klebsiella oxytoca* (*35*), an abundant organism in pulp and paper waste; 2) *Erwinia* which cause soft-rot of plant tissue (*39*); and 3) *Bacillus* (*40*). All three organisms utilize cellobiose and thus do not require supplemental ß-glucosidase. The latter two organisms also secrete endoglucanases which can assist cellulose hydrolysis. Most of our published studies have focussed on *K. oxytoca* P2 in which the ethanol pathway genes have been chromosomally integrated (*41*). In this organism and in plasmid-bearing derivatives of *Erwinia*, ethanol is produced from solubilized sugars at greater than 90% of theoretical yield. Other studies indicate that no major barriers exist to prevent the transfer of the ethanol pathway to *Bacillus,* although optimal strains have not been reported (*40*).

Optimization of SSF using *K. oxytoca* P2. Commercial fungal cellulases produced by *Trichoderma* are most active under conditions too extreme (45-50°C, pH 4.0-5.0) for the growth of strain P2. A series of experiments was conducted with Sigmacell 50 (crystalline cellulose; Sigma Scientific Company, St. Louis, MO) to identify the optimal conditions for an SSF process with this biocatalyst (*37*). Spezyme CE and Spezyme CP cellulases were generously provided by Genencor International (South San Francisco, CA). Temperature was varied from 30°C-40°C at pH 5.0-pH 6.0. Surprisingly, ethanol yields exceeded 70% of the theoretical maximum (0.568 g ethanol/g cellulose) in most cases. The highest yield and rates of ethanol production were obtained at pH 5.2 and 35°C. Under these conditions, cellulase enzymes from *Trichoderma* are very stable and continue to be active for many days. Further studies were conducted to examine the dose dependence of cellulase enzymes. With 100 g cellulose/L, a loading of 1000 filter paper units of cellulase (FPU)/L (ie., 10 FPU/g cellulose) approached saturation. Under these conditions, the overall efficiency for fermentation plus saccharification was 72% of the theoretical maximum (Figure 4).

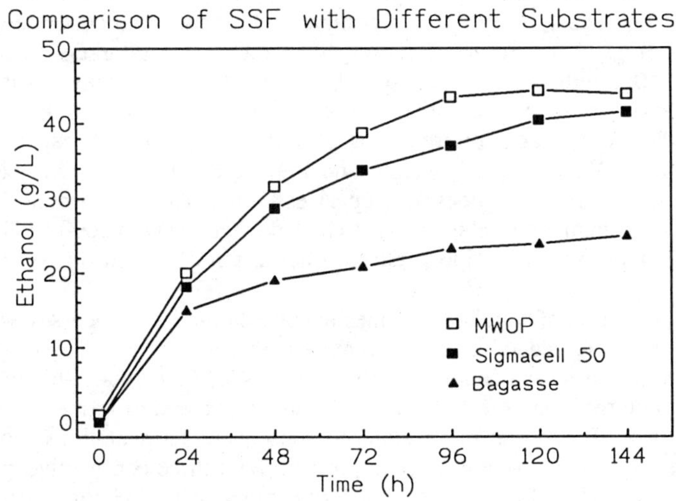

Figure 4. Production of ethanol from different substrates by SSF using *K. oxytoca* strain P2. Fermentations contained 100 g substrate/L and 1,000 FPU cellulase/L.

Further studies were conducted using sugar cane bagasse which had been treated with dilute acid to remove hemicellulose (Figure 4) (*42*). Pretreatment was essential for saccharification and fermentation. However, this material was much less digestible than Sigmacell 50 and required approximately twice as much cellulase enzyme to achieve high yields. Partial saccharification of acid-treated bagasse (pH 4.8 and 48°C) for 12 h with enzymes alone (without biocatalysts) improved mixing and fermentability with a modest benefit to yield. In many cases, SSF with acid treated bagasse stopped after less than 50% of the cellulose had been digested. A brief heat treatment (followed by re-inoculation) was found to rejuvenate the saccharification process. Although the basis for this effect is not understood, it is possible that brief heating to around 60°C allows cellulases which are bound at nonproductive sites to be released. Subsequent binding to at new sites may allow saccharification to resume.

Some of our most successful studies have been conducted with mixed office waste paper (*38*). This substrate is highly digestible with commercial cellulase (Figure 4). Enzyme loadings of around 8.3 FPU/g cellulose approach saturation with Spezyme CP. Over 40 g ethanol/L was produced after 96 h. As with bagasse, partial saccharification prior to inoculation improved mixing but was of little benefit to ethanol yield. Dilute acid pretreatment of mixed waste office paper solubilized approximately 10% of the dry weight and improved mixing in SSF experiments. However, this pretreatment does not appear essential. Yields of around 80% of the theoretical maximum were achieved in batch fermentations with 1000 FPU/L and 100 g MWOP/L (approximately 530 L ethanol/metric ton of mixed office waste paper) (Table III).

The re-utilization of cellulase enzymes in consecutive SSF processes with MWOP allowed a dramatic reduction in the requirement for fungal cellulase (Figure 5). Fungal endoglucanase and cellobiohydrolase have specific cellulose-binding domains which facilitate recycling (*38*). Cellulosic residue at the end of fermentation contains bound cellulase. By adding this residue to subsequent fermentations, both product yield and enzyme effectiveness were improved. With three consecutive recycles, 40 g ethanol was produced after each 80-h fermentation with 83% of theoretical yield using an average of only 570 FPU/L of fermentation broth. For this substrate, the estimated cost of cellulase enzyme produced on site is $0.085 per liter of ethanol, $0.32 per gallon of ethanol. Approximately 539 liters of ethanol per metric ton are projected using this approach (Figure 6).

Figure 6 shows a comparison of results from SSF fermentations with *K. oxytoca* P2 and yeasts. Further detail is provided in Table III. Two important parameters are highlighted in this comparison, ethanol production per 1000 FPU commercial cellulase and ethanol yield per metric ton of feedstock. Both represent major cost factors for a commercial process. This comparison illustrates the benefit of utilizing *K. oxytoca* P2 containing a cellobiose uptake system for batch fermentations for recycling. SSF processes with P2 allow the maintenance of high ethanol yields with a fraction of the cellulase enzyme needed with yeast as a biocatalysts. However, the cost of fungal cellulase remains substantial and further efforts should be made to reduce the levels of these enzymes needed in bioconversion.

Opportunities for Synergy in Biomass to Ethanol Processes

The production of ethanol from lignocellulose offers many potential opportunities for

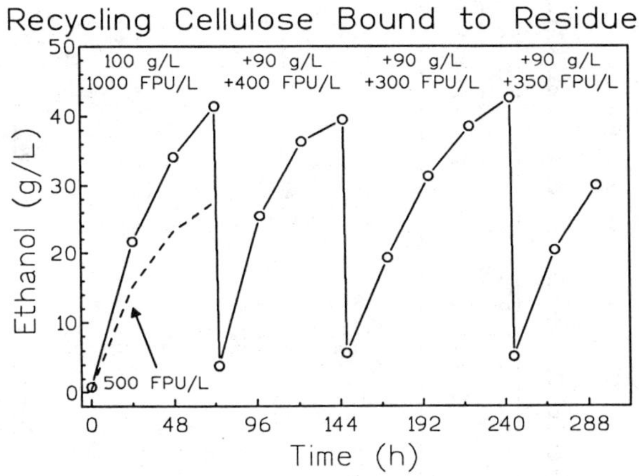

Figure 5. Recycling of cellulase activities during SSF. Fermentations were conducted with mixed waste office paper and *K. oxytoca* strain P2. The initial SSF contained 100 g acid-treated MWOP/L and 1,000 FPU Spezyme CP cellulase (Stage 1). At the end of Stage 1, residue containing bound cellulase was harvested by centrifugation and resuspended in fresh medium to which 90 g MWOP and 400 FPU cellulase/L were added. This Stage 2 SSF was re-inoculated and allowed to ferment. The residue from the Stage 2 SSF was harvested, resuspended in fresh medium containing 90 g MWOP and 300 FPU cellulase/L, re-inoculated, and allowed to ferment (Stage 3). The residue from the Stage 3 SSF was harvested, resuspended in fresh medium containing 90 g MWOP/L and 350 FPU cellulase/L, re-inoculated, and allowed to ferment (Stage 4). Although this fourth SSF was prematurely terminated, it serves to illustrate that cellulase enzymes can continue to be recycled. For comparison, a dashed line has been added which represents ethanol production during a simple batch SSF containing 120 g MWOP/L and 500 FPU/L cellulase (without recycling).

Table III. Comparison of SSF Results for Different Substrates and Biocatalysts

Biocatalyst	Substrate (g/L)	Cellulase (FPU/L)	Cellobiase (IU/L)	Ferm. Time (h)	Ethanol (g/L)	Yield gE/gS	Yield gE/1000 FPU	Yield LE/t	Reference
1. Sigmacell 50 (purified crystalline cellulose)									
K. oxytoca P2	100 g	1,000	none	96	37.0	0.370	37.0	468	37
				168	42.1	0.421	42.1	533	
K. oxytoca P2	100 g	2,500	none	96	41.4	0.414	16.6	524	37
				168	46.8	0.468	18.7	592	
B. custersii	75 g	1,950	none	72	32.0	0.427	16.4	541	43
B. custersii	100 g	1,900	none	240	40.0	0.400	21.1	506	43
2. Office Paper (80% cellulose, 10% hemicellulose)									
S. cerevisiae	200 g	3,440	387	72	55.8	0.279	16.2	353	44
				96	59.3	0.297	17.2	376	44
S. cerevisiae	200 g	6,880	773	72	63.0	0.315	9.2	395	44
				96	65.8	0.329	9.6	415	44
K. oxytoca P2	120 g	1,000	none	96	40.8	0.340	40.8	430	38
3. Office Paper Treated with Steam (ST) (210°C) or Acid (AT) (hemicellulose removed)									
S. cerevisiae	100 g ST	2,649	274	48	29.4	0.294	11.1	372	45
S. cerevisiae	100 g AT	2,649	274	48	35.6	0.356	13.4	451	45
K. oxytoca P2	93 g AT	567	none	80	39.6	0.426	69.8	539	38
4. Acid-Treated Office Paper (90% cellulose, hemicellulose removed)									
K. oxytoca P2	100 g	1,000	none	96	43.5	0.435	43.5	550	38
K. oxytoca P2	100 g	1,000	none	72	41.4	0.414	41.4	524	38

5. Refuse-Derived Fiber									
S. cerevisiae	100 g	2,408	275	72	25.0	0.250	10.4	316	46
S. cerevisiae	100 g	2,408	275	72	22.7	0.227	9.4	287	46
6. Acid-Treated Bagasse (63% cellulose, hemicellulose removed)									
K. oxytoca P2	100 g	1,000	none	96	23.3	0.233	23.3	295	42
K. oxytoca P2	160 g	1,600	none	96	26.3	0.164	16.4	208	42
K. oxytoca P2	160 g	3,200	none	96	32.5	0.203	10.2	257	42
				168	38.6	0.241	12.1	305	
K. oxytoca P2	160 g	1,600	none	168+72	39.2	0.245	24.5	310	42
(includes second heat treatment and SSF)									
7. Acid-Treated Hybrid Poplar (69.5% cellulose, hemicellulose removed)									
S. cerevisiae + *B. clausenii*	130 g	1,710	13,680	144	42.0	0.323	24.6	409	47
S. cerevisiae + *B. clausenii*	108 g	975	1,950	192	35.3	0.327	36.2	414	47
S. cerevisiae	130 g	1,710	13,680	72	43.0	0.331	25.1	419	47
S. cerevisiae	108 g	975	1,950	192	35.7	0.331	36.6	419	47
8. SO_2-Treated Willow (60% cellulose, hemicellulose removed)									
S. cerevisiae	100 g	1,800	225	72	29.3	0.293	16.3	371	48
S. cerevisiae	150 g	2,700	338	72	39.1	0.261	14.5	330	48

Abbreviations: gE/gS, grams ethanol produced per gram of substrate; gE/1,000 FPU, grams of ethanol produced per 1,000 FPU cellulase; LE/t, liters of ethanol per metric ton (dry weight) of substrate (ethanol = 0.790 kg/L).

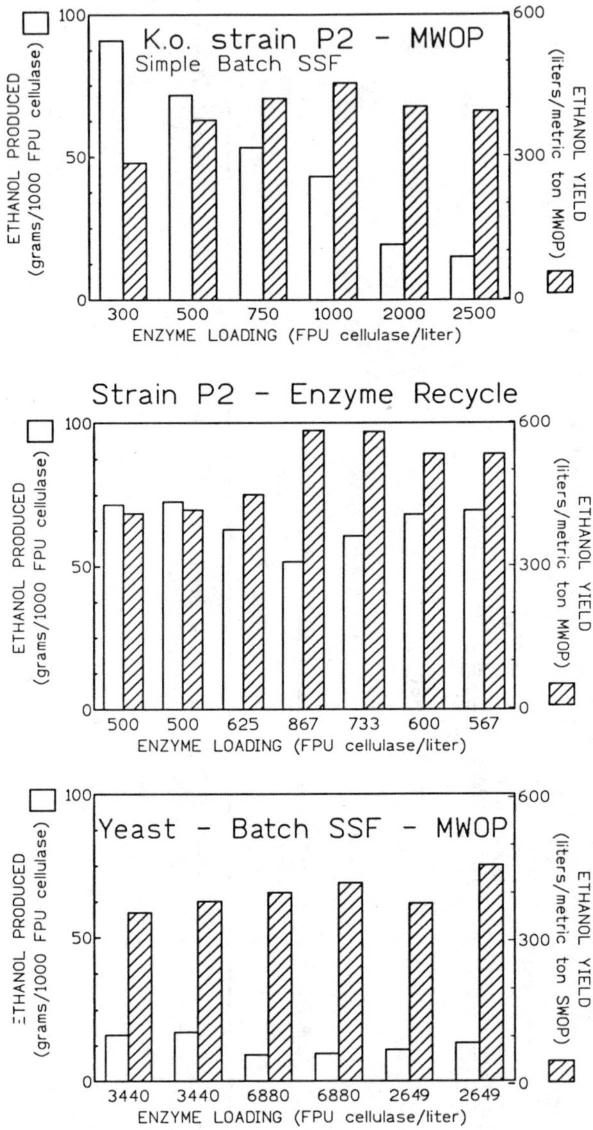

Figure 6. Comparison of ethanol production from waste office paper (MWOP) by *K. oxytoca* strain P2 and yeast. More detail concerning fermentation conditions and references are provided in Table III.

synergy with other types of manufacturing processes. Perhaps the simplest of these would involve a combination of electricity with boilers fired from the residues after hemicellulose hydrolysis and ethanol production from hemicellulose syrups. This process could reduce heavy metal contamination of boiler feeds derived from municipal landfill waste (metals precipitated with the gypsum) and upgrade thermal value. Low grade steam could be used for ethanol purification and other processes.

Cane sugar production plants could also benefit from increased ethanol yields by fermenting hemicellulose syrups, analogous to the conversion of bagasse to furfural plant in south Florida. Although bagasse is burned to power sugar refining, excess bagasse typically accumulates. Hemicellulose could be stripped by dilute acid hydrolysis and fermented to ethanol, leaving sufficient residue as a boiler fuel for both processes.

Ethanol production from grain and cane sugar offer an extremely attractive opportunity for synergy. Spent yeasts from these processes could be recycled as a nutrient source for hemicellulose or cellulose fermentations. Corn fiber residues, corn cobs, corn stover, bagasse or other lignocellulosic residues could serve as a feedstock for the biomass to ethanol process. Again, undigested residues could be burned to provide the energy.

Acknowledgments

This research was supported in part by the Florida Agricultural Experimental Station (publication number R-05323) and by grants from the U.S. Department of Agriculture, National Research Initiative (95-37308-1843) and the U.S. Department of Energy, Office of Basic Energy Science (DE-FG02-96ER20222).

Literature Cited

1. Wyman, C.E. In *Enzymatic Degradation of Insoluble Carbohydrates*; Saddler, J.N.; Penner, M.H., Eds.; ACS Symposium Series No. 618, American Chemical Society: Washington, DC, **1995,** pp 272-290.
2. Wyman, C.E. *Bioresource Technol.* **1994**, *50*, 3-16.
3. Lynd, L.R.; Cushman, J.H.; Nichols, R.J.; Wyman, C.E. *Science* **1991,** 251, 1318-1323.
4. Sheehan, J.J. In *Enzymatic Conversion of Biomass for Fuels Production*, Himmel, M.E.; Baker, J.O.; Overend, R.P., Eds.; ACS Symposium Series No. 566, American Chemical Society: Washington DC, **1994,** pp 1-53.
5. du Preez, J.C. *Enzyme Microb. Technol.* **1994**, *16*, 944-956.
6. Grohmann, K. In *Bioconversion of Forest and Agricultural Residues*, Saddler, J.N., Ed.; CAB International, Wallingford, U.K., **1993**, pp 183-209.
7. Hahn-Hagerdal, B.; Hallborn, J.; Jeppsson, H.; Olsson, L.; Skoog, K.; Walfridsson, M. In *Bioconversion of Forest and Agricultural Residues*, Saddler, J.N., Ed.; CAB International, Wallingford, U.K., **1993**, pp 231-290.
8. McMillan, J.D. In *Enzymatic Conversion of Biomass for Fuels Production*, Himmel, M.E.; Baker, J.O.; Overend, R.P., Eds. ACS Symposium Series No. 566, American Chemical Society, Washington DC, **1994**, pp 293-324.
9. McMillan, J.D. In *Enzymatic Conversion of Biomass for Fuels Production*,

Himmel, M.E.; Baker, J.O.; Overend, R.P., Eds. ACS Symposium Series No. 566, American Chemical Society, Washington DC, **1994**, pp 411-437.
10. Philippidis, G.P. In *Enzymatic Conversion of Biomass for Fuels Production*, Himmel, M.E.; Baker, J.O.; Overend, R.P., Eds. ACS Symposium Series No. 566, American Chemical Society, Washington DC, **1994**, pp 188-217.
11. Philippidis, G.P.; Smith, T.K.; Wyman, C.E. *Biotechnol. Bioeng.* **1992**, *41*, 846-853.
12. Ramos, L.P.; Saddler, J.N. In *Enzymatic Conversion of Biomass for Fuels Production*, Himmel M.E.; Baker, J.O.; Overend, R.P., Eds. ACS Symposium Series No. 566, American Chemical Society, Washington DC, **1994**, pp 325-341.
13. von Sivers, M.; Zacchi, G. *Bioresource Technol.* **1995**, *51*, 43-52.
14. Ingram, L.O.; Conway, T.; Clark, D.P.; Sewell, G.W.; Preston, J.F. *Appl. Environ. Microbiol.* **1987**, 53, 2420-2425.
15. Ingram, L.O.; Alterthum, F.; Conway, T.; Ohta, K.; Shanmugam K.T. Ethanol Production by *Escherichia coli* Strains Co-expressing *Zymomonas pdc* and *adh* Genes. **1991**, U.S. Patent No. 5,000,000.
16. Conway, T.C.; Sewell, G.W.; Osman, Y.A.; Ingram, L. O. *J. Bacteriol.* **1987**, *169*, 2591-2597.
17. Conway, T.; Osman, Y.A.; Ingram, L.O. *J. Bacteriol.* **1987**, *169*, 949-954.
18. Ingram, L.O.; Conway, T. *Appl. Environ. Microbiol.* **1988**, *54*, 397-404.
19. Alterthum, F.; Ingram L.O. *Appl. Environ. Microbiol.* **1989**, *55*, 1943-1948.
20. Ohta, K.; Alterthum, F.; Ingram, L.O. *Appl. Environ. Microbiol.* **1990**, *56*, 463-465.
21. Ohta, K.; Beall, D.S.; Shanmugam, K.T.; Ingram, L.O. *Appl. Environ. Microbiol.* **1991**, *57*, 893-900.
22. Beall, D.S.; Ohta, K.; Ingram, L.O. *Biotechnol. Bioengin.* **1991**, *38*, 296-303.
23. Guimaraes, W.V.; Dudey, G.L.; Ingram, L.O. *Biotechnol. Bioengin.* **1992**, *40*, 41-45.
24. Lindsay, S.E.; Bothast, R.J.; Ingram, L.O. *Appl. Microbiol. Biotechnol.* 1995, *43*, 70-75.
25. Grohmann, K.; Baldwin, E.A.; Buslig, B.S.; Ingram, L.O. *Biotechnol. Lett.* **1994**, *16*, 281-286.
26. Asghari, A.; Bothast, R.J.; Doran, J.B.; Ingram, L.O. *J. Indust. Microbiol.* **1996**, *16*, 42-47.
27. Beall, D.S.; Ingram, L.O.; Ben-Bassat, A.; Doran, J.B.; Fowler, D.E.; Hall, R.G.; Wood, B.E. *Biotechnol. Lett.* **1992**, *14*, 857-862.
28. Guimaraes, W.V.; Ohta, K.; Burchhardt, G.; Ingram, L.O. *Biotechnol. Lett.* **1992**, *14*, 415-420.
29. Barbosa, M. de F.S.; Beck, M.J.; Fein, J.E.; Potts, D.; Ingram, L.O. *Appl. Environ. Microbiol.* **1992**, *58*, 1382-1384.
30. Burchhardt, G.; Ingram, L.O. *Appl. Environ. Microbiol.* **1992**, *58*, 1128-1133.
31. York, S.W.; Ingram, L.O. *Biotechnol. Lett.* **1996**, *18*, 683-688.
32. York, S.W.; L.O. Ingram. *J. Indust. Microbiol.* **1996**, *16*, 374-376.
33. Gauss, W. F.; Suzuki, S.; Takagi, M. Manufacture of alcohol from cellulosic materials using plural ferments. **1976**, U.S. Patent No. 3,990,944.

34. Emert, G.H.; Katzen, R.; Fredrickson, R.E.; Kampish, K.F.; Yeats, C.E. *J. Appl. Poly. Sci.* **1983**, 37, 787-795.
35. Ohta, K.; Beall, D.S.; Mejia, J.P.; Shanmugam, K.T.; Ingram, K.T. *Appl. Environ. Microbiol.* **1991**, 57, 2810-2815.
36. Ingram, L.O.; Beall, D.S.; Burchhardt, G.F.H.; Guimaraes, W.V.; Ohta, K.; Wood, B.E.; Shanmugam, K.T. Ethanol Production by Recombinant Hosts. **1995,** U.S. Patent No. 5,424,202
37. Doran, J.B.; Ingram, L.O. *Biotechnol. Progr.* **1993,** *9*, 533-538.
38. Brooks, T.A.; Ingram, L.O. *Biotechnol. Progr.* **1995**, *11*, 619-625.
39. Beall, D.S.; Ingram, L.O. *J. Indust. Microbiol.* **1993**, *11*, 151-155.
40. Barbosa, M. de F.S.; Ingram, L.O. *Cur. Microbiol.* **1994**, *28*, 279-282.
41. Wood, B.E.; Ingram, L.O. *Appl. Environ. Microbiol.* **1992**, *58*, 2103-2110.
42. Doran, J.B.; Aldrich, H.C.; Ingram, L.O. *Biotechnol. Bioengin.* **1994**, *44*, 240-247.
43. Spindler, D.D.; Wyman, C.E.; Grohmann, K.; Philippidis, G.P. *Biotechnol. Lett.* **1992**, *14*, 403-407.
44. Wayman, M.; Chen, S.; Doan, K. *Process Biochem.* **1992**, *27*, 239-245.
45. Capek-Menard, E.; Jollez, P.; Chornet, E.; Wayman, M.; Doan, K. *Biotechnol. Lett.* **1992**, *14*, 985-988.
46. Wayman, M.; Doan, K. *Biotechnol. Lett.* **1992**, *14*, 335-338.
47. Spindler, D.D.; Wayman, C.E.; Grohmann, K. *Appl. Biochem. Biotechnol.* **1991**, *28/29*, 773-786.
48. Eklund, R.; Zacchi, G. *Appl. Biochem. Biotechnol.* **1995**, *34/35*, 105-108.

Chapter 4

Fuel Ethanol Production from Lignocellulosic Sugars

Studies Using a Genetically Engineered *Saccharomyces* Yeast

M. S. Krishnan[1,2], Y. Xia[1], N. W. Y. Ho[1], and G. T. Tsao[1,2]

[1]Laboratory of Renewable Resources Engineering and [2]School of Chemical Engineering, Purdue University, West Lafayette, IN 47907

Fermentation studies of ethanol production from lignocellulosic sugars using the genetically engineered *Saccharomyces* yeast 1400 (pLNH33) and its parent *Saccharomyces* yeast strain 1400 are reported. While the parent strain 1400 is unable to ferment xylose, the recombinant yeast 1400 (pLNH33) ferments xylose and mixtures of glucose and xylose. High ethanol yields upto 84% were obtained by fermentation of glucose-xylose mixtures using the recombinant yeast. The kinetics of ethanol inhibition of yeast cell growth on glucose and xylose are presented. Results of ethanol production from corn fiber and corn cob by the simultaneous saccharification and fermentation (SSF) process are also reported.

Ethanol has received attention recently as an octane booster and a transportation fuel. The economics of fuel ethanol production are significantly influenced by the cost of the raw materials used in the production process. Lignocellulosic materials such as agricultural residues and municipal waste paper have been identified as potential feedstocks, in view of their ready availability and low cost (*1*). These lignocellulosic hydrolyzates that are produced either chemically or enzymatically contain both pentoses and hexoses. The pentoses are comprised of D-xylose and L-arabinose while the major hexose is D-glucose (*2*). While the glucose is readily fermented by using *Saccharomyces* yeasts, few microorganisms have the ability to ferment xylose. For the economics of the biomass to ethanol process, it is necessary to convert the xylose to ethanol as well. *Pichia stipitis* and *Candida shehatae* are the best wild type xylose fermenting yeasts that have been reported in the literature (*3*). Recent advances in molecular biology techniques have led to the development of genetically engineered microorganisms for xylose fermentation. These include recombinant bacterial strains of *E. coli* (*4*), *Klebsiella oxytoca* (*5*) and *Zymomonas mobilis* (*6*).

Although these strains show good xylose fermentation performance, the low ethanol tolerance of these microorganisms is a limiting factor in the process. *Saccharomyces* yeasts have a relatively higher ethanol tolerance and hence attempts have been made to develop recombinant *Saccharomyces* yeasts that can ferment xylose (*7,8*). However, the ethanol yields and productivities are low. This has been attributed to the cofactor imbalance and an insufficient capacity for xylulose conversion through the pentose phosphate pathway.

A recombinant yeast denoted 1400 (pLNH33) has been developed by Nancy Ho and co-workers at the Laboratory of Renewable Resources Engineering, Purdue University (*9,10*). This strain was developed using the high ethanol tolerance *Saccharomyces* yeast 1400 (*11*) as the host and cloning the xylose reductase, xylitol dehydrogenase genes (both from *Pichia stipitis*) and xylulokinase gene (from *S. cerevisiae*) into yeast 1400. The recombinant yeast ferments glucose and xylose simultaneously to ethanol in high yields.

In this paper, we report the fermentation studies conducted on glucose, xylose and their mixtures using this recombinant yeast. Ethanol tolerance is a key factor influencing process economics, motivating us to investigate the kinetics of ethanol inhibition on these genetically engineered yeasts. Results of the simultaneous saccharification and fermentation (SSF) process using corn fiber and corn cob as model feedstocks are also presented.

Methods

Microorganisms. *Saccharomyces* yeasts 1400 and genetically engineered 1400(pLNH33) were used in the experimental work. The yeast 1400 (pLNH33) was obtained from Dr. Nancy Ho at LORRE.

The yeast strain 1400 (*11*) is a protoplast fusion product of *Saccharomyces diastaticus* and *Saccharomyces uvarum.*

Culture Conditions. The recombinant yeast 1400 (pLNH33) was maintained on YEPX seed cultures. The composition of the seed culture media per liter of distilled water is as follows: D-xylose 20g, yeast extract 10g, Bactopeptone 20g. The yeast was grown to an OD of 400-450 Klett Unit (measured by a Klett-Summerson colorimeter) and then maintained at 4°C. The medium for the preparation of the inoculum was the YEPX medium described above. 1 ml of the seed culture was added to a sterilized 250 ml Erlenmeyer flask with silicone sponge closure, containing 50 ml of medium. The inoculum was incubated at 30°C in a floor shaker at 150-200 rpm for 18-20 hours (when the cells were in the late exponential phase) before being used to inoculate the fementation medium.

The parent yeast strain 1400 was maintained on YEPD agar plates. The composition of the plate media per liter of distilled water is as follows: glucose 20g, yeast extract 10g, Bactopeptone 20g and agar 20g. The yeast was inoculated on the plate medium at 30°C for 48 hours and then maintained at 4°C. For the preparation of the inoculum, yeast extract-Bactopeptone medium with 20 g/L glucose was used. A loopful of yeast cells were transferred from the agar plate into 50 ml sterilized medium in a 250 ml Erlenmeyer flask.

Fermentation Conditions. The fermentation was performed in 250 ml Erlenmeyer flasks with silicone sponge closures, containing 100 ml sterilized medium. The fermentation medium consisted of 20 g/L Bactopeptone, 10 g/L yeast extract and appropriate concentrations of glucose and/or xylose. The inoculum sizes used were in the range of 0.1 g/L-2.5 g/L. The fermentation conditions were same as those indicated earlier for the inoculum preparation.

Pretreatment and Hydrolysis of Corn Fiber. Corn fiber (corn hull from A. E. Staley, Lafayette, IN) was pretreated with 0.5% dilute hydrochloric acid at 120°C for 45 minutes. Enzymatic hydrolysis of the pretreated corn fiber was performed at 45°C using Iogen cellulase having an activity of 154 FPU/ml. The pretreated corn fiber was thoroughly washed, following which cellulase was added to the glucose free medium.

Simultaneous Saccharification and Fermentation (SSF) of Corn Fiber. Dry pretreated corn fiber (25% w/v) was added into a 250 ml side-arm Erlenmeyer flask. Yeast extract (10 g/L) and Bactopeptone (20 g/L) were also added and the pH was adjusted to 5. The yeast was inoculated from the seed culture to give an initial cell concentration of 0.5 g/L. Iogen cellulase was added to the medium to give an activity of about 5-10 FPU per gram of cellulose from corn fiber. This fermentation medium was diluted using deionized water to give the desired solid fraction. The SSF of pretreated corn fiber was conducted at 30°C in a shaker at 150-200 rpm.

Analytical Methods.

Cell Mass. A spectrophotometer (Coleman model 55, Perkin-Elmer, Maywood, IL) was used to measure the absorbance of the samples at a wavelength of 600 nm that is in the visible region. Samples were diluted as required to assure absorbences of less than 0.5. In this region the calibration curve was linear with a slope of 0.65 g dry weight per unit absorbance.

HPLC. Hitachi HPLC (Hitachi Ltd., Tokyo, Japan) with RI detector was used to analyze the concentrations of glucose, xylose, xylitol, ethanol and glycerol. A BioRad HPX-87H Ion-Exclusion column was used. The mobile phase was 0.005M H_2SO_4 at a flow rate of 0.4 mL/min.

Glucose Analyzer. YSI 2700 Select Biochemistry Analyzer (YSI Inc., Yellow Springs, OH) equipped with glucose membrane was used for rapid analysis of glucose concentration in the fermentation media.

Results and Discussion

Fermentation Studies on Glucose and Xylose. The fermentation performance of the parent yeast and the genetically engineered yeast was studied on glucose and xylose separately. In both these experiments the substrate concentrations selected were well above the growth limiting concentrations and not inhibitory to growth of the yeasts.

The results of the glucose fermentation study using yeasts 1400 and 1400 (pLNH33) is shown in Figures 1a and 1b, respectively. These experiments were performed using approximately 100 g/L glucose as the substrate with an initial cell density in the range of 0.05-0.10 g/L. In both cases, a maximum cell density in the range of 11-11.5 g/l was obtained after 14 hours. The specific growth rates of the yeast was calculated from the slope of a semi-log plot of cell dry weight versus time. For the yeast 1400, the specific growth rate under the experimental conditions was calculated to be 0.49±0.02 hr^{-1} while for the recombinant yeast 1400 (pLNH33), the specific growth rate was calculated to be 0.48±0.02 hr^{-1} . Thus the specific growth rates of both yeasts on glucose are identical. The specific glucose utilization rates and ethanol yields in both cases were 2.3 g/g-hr and 0.49, respectively. Based on a theoretical yield of 0.51 g/g, these yields correspond to 96 % of the theoretical yield.

The results of the xylose fermentation study using the above yeasts are shown in Figures 2a and 2b. These experiments were performed using approximately 50 g/L xylose as the substrate with an initial cell density of 0.05-0.10 g/L. As seen in Figure 2a, the yeast 1400 is unable to grow and ferment xylose to ethanol. In comparison, the recombinant yeast 1400 (pLNH33) grows to a cell density of 9.4 g/L, and produces 20.45 g/L ethanol from 52.06 g/L initial xylose after 36 hours. Glycerol and xylitol are the by-products that are produced in minor amounts, to the extent of 1.73 g/L and 2.85 g/L respectively. The specific growth rate of the recombinant yeast was calculated, as explained earlier, to be 0.19±0.02 hr^{-1}. The specific xylose utilization rate was calculated to be 0.30 g/g-hr. Based on the xylose consumed, the ethanol yield was calculated to be 0.40 g/g. This corresponds to 78 % of the theoretical yield (based on a theoretical yield of 0.51 g/g). These results indicate that the specific xylose utilization rate and ethanol yield from xylose are lower than those from glucose.

Fermentation Study on a Glucose-Xylose Mixture. From the above studies on single substrates, it is clear that the fusion product 1400 lacks the ability to ferment xylose. On the other hand, the recombinant yeast 1400 (pLNH33) shows a good fermentation performance on xylose. Typically, both glucose and xylose are present in lignocellulosic hydrolysates. This requires that the fermentative microorganism be able to ferment xylose in presence of glucose. The recombinant yeast 1400 (pLNH33) has been genetically designed to ferment both glucose and xylose present in the same medium (*9*). To demonstrate this, fermentation of a 1:1 mixture of glucose and xylose was performed. The composition of the sugar mixture was 52.8 g/L glucose and 56.3 g/L xylose and the initial cell density was 2.3 g/L. As shown in Figure 3, the recombinant yeast ferments glucose and xylose simultaneously to ethanol. However, the glucose utilization rate at 8.11 g/L-hr is relatively higher than the xylose utilization rate at 1.77 g/L-hr. A final cell density of 11.5 g/L and an ethanol concentration of 47.9 g/L were achieved after 48 hours. Based on a theoretical yield of 0.51 g/g, the ethanol yield from this experiment was calculated to be 80%.

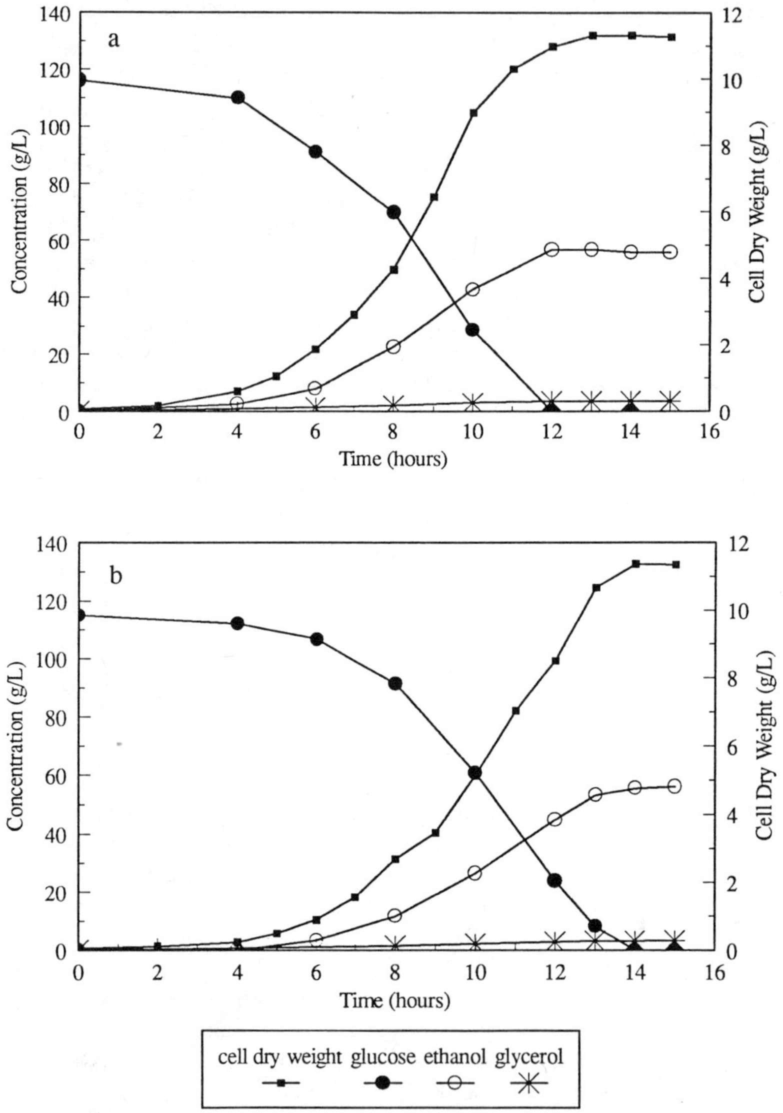

Figure 1. Growth and fermentation performance of a) yeast strain 1400 and b) recombinant yeast strain 1400 (pLNH33) on glucose.

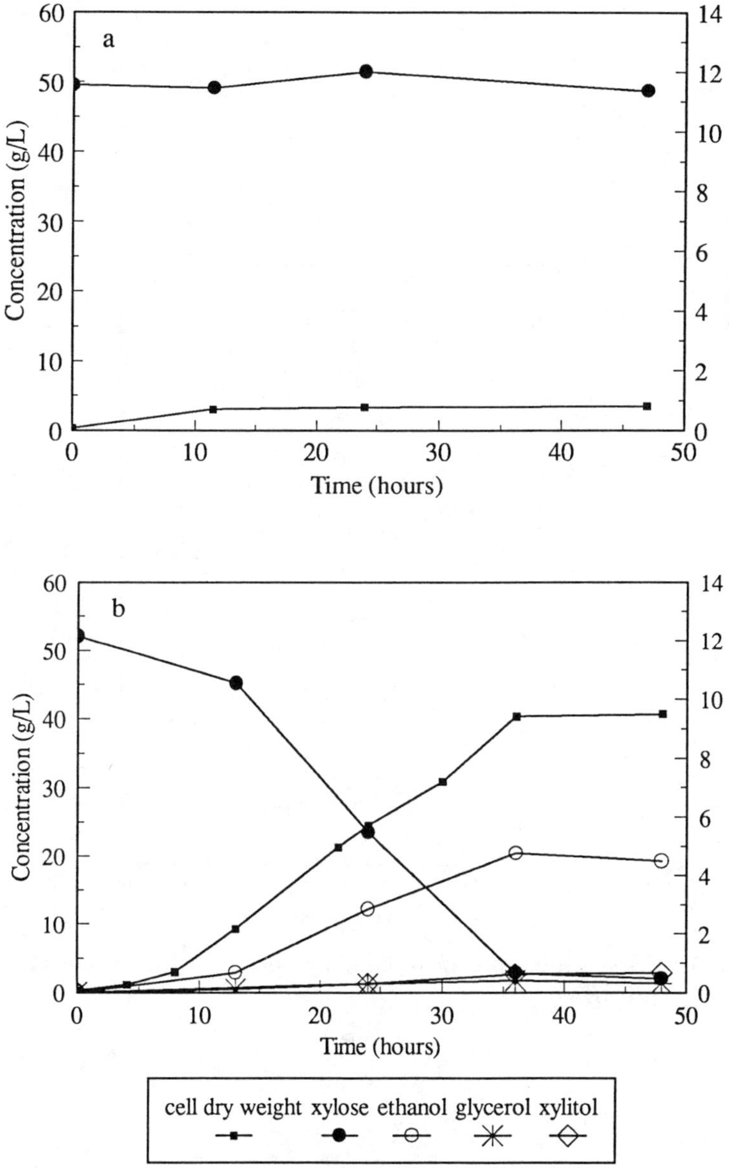

Figure 2. Growth and fermentation performance of a) yeast strain 1400 and b) recombinant yeast strain 1400 (pLNH33) on xylose.

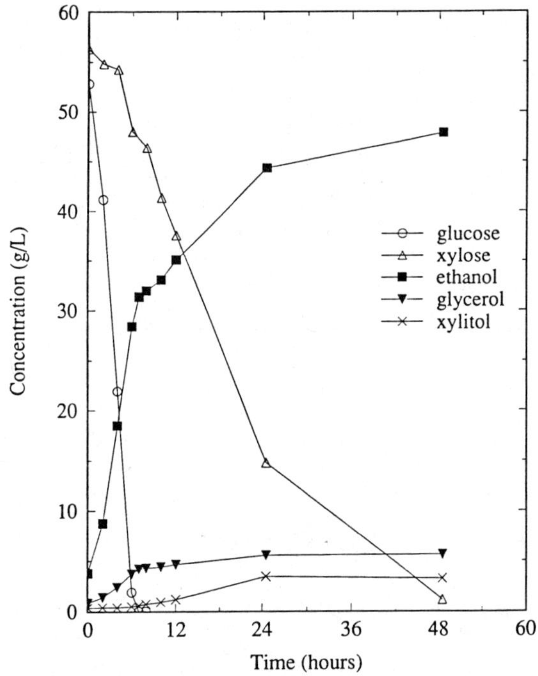

Figure 3. Fermentation of a glucose-xylose mixture by recombinant yeast strain 1400 (pLNH33).

Fermentation Kinetics. The effects of substrate and product inhibition on glucose and xylose fermentation were studied in order to determine the optimal fermentation conditions for achieving highest ethanol yields and productivities.

Substrate Inhibition. These experiments were conducted by using different initial glucose or xylose concentrations in the fermentation media. For the glucose fermentation, significant lag times are observed in experiments with an initial concentration greater than 200 g/L (*12*). This can be attributed to the effect of glucose inhibition on the yeast growth, as all other conditions such as temperature and availability of nutrients are favorable. Similar studies have also been conducted for the xylose fermentation using recombinant yeast 1400 (pLNH33). These studies indicate that the effect of xylose inhibition for yeast growth is significant above 70 g/L (data on specific growth rates not included).

A classical process methodology for concentrated sugar fermentations is to feed the substrate into the fermentation medium in steps, instead of feeding all of it initially (*11,12*). To achieve higher ethanol concentrations from glucose fermentation, this modified feeding scheme was used as shown in Figure 4. The fermentation was begun with 106 g/L glucose initially, followed by four successive additions of glucose with nutrients later during the fermentation. The arrows in the figure indicate the times at which glucose with nutrients was added into the fermentation medium. The addition of nutrients along with glucose helped in maintaining a high cell viability throughout the fermentation. At the end of 67 hours, an ethanol concentration of 133 g/L was obtained at a productivity of 1.99 g/L-hr. This reflects the high ethanol tolerance of the *Saccharomyces* strain 1400. Since ethanol tolerance significantly influences the economics of ethanol recovery, this high ethanol tolerance yeast has potential applications in commercial ethanol production.

This fed-batch experimental methodology was also used for the xylose fermentation using the recombinant yeast 1400 (pLNH33), as shown in Figures 5a and 5b. An ethanol concentration of 50.34 g/L was obtained after 193 hours of fermentation, giving a productivity of 0.26 g/L-hr. Thus in comparison to the glucose fed-batch fermentation, relatively lower ethanol concentrations and productivities were achieved. Ethanol inhibition of the xylose fermentation may be one of the several reasons possible for this observation.

Product Inhibition. In order to determine the effect of ethanol on the cell growth rates and the fermentation rate, experiments with a range of initial ethanol concentrations in the fermentation media were performed. This study was carried using both glucose and xylose as substrates. The sugar concentrations in both experiments were initially 50 g/L, which is well in excess of the saturation constant. This sugar concentration was selected in order to separate ethanol inhibition effects from those of substrate or nutrient limitation. The effects of ethanol concentration as a single independent variable can be clearly discerned using this method as compared to batch studies with produced ethanol.

Figure 6 shows the experimental data of variation in specific growth rate of yeast 1400 on glucose as a function of the initial ethanol concentration. When there is no initial ethanol in the medium, the highest specific growth rate of 0.6 hr^{-1} is

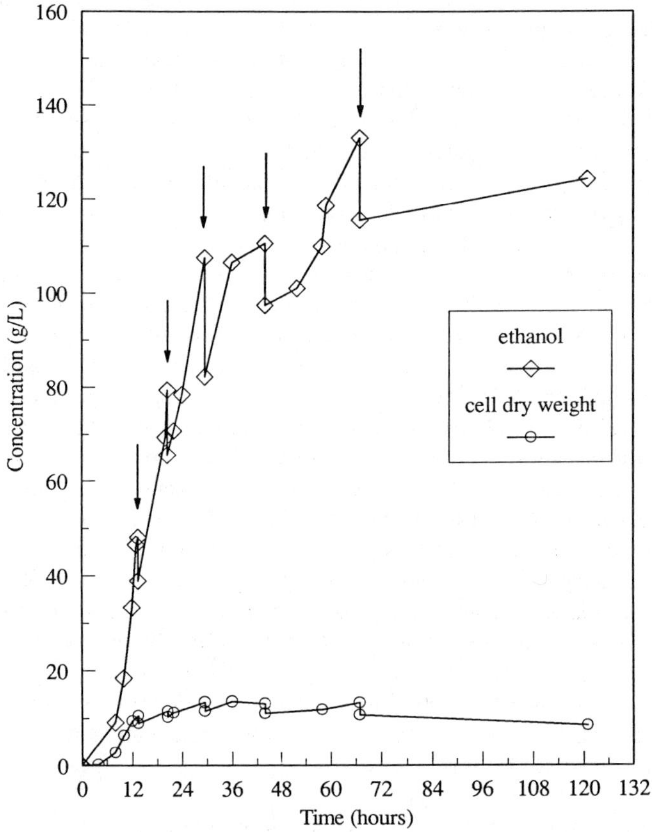

Figure 4. Ethanol and cell density profiles during a glucose fed-batch fermentation using the yeast strain 1400.

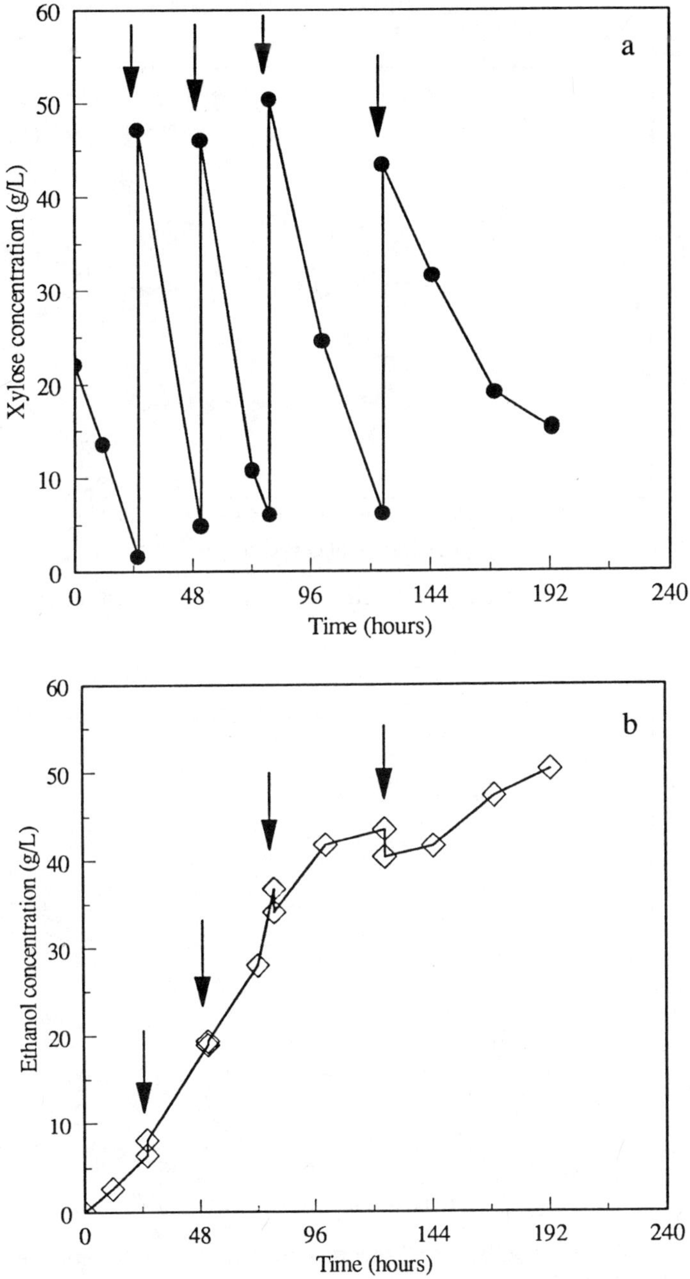

Figure 5. a) Xylose and b) ethanol concentration profiles during a xylose fed-batch fermentation using the recombinant yeast strain 1400 (pLNH33).

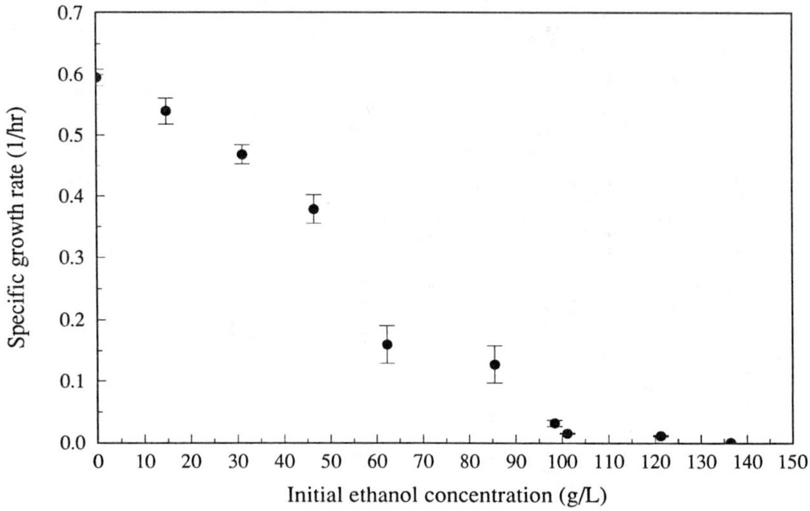

Figure 6. Effect of ethanol on specific growth rate of yeast strain 1400 on glucose.

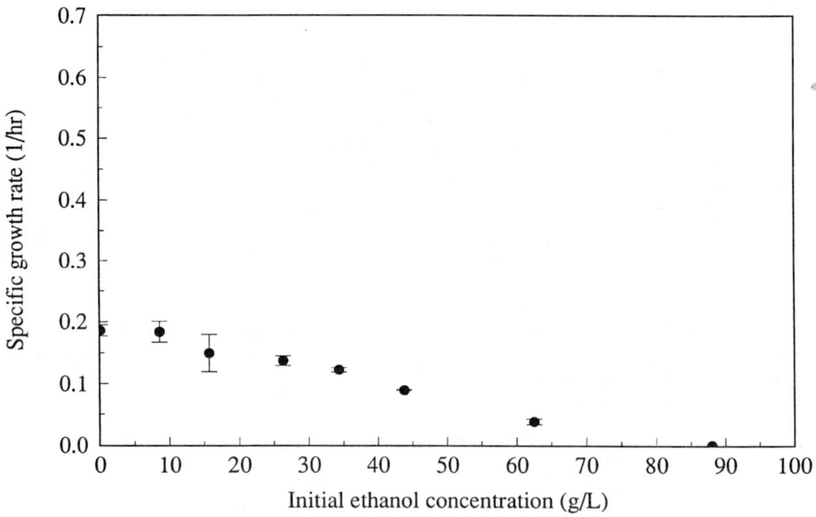

Figure 7. Effect of ethanol on specific growth rate of recombinant yeast strain 1400 (pLNH33) on xylose.

obtained. There is a decline in the specific growth rate as the initial ethanol increases. The effect of ethanol inhibition becomes significant beyond a concentration of ~100 g/L. The cell growth stops completely above an ethanol concentration of 136 g/L.

The above experiment was also performed with the recombinant yeast using xylose as the fermentation substrate. The slower growth rate on xylose is evident from the highest specific growth rate of 0.19 hr^{-1}, which is about 3 times lower than the highest specific growth rate on an equal concentration of glucose. The inhibitory effect of ethanol on the cell growth on xylose is also stronger, in comparison to cell growth on glucose. The cell growth on xylose is inhibited strongly beyond an ethanol concentration of 60 g/L, as seen in Figure 7. However, the genetically engineered yeast has been designed to overcome this ethanol inhibition by being able to use glucose for growth, and then cofermenting the rest of the glucose and xylose to ethanol.

Fermentation Performance of the Recombinant *Saccharomyces* 1400 (pLNH33) on Lignocellulosic Hydrolyzates. The results which have been discussed above were obtained from studies conducted on pure sugars. These studies are useful, as they provide insight into the fermentation kinetics. In order to study the fermentation performance of this recombinant yeast on "real" substrates, some experiments conducted using corn fiber and corn cob as model feedstocks are presented.

The pretreatment of the lignocellulosic material is an important process step to achieve higher ethanol yields. Various physical and chemical pretreatment methods have been reported in the literature. These include physical treatments, chemical treatment using strong acids and bases, steam explosion and the low temperature ammonia fiber explosion (AFEX) process (*13*). Pretreatment studies in our laboratory have resulted in the development of two pretreatment techniques: dilute acid hydrolysis and ammonia steeping followed by dilute acid hydrolysis. Pretreatment of the corn fiber was accomplished by acid hydrolysis using 0.5% HCl. The corn cob which contains a higher fraction of lignin was pretreated by the ammonia steeping process followed by the dilute acid hydrolysis.

Corn Fiber Studies. Corn based starch that is mainly obtained by corn wet milling is a predominant feedstock for fuel ethanol production. A low value side stream called "corn fiber" is generated in the corn wet milling process. This stream contains the hulls, fine fibers and residual starch granules from washing the starch. About 9 to 10% (w/v) of the original dry weight of the corn is recovered in the corn fiber stream. Thus corn fiber is an attractive starting material for ethanol production, in view of its availability, limited use and low cost.

The typical composition of corn fiber used in these studies is as follows: 25% starch, 25% hemicellulose and 15% cellulose. Pretreatment of corn fiber was accomplished by hydrolysis using 0.5% (w/v) HCl at 120°C for 45 minutes. This pretreatment by dilute acid hydrolysis readily hydrolyses the starch to glucose, and hemicellulose to primarily xylose. The cellulosic residue is then treated with the cellulase enzyme to hydrolyze it to glucose.

In the simultaneous saccharification and fermentation (SSF) studies, the recombinant yeast 1400 (pLNH33) and the Iogen cellulase are added together to the medium containing the pretreated corn fiber. Figure 8 shows the result of a batch SSF process using 10% w/v dry pretreated corn fiber with 0.46 FPU/ml (about 6 FPU per gram of cellulose from corn fiber) Iogen cellulase enzyme in 50 ml medium. As seen from the substrate profiles, there is a simultaneous utilization of glucose and xylose by the recombinant yeast. An ethanol concentration of 25.1 g/L was obtained at the end of 72 hours of SSF. Based on 15 g/L of glucose released from cellulose, 24.6 g/L glucose and 16 g/L xylose from the pretreated corn fiber, the total fermentable sugar is 55.6 g/L. The ethanol yield obtained in this experiment was 0.45 g/g, that corresponds to 88 % of the theoretical yield at 0.51 g/g.

The fed-batch SSF process is an effective process methodology for achieving high ethanol productivities and reducing enzyme costs by lower cellulase loadings. Figure 9a shows a fed-batch SSF process starting with 10% (w/v) pretreated corn fiber, followed by an identical feeding to give a total of 20% (w/v) solids. Ethanol concentrations of 38.9 g/L and 41 g/L were obtained after 72 and 96 hours, respectively. Instead of a single feeding, multiple feeds can also be used in the fed-batch SSF process. Figure 9b shows the result of such an experiment, in which the SSF is started with 10% (w/v) pretreated corn fiber as before, followed by two identical feedings at 12 hour intervals to give a total of 30% (w/v) solids. Ethanol concentrations of 44.4 g/L and 48.7 g/L were obtained after 72 and 96 hours, respectively. Based on these results, the ethanol productivities during SSF of pretreated corn fiber lie in the range of 0.44-0.62 g/L-hr.

Corn Cob Studies. Corn cob is a low value agricultural residue having limited use. A typical composition of corn cob used in these studies is as follows: 41.1% cellulose, 36% xylan, 6.8% lignin and 3.2% acetate. Compared to corn fiber, the percentage of xylose in sugars obtained from corn cob will be typically higher. This requires the use of a xylose fermenting microorganism for effectively fermenting xylose to ethanol.

Recently, an effective pretreatment process for lignocellulosic biomass has been developed (Cao, N. J., personal communication). This process involves steeping the raw material in 10% ammonium hydroxide solution at ambient temperatures for 24 hours. It has been determined that this ammonia steeping process efficiently removes lignin, acetate and extractives. Following this treatment, a dilute acid hydrolysis is used to hydrolyze the hemicellulose fraction. The residual fraction is primarily comprised of cellulose, that can be subjected to enzymatic hydrolysis. This process methodology systematically separates lignin, hemicellulose, cellulose and enables separate processing of each fraction. Since the lignin is removed in the initial stages of the process, the adsorption of cellulase on the lignin is minimized. Thus this pretreatment method also allows lower cellulase loadings in the SSF process. By coupling this pretreatment method with the use of the recombinant *Saccharomyces* 1400 (pLNH33) in the fermentation, promising results have been obtained (Cao, N. J., et al. *Biotechnol. Lett.*, in press).

The ammonia steeping pretreatment removes almost all the acetate from the raw material. This is a key step in the process, as acetic acid has been determined to be inhibitory for xylose fermentation (*14*). As discussed earlier, a dilute HCl pretreatment hydrolyses the hemicellulose and a xylose rich hydrolyzate is obtained,

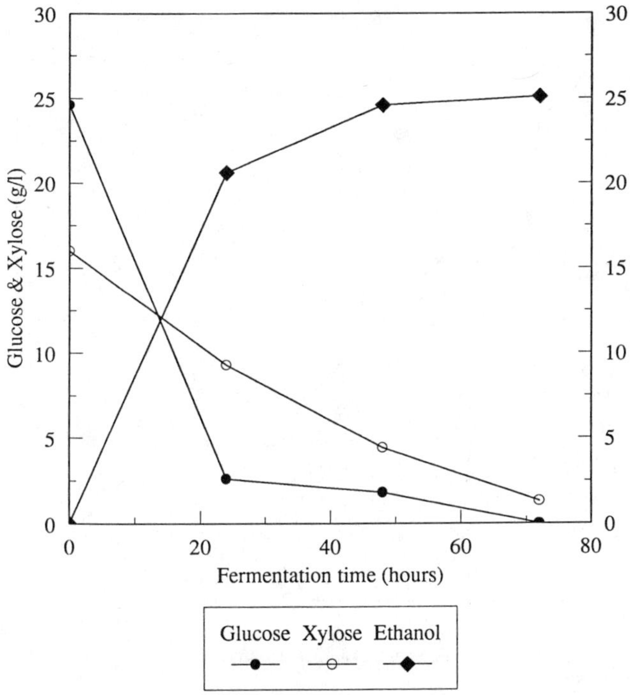

Figure 8. Simultaneous saccharification and fermentation (SSF) of 10% (w/v) pretreated corn fiber using the recombinant yeast strain 1400 (pLNH33) and Iogen cellulase.

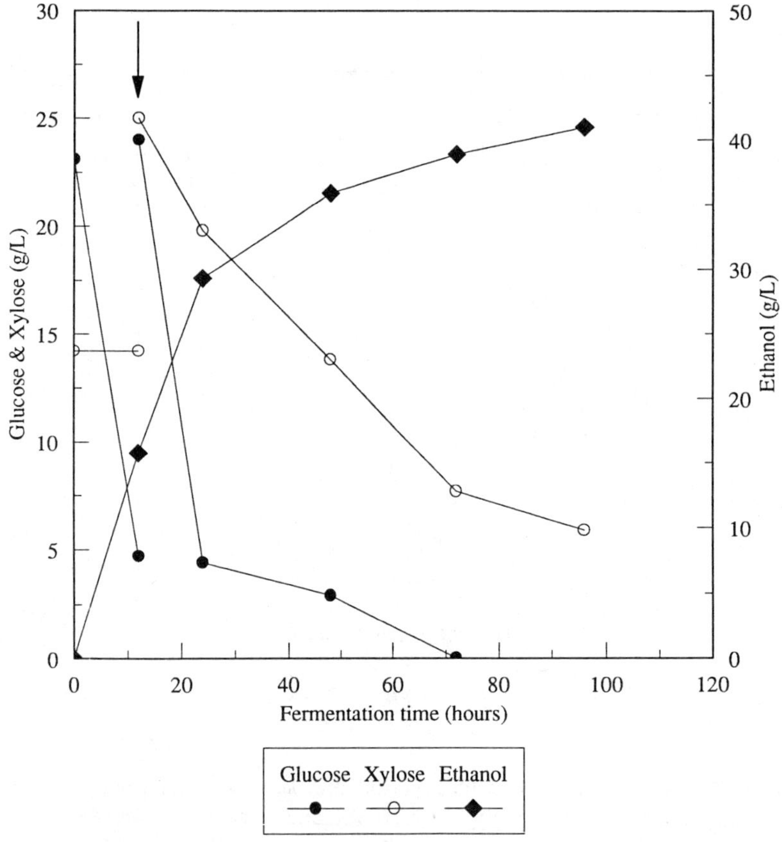

Figure 9a. Single stage fed-batch SSF process of pretreated corn fiber using the recombinant yeast strain 1400 (pLNH33) and Iogen cellulase.

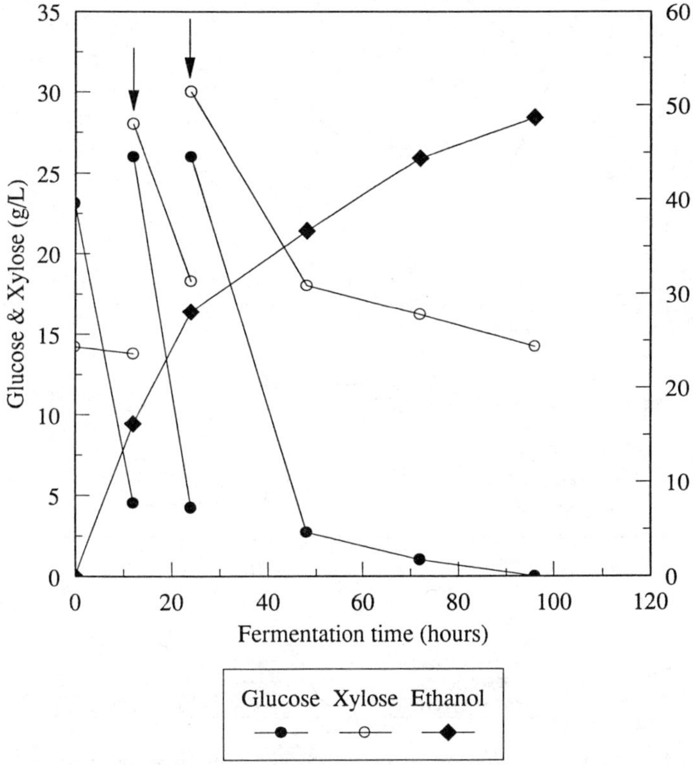

Figure 9b. Two stage fed-batch SSF process of pretreated corn fiber using the recombinant yeast strain 1400 (pLNH33) and Iogen cellulase.

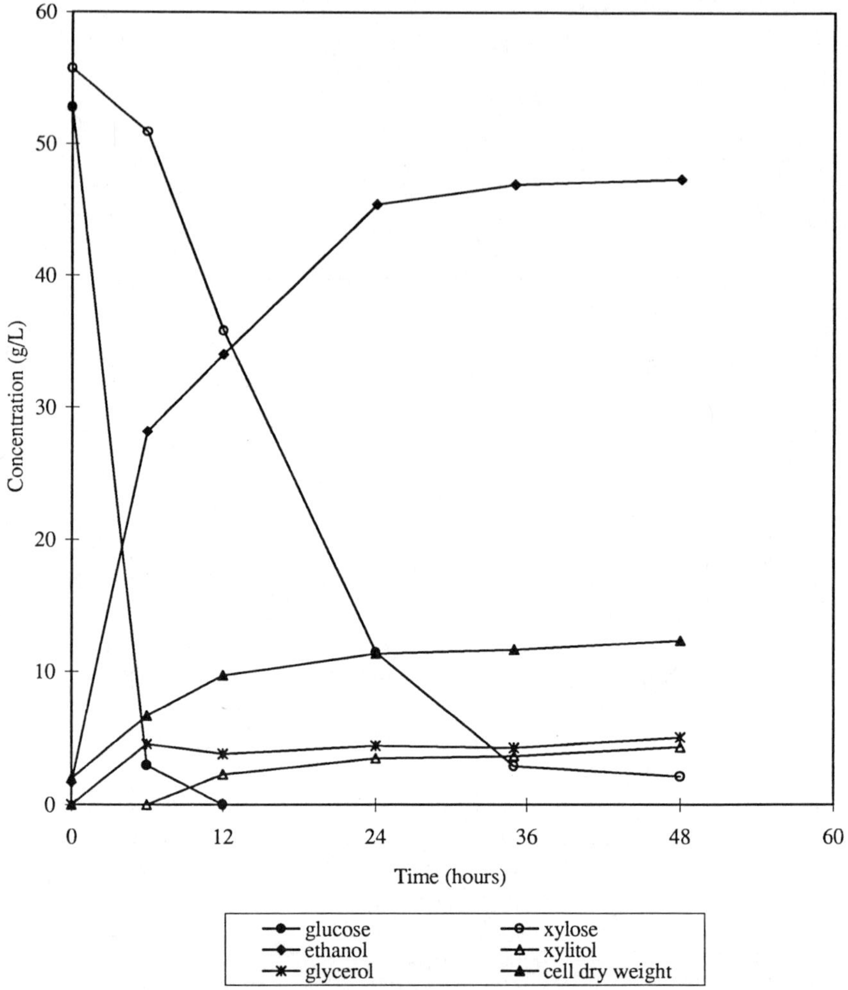

Figure 10. Batch fermentation of cellulose-hemicellulose hydrolyzate from corn cob using the recombinant yeast strain 1400 (pLNH33).

that is treated with a weak base anion exchange resin. Iogen cellulase was used to hydrolyze the cellulosic residue, which results in a glucose rich hydrolyzate. Both the hemicellulose and cellulose hydrolyzates were mixed and fermented after adjusting the pH to 5. Figure 10 shows the result of a batch fermentation of this mixed hydrolyzate containing 52.8 g/L glucose and 55.7 g/L xylose using the recombinant yeast. An ethanol concentration of 46.9 g/L was obtained within 36 hours, giving a high yield of 84% (based on a theoretical yield of 0.51 g/g).

Results of the SSF process of cellulosic residue (pretreated by different methods) obtained from 20 g corn cob using the Iogen cellulase and the recombinant

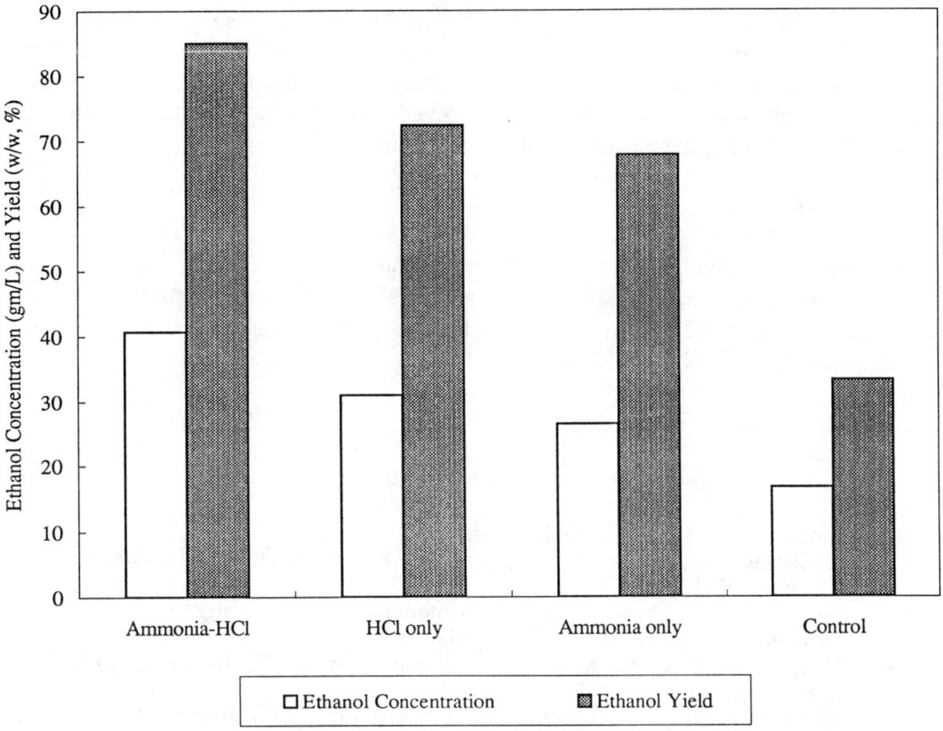

Figure 11. Effectiveness of different pretreatment methods for the SSF of cellulosic residue from corn cob using the recombinant yeast strain 1400 (pLNH33) and Iogen cellulase.

yeast 1400 (pLNH33) at 35°C are shown in Figure 11. The best results were obtained with the pretreatment involving ammonia steeping followed by dilute acid hydrolysis. An ethanol concentration of 40.7 g/L was obtained after 48 hours with a yield of 86% based on dry cellulose from corn cob. Application of the pretreatment method allowed a low cellulase loading (8.5 IFPU/g corn cob) to be used in the SSF process.

Conclusions

Fermentation studies using the recombinant yeast 1400 (pLNH33) show good results with high ethanol yields achieved on glucose, xylose and their mixtures. The simultaneous sugar utilization pattern exhibited by this yeast is beneficial for the SSF process, as the xylose fermentation time is reduced significantly. Moreover, this recombinant yeast retains the high ethanol tolerance of its parent (the fusion yeast strain 1400). This improves the economics of ethanol recovery greatly, as high final ethanol concentrations can be achieved in the fermentation broth.

Promising results with the yeast 1400 (pLNH33) have also been obtained on lignocellulosic feedstocks like corn fiber and corn cob. Significant process improvements have also been made in the biomass pretreatment technology. The use of this recombinant yeast coupled with improved pretreatment techniques provides a firm base for developing a mature biomass to ethanol technology.

Acknowledgments

This work was funded by Amoco Corporation and the Consortium for Plant Biotechnology Research. The corn cob studies were conducted by M.S. Krishnan with N. J. Cao and C. S. Gong. We acknowledge the assistance of Jean Lu and Xuezhi Yu for the HPLC analysis.

Literature Cited

1. Lynd, L. R.; Cushman, J. H., Nichols, R. J., Wyman, C. E., *Science*, **1991**, *251*,1318-1322.
2. Schneider, H., *Crit. Rev. Biotechnol.*, **1989**, *9(1)*, 1-40.
3. Hahn-Hagerdal, B.; Jeppson, H., Skoog, K., Prior, B. A., *Enzyme Microb. Technol.*, **1994**, *16*, 933-943.
4. Ohta, K.; Beall, D. S., Mejia, J. P., Shanmugam, K. T., Ingram, L. O., *Appl. Environ. Microbiol.*, **1991**, *57*, 893-900.
5. Ohta, K.; Beall, D. S., Mejia, J. P., Shanmugam, K. T., Ingram, L. O., *Appl. Environ. Microbiol.*, **1991**, *57*, 2810-2815.
6. Zhang, M.; Eddy, C., Deanda, K., Finkelstein, M., Picataggio, S., *Science*, **1995**, *267*, 240-243.
7. Kotter, P.; Amore, R., Hollenberg, C. P., Ciriacy, M., *Curr. Genet.*, **1990**, *18*, 493.
8. Tantirungkij, M.; Nakashima, N., Seki, T., Yoshida, T., *J. Ferment. Bioeng.*, **1993**, *75*, 83.
9. Ho, N. W. Y.; Tsao, G. T., "Recombinant Yeasts for Effective Fermentation of Glucose and Xylose", PCT Patent No. W095/13362, 1995.
10. Ho, N. W. Y.; Chen, Z. D., Brainard, A., In *Proceedings of Tenth International Conference on Alcohol Fuels*, Colorado Springs, CO, 1993, pp. 738.
11. D' Amore, T.; Panchal, C. J., Russell, I., Stewart, G. G., *Crit. Rev. Biotechnol.*, **1990**, *9(4)*, 287-305.
12. Krishnan, M. S.; Xia, Y., Tsao, G. T., Kasthurikrishnan, N., Srinivasan, N., Cooks, R. G., *Appl. Biochem. Biotechnol.*, **1995**, *51/52*, 479-493.
13. McMillan, J. D., In *Enzymatic Conversion of Biomass for Fuels Production*; Himmel, M. E., Baker, J. O., Overend, R. P., Eds.; ACS Symposium Series 566; American Chemical Society: Washington, DC, 1994, pp. 292-324.
14. Mohandas, D. V.; Whelan, D. R., Panchal, C. J., *Appl. Biochem. Biotechnol.*, **1995**, *51/52*, 307-318.

Chapter 5

Carbon Dioxide Effects on Fuel Alcohol Fermentation

Daniel W. Karl[1], Kris M. Roth[2,3], Frederick J. Schendel[4], Van D. Gooch[5], and Bruce J. Jordan[2]

[1]Daniel Karl Scientific Consulting, 430 Saratoga Street South, St. Paul, MN 55105
[2]Morris Ag-Energy, Inc., P.O. Box 111, Morris, MN 56267
[3]Department of Chemistry, University of Minnesota, Morris, MN 56267
[4]ENCORE Technologies, 111 Cheshire Lane, Suite 500, Minnetonka, MN 55305
[5]Department of Biology, University of Minnesota, Morris, MN 56267

High levels of carbon dioxide are known to be inhibitory to yeast growth, at least at the low temperatures prevailing in the brewing industry, and have also been suggested to favor increased diversion of carbon to glycerol. Since it was not clear whether the inhibitory effects depend on the bulk concentration of CO_2 or on its partial pressure, it was not clear whether the same results would be obtained under the higher temperatures employed in fuel alcohol fermentation. We first determined the conditions prevailing in an industrial corn-to-ethanol fermentation plant employing relatively small fermentors, then carried out laboratory fed-batch fermentations with glucose feed with CO_2 partial pressures of 0.5, 1.5, 2.5, and 3.5 atm absolute. Elevated carbon dioxide slowed the fermentation, particularly at the later stages, decreased the maximum number of viable cells obtained and increased cell death rates slightly. High carbon dioxide also decreased overall glycerol production. Low-level aeration also decreased glycerol productivity on a per-cell basis but stimulated cell growth to a compensating extent so that the final level was comparable to the control.

Carbon dioxide has both stimulatory and inhibitory effects on the metabolism of living cells. It is known to be required, at relatively low concentrations, by several essential biochemical pathways. For yeast, carbon dioxide concentrations up to 5% in the gas phase have been found to be stimulatory *(1,2)*. Inhibition of various functions begins in at higher concentrations. Aerobic metabolism is significantly inhibited at 0.5 atm CO_2 *(3)*, but fermentation *per se* is not inhibited at 3.5 atm *(4)* and only begins to be inhibited at 10 atm *(5)*. Anaerobic yeast growth is inhibited by lower concentrations, with effects apparent as low as 1.5 atm, depending on the temperature at which the yeast are growing and the strain of yeast *(6-9)*; some selected strains have been propagated at elevated CO_2 concentrations *(10)*. Both rate and extent of growth are affected by inhibitory levels of CO_2 under conditions of the brewing industry; the presence of abortive buds and enlargement of the cells suggests interference at specific steps in the cell cycle. Rice *et al.* present evidence that it is the concentration of dissolved carbon dioxide, and not its partial pressure, which determines the extent of inhibition *(6)*. Thus a given partial pressure of CO_2 became less inhibitory as the temperature increased, within the temperature range encountered in brewing. Whether this trend continues into the substantially higher temperature range employed in fuel alcohol production is not known.

The mechanisms involved in CO_2 inhibition are unclear, although there are many candidates *(11)*. Carbon dioxide is believed to partition freely into and through biological membranes *(12,13)*, so a purely osmotic mechanism seems unlikely. Yeast cells employ ion-transport mechanisms to maintain their internal pH in spite of the perturbing effect of membrane-permeating weak acids such as CO_2, acetic acid, or propionic acid; this is effective in limiting the intracellular pH change to about one unit for a four-unit change in the external pH *(14)*, but carries a cost in energy expenditure. It is not known whether yeast cells have a mechanism for expelling the resulting bicarbonate ion; if they do not, intracellular bicarbonate concentrations could become high enough to inhibit cytoplasmic enzymes *(15)*. Carbon dioxide is similar to other weak acids in inducing potassium uptake by yeast *(14)*. Action at or within the plasma membrane may also account for some or all of the inhibition, similar to mechanisms postulated for ethanol inhibition *(16,17)*.

Oura in 1977 argued that carbon dioxide can also have a substantial influence to increase production of the fermentation coproducts glycerol and succinic acid *(18)*. Glycerol, which is produced to maintain redox balance within the cell, can account for a substantial diversion of carbon away from ethanol production. A major part of the glycerol production occurs to correct a redox imbalance due to production of succinic acid. Decreased carbon dioxide partial pressure and increased available nitrogen were suggested as means of minimizing succinate-associated glycerol production. The available evidence, however, suggests that high carbon dioxide partial pressure decreases rather than increases glycerol production, at least under semi-aerobic conditions in continuous fermentation *(19,20)*.

Although hydrostatic pressures of a few atmospheres have no detectable effects on yeast, hydrostatic pressure increases the saturation concentration of carbon dioxide. Tall fermentors may engender hydrostatic pressures of two atmospheres or more; adding atmospheric pressure and supersaturation may result in local CO_2 partial pressures of 3.5 atmospheres near the bottom of a tall industrial fermentor. Indeed, adoption of the use of tall tanks prompted much of the brewing industry's interest in carbon dioxide effects. We sought to determine whether carbon dioxide effects on fermentation and carbon diversion to glycerol under conditions of the fuel alcohol industry should be a consideration in fermentor design and plant operation.

This investigation had three parts. First, we determined the conditions prevailing during ethanol fermentation in a commercial ethanol plant. Second, we conducted controlled laboratory fermentations at various carbon dioxide pressures. We did not attempt an exact duplication of the industrial process, but rather to simulate certain of the biologically relevant conditions in a more controllable fashion. Thus we used a steady glucose feed to simulate the continuous release of glucose from starch in the industrial fermentation, and used yeast extract to provide the complex nitrogen compounds provided in the industrial proceed by recycled stillage (backset). The remainder of the medium was based on a well known defined medium to insure nutritional adequacy. Finally, based on what we observed in the laboratory runs, we attempted to decrease carbon dioxide levels and control glycerol production in the industrial process by operation at reduced pressure or with slow air sparge.

Plant and Operation

Morris Ag-Energy operates a dry-milling ethanol plant at Morris, Minnesota. At the time of these experiments the capacity of this plant had been increased from 4.5 million to about 6 million gpa, primarily through debottlenecking and improved operation. The plant employs conventional batch fermentation and primary distillation systems and a molecular-sieve dehydration system. The fermentors are relatively small, of shallow-tank design, about 14,500 gallons (58000 l) capacity with a working depth of about 8 ft (2.5 m). They are fitted with top-drive slow-speed agitators and internal cooling coils. One fermentor was used for all the in-plant runs; it was modified for pH monitoring and sparging with air or other gasses. A bilobe rotary compressor (Roots blower) of about 60 cfm free air capacity was used for vacuum or headspace gas recirculation.

In this operation, corn is ground in a hammermill, then mixed with water (48 gpm, 182 l/min) and recycled thin stillage (39 gpm, 148 l/min) to make a mash of 20 ^{o} Brix. The starch is gelatinized and digested at 90 ^{o}C with 120 ml/min of commercial bacterial alpha amylase (IBIS) in a series of stirred tanks; the pH is controlled at 6.5 with ammonia. Further saccharification with glucoamylase (Alltech, 1 volume per 2900 volumes mash) occurs in the fermentor after cooling to 32 ^{o}C and pH adjustment. Yeast is produced continuously in a semiaerobic yeast propagator fed the same mash; all of the glucoamylase is added through the yeast propagator so that the yeast see a high initial glucose concentration. The yeast propagator operates at pH 3.5 (adjusted with sulfuric acid) with a cell count typically 0.5×10^7 to 1.5×10^7 ml^{-1} and viability 75%. The metabolism of the yeast in the propagator is primarily fermentative but they retain the ability to quickly consume added oxygen. For these experiments the yeast suspension from the propagator was 1/8 the total fermentor charge, resulting in an initial pH of 5.6 which declines during the course of the fermentation to a limiting value of about 3.8. After fermentation the beer is pumped to a series of beer wells where fermentation is completed and then to the distilling column. Although the normal residence time in the fermentor is 40 h, some of the experimental runs were kept at least 48 h to monitor the completion of the fermentation.

Methods

Analytical. Cell viability was determined using methylene blue *(21)* , a microscope, and a hemocytometer; total and viable cell counts were determined from the same data. Glucose and glycerol were determined enzymatically using prepared commercial reagents (Sigma 315-100 and 337-40A). Total glucose was determined after acid hydrolysis (0.25M H_2SO_4, 30 min, 100 ^oC); for most runs this analysis was employed only as a check of the total conversion. FAN (free alpha-amino nitrogen) was determined by the EBC ninhydrin method *(22)*. Ammonia was determined by a modified Berthelot reaction *(23)*. Ethanol was determined by gas chromatography using a Hayesep R column (Alltech Instruments). Ethanol values from the industrial runs were considered as relative values only and no interpretation was made of the absolute levels, due to possible handling losses. Carbon dioxide was determined by a modification of the Martin manometric method *(24)* using a commercial differential pressure sensor (Omega PX26-005DV) instead of a manometer. The reference side of the pressure sensor was connected to the vacuum pump via a 2 l flask which acted as ballast. Linear calibration curves were obtained through at least 60 mM CO_2. Concentration was related to partial pressure by the general approach of Schumpe *(25)*; the effect of ethanol was specifically included based on literature data at low concentration *(26)*. The effect of ethanol on CO_2 solubility was judged to be sufficiently linear and reproducible to permit use of this approach over the limited range of ethanol concentrations encountered in these experiments. Fermentor pH was determined *in situ* with commercial instruments (Omega) and electrodes (Phoenix).

Laboratory Fermentations. These fed-batch runs employed a Biolafitte Fermentor at the BioProcess Institute, University of Minnesota, St. Paul. The medium is listed in Table I; this was based originally on the medium of Oura *(27)* with NH_4Cl and yeast extract added to simulate the ammonia and FAN levels prevailing in the industrial fermentation, but it was necessary to increase the yeast extract to even approach the rates and cell counts prevailing in the industrial fermentation. Concentrations were figured on the basis of a 16 l final volume. The initial glucose concentration was 60 g/l; additional glucose was added to the fermentor as a concentrated solution during the run. The inoculum was Alltech alcohol-production yeast grown in 1 l of the same medium in a 2 l unsealed flask shaken at 200 rpm. This is the same yeast source employed by the plant to inoculate their yeast propagator. The fermentor was agitated vigorously and continuously sparged with a mixture of nitrogen and carbon dioxide. Gas mixtures were prepared by continuous metering through rotameters (Cole Parmer), using the manufacturer's calibration graphs. A back pressure regulator was used to maintain constant pressure and thus constant CO_2 partial pressure, calculated from the back pressure and the CO_2 content of the sparge gas. Volumetric productivity of ethanol and glycerol was calculated as the increase in concentration since the previous point, divided by the intervening time. Productivity per cell was calculated by dividing the volumetric productivity by the geometric mean of the cell counts at the beginning and end of the interval. Exponential growth and death rates were calculated from those parts of the growth curve which were linear on semi-logarithmic plots. In the air-supplemented laboratory run, oxygen uptake was monitored by continuous mass spectrometry of the inlet and outlet gas streams. An absolute calibration was not performed but the data serve to estimate the relative level and percentage consumption of supplied oxygen. Total oxygen consumption was calculated from these data and the known flow rates.

Table I. Medium Ingredients for Laboratory Reactor Runs

Ingredient	Concnetration
Yeast Extract	11.00 g/l
Tween 80 (d=1.06 g/ml)	5.51 g/l
Monopotassium Phosphate	2.20 g/l
Ammonium Sulfate	0.65 g/l
Magnesium Chloride (hexahydrate)	0.52 g/l
Inositol	0.12 g/l
Glucose (Initial)	60.00 g/l
Calcium Chloride (dihydrate)	90.00 mg/l
Ergosterol	51.00 mg/l
Ferric Ammonium Sulfate (hexahydrate)	35.00 mg/l
Manganese Sulfate (monohydrate)	3.80 mg/l
Copper Sulfate (pentahydrate)	0.50 mg/l
Sodium Molybdate (dihydrate)	3.30 ug/l
Zinc Sulfate (hepthydrate)	2.30 ug/l
Cobalt Sulfate (hexahydrate)	2.30 ug/l
Potassium Iodide	1.70 ug/l

Results

Normal Fermentation. Figure 1 shows a normal industrial batch fermentation as described above. Similar patterns were seen in other runs, with some differences of rate. The free glucose is seen to decrease smoothly from an initial level of about 60 g/l. This level reflects the balance between glucose production by the glucoamylase and its consumption by the yeast. Ammonia and FAN decrease rapidly; both are largely consumed by 8 h and exhausted by 12 h (Figure 1a). A portion of the FAN is not utilizable. The increase in viable cell count is essentially complete by 9h (Figure 1b), the rate of glucose utilization slows at his point, and the general pattern is consistent with nitrogen limitation. The initial cell viability is low, typically 60-70%; this increases to the 88-93% range by 4 h and remains nearly constant within that range through 40 h (viability data are not plotted on the graph). Fermentation, as reflected by ethanol production, was 82% complete at 20 h and 94% complete at 30h. Essentially all of the starch was converted to glucose and fermented by 37 h. Glycerol production in the industrial fermentation appears to be largely growth-

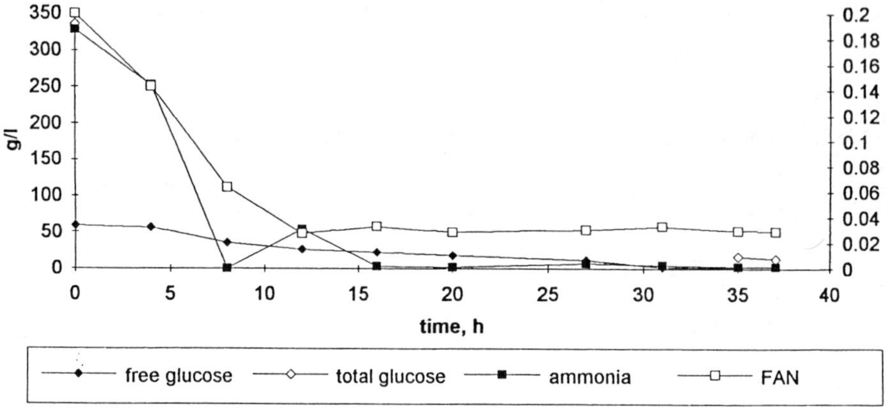

Figure 1a. Substrate Levels, Control Run

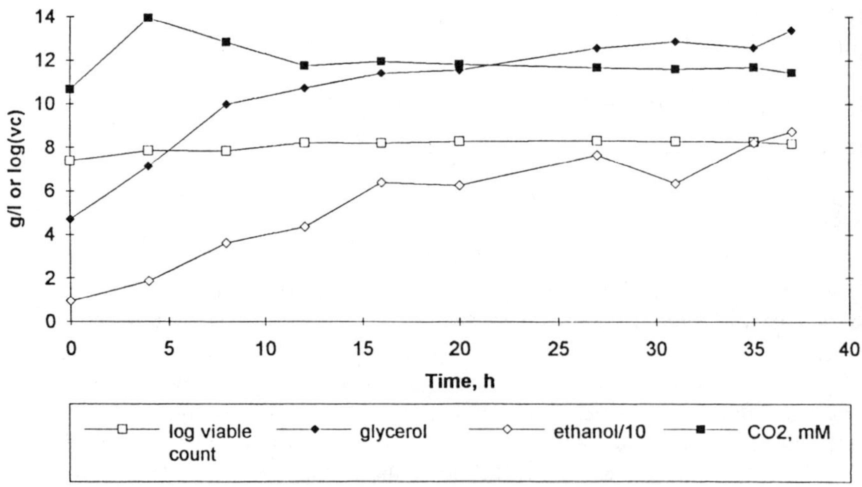

Figure 1b. Product Levels, Control Run

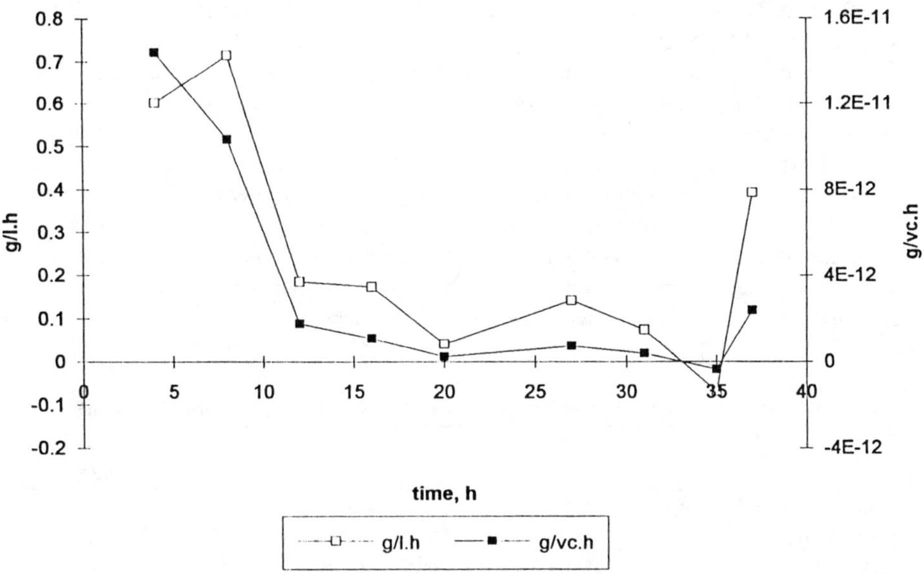

Figure 1c. Glycerol Productivity, Control Run

associated; the productivity on a per-cell basis drops abruptly after 8 hours (Figure 1c), at the same time the N sources are exhausted. The high starting glycerol level is due to the recycle of thin stillage and glycerol produced in the yeast propagator; there is a further increase in glycerol concentration in the beer well and still (not shown), possibly due to breakdown of cell components. The dissolved carbon dioxide concentration in a control run is shown in figure 1b; it increased from 27 mM at the start to 35 mM at the time of peak fermentation, then settled back to about 29 mM. These concentrations correspond to partial pressures of 1.5, 1.7, and 1.3 atm (solubility calculated separately for the estimated medium composition at each point), confirming that supersaturation is significant in this system.

Laboratory Fermentations. We set up a series of laboratory fermentations to test the effects of CO_2. The conditions of inoculum, pH (4.0), temperature (32.5 ^oC), ammonium and FAN and the initial glucose concentration were initially set to simulate the industrial process, and the glucose feed rate was set to simulate the rate of release of free glucose by glucoamylase in the industrial process. In one run, a low level of oxygen was added as air along with 1.5 atm CO_2. The results of this series of experiments are summarized in Table II. Peak viable cell counts ($5x10^7$ to 10^8 cells/ml) were lower than the levels seen in the plant but peak viability was typically 95-96%. The limiting factor or factors were not identified. Ammonia was not depleted and FAN was not reduced to the levels seen in the plant, so usable N was not limiting. Unlike the industrial fermentation, the lab runs showed decreasing viability after the peak cell count was reached.

Table II. Effect of CO_2 on the Production of Cells, Glycerol, and Ethanol

CO_2, atm	0.5	1.5	2.5	3.5	1.5, + air supplement
peak count, 10^6 cells/ml	85	87	68	52	98
growth rate, hr^{-1}	0.242	0.198	0.199	0.198	0.248
death rate, hr^{-1}	0.016	biphasic	0.016	0.021	0.019
glycerol yield, g	81	68	64	46	71
g glycerol/g glucose	0.021	0.019	0.018	0.013	0.019
mean productivity 10-60 h, pg/vc.h	2.25	2.29	1.25	2.07	1.53
ethanol yield, kg	1.96	1.91	1.86	1.52	1.42
g ethanol/g glucose	.51	.54	.51	.42	.38

Carbon Dioxide Effects. Carbon dioxide at 3.5 atm was somewhat inhibitory to cell growth (Table II). The peak cell count was decreased relative to the lower CO_2 levels, and the rate of cell death was slightly increased compared with the runs at 0.5 and 2.5 atm (cell death in the 1.5 atm run was biphasic and cannot be readily compared with the others). Peak viable count was also somewhat reduced at 2.5 atm compared with the lower CO_2 levels. An increase in cell size was noticed at

2.5 and 3.5 atm, but no unusual budding was observed. The fermentation rate decreased with increasing carbon dioxide concentrations (Table III); this effect was confined to roughly the last half of the fermentation. There were substantial differences between treatments at 30 and 45 h, but by 65 h ethanol had reached comparable levels in all the runs except possibly the run at 3.5. The ethanol production measured for the 3.5 atm run does not account for all the glucose apparently consumed; we have no ready explanation for this discrepancy but the fermentation rate is still depressed when measured as glucose consumption.

Table III. Effect of CO_2 on the Percentage Completion[a] of the Fermentation

Time	30h	45h
CO_2atm.	% completion	
0.5	80	100
1.5	71	90
2.5	46*	71
3.5	65	81

[a]Percentage completion was calculated as the increase in alcohol concentration through the indicated time divided by the total increase in alcohol concentration when all glucose was consumed. *glucose feed was started late, at 20h, in the 2.5 atm experiment, contributing to the low completion at 30 h.

Glycerol and Ethanol Production Kinetics. Both glycerol and ethanol accumulated most rapidly while the cell number was still increasing; volumetric productivity declined after the peak cell number was reached (Figure 2). Although this would suggest that the production of both glycerol and ethanol was largely growth associated, a more detailed analysis disputes that conclusion. When the glycerol productivity is calculated on a per-viable-cell basis, most of the runs show a similar pattern: Initially high per-cell productivity, about $7x10^{-12}$g/(cell.hour) declines within the first 6 hours to a plateau which persists through the remainder of the growth phase, stationary phase, and death phase (Figure 2a-c). Since the cell number is still increasing at 6 h, the higher glycerol productivity in the early periods is more specifically related to rapid growth than to growth *per se*. The bulk of the glycerol accumulation occurs later when the per-cell accumulation rate is constant, and cell growth has ceased; thus it is not growth-associated. The glycerol accumulation rate in this period is greater than that in the comparable period of the industrial fermentation. Overall glycerol yield was greatest at 0.5 atm CO_2 and least at 3.5 atm CO_2 among the non-aerated runs. An exception to this pattern was the air-supplemented run. Besides eliminating the initial spike air supplementation decreased the per-cell productivity during the plateau phase. However, the cell count in this run was enough higher to counteract this decrease so that the overall glycerol yield was little different than the control. The total oxygen uptake

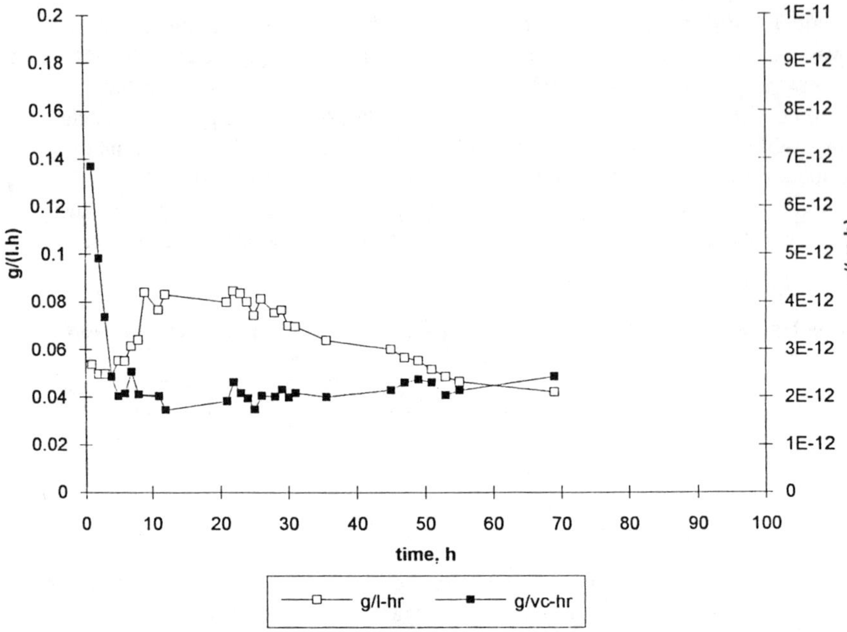

Figure 2a. Glycerol Productivity, 3.5 Atm CO2.

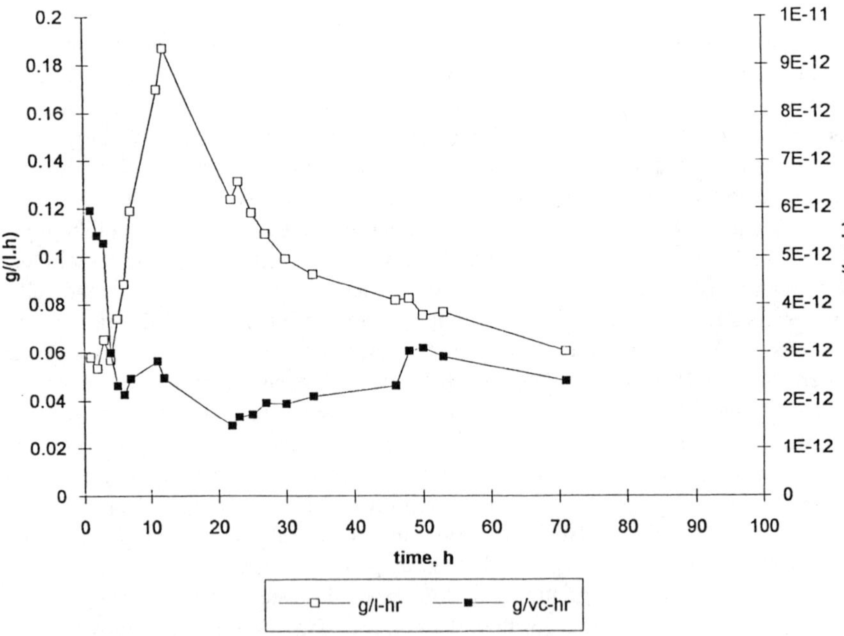

Figure 2b. Glycerol Productivity, 1.5 Atm CO2.

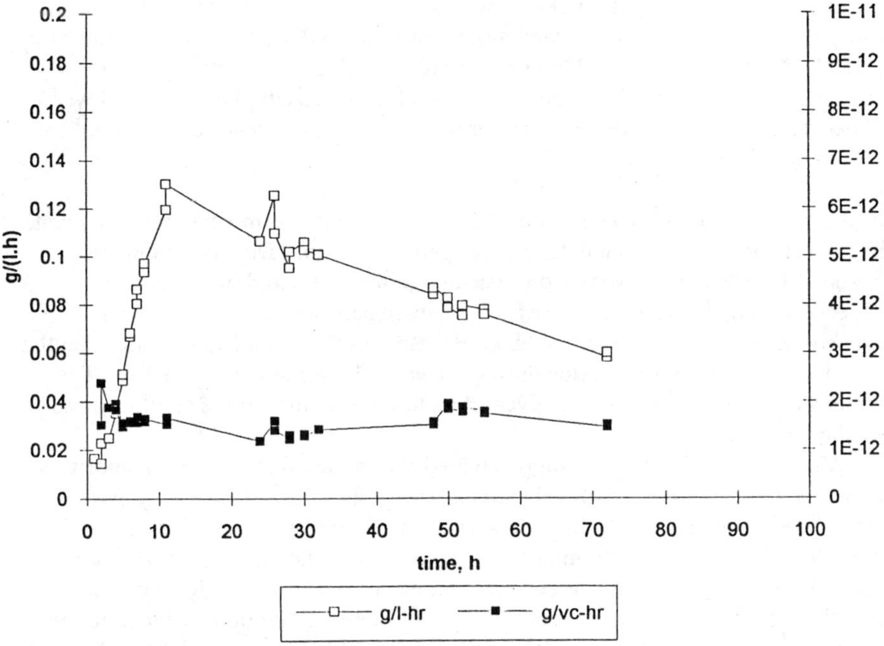

Figure 2c. Glycerol Productivity, 1.5 Atm CO2, Air Supplemented

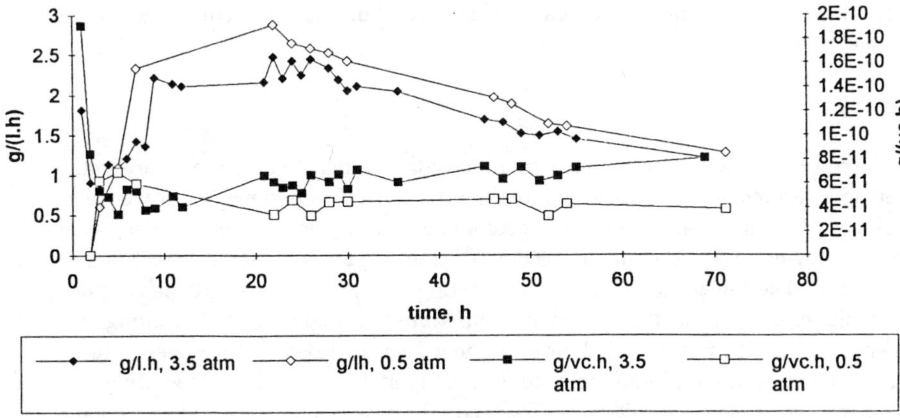

Figure 2d. Ethanol Productivity

amounted to 5.8 mmol/l based on the flow rate and composition of the gas streams entering and leaving the fermentor.

Ethanol productivity did not follow as simple a pattern and declined more slowly (Figure 2d). but also showed more rapid accumulation in the early part of the run and a plateau of decreased but continuing per-cell productivity as the run progressed (Figure 2d). The changes in per-cell productivity during the run were influenced by the CO_2 level. Ethanol production was judged not to be strongly growth-associated.

In-plant Experiments. We attempted to manipulate the fermentation in the plant, first by carrying out a fermentation under partial vacuum. Due to the limited capacity of the Roots blower used as vacuum pump, we could only achieve appreciable vacuum after the rate of CO_2 release had slowed. We ultimately reduced the headspace pressure to about 0.6 atm absolute, but had no impact on the level of dissolved carbon dioxide during the critical middle part of the run. The fermentation timecourse and the glycerol yield were similar to the control run (data not shown).

We also carried out an air-supplemented run in the plant. The fermentor was aerated at 20 cfm through sintered metal spargers, and the headspace gas was recirculated at about 40 cfm through separate perforated-pipe spargers. Figure 3c shows that there was a small impact of CO_2 levels, as shown by the reversal of the CO_2 trace when the sparge was shut off. However, the lowered CO_2 concentration was still at the level seen in the control run and there was no improvement in the fermentation kinetics. As in the lab run, aeration led to a slightly higher cell count and altered the pattern of glycerol accumulation. However this time the early growth-associated glycerol production was not eliminated, though it was reduced on a per-cell basis (Figure 3c). Later in the run, second wave of glycerol production occurred which brought the overall glycerol yield up to a level comparable to the control run.

Discussion

The results of the laboratory fermentations confirm that CO_2 is inhibitory to the ethanol fermentation. This inhibition manifests itself in two ways: a lower rate of glucose consumption and ethanol accumulation at higher CO_2 levels, associated with a lower peak cell count at 2.5 and 3.5 atm and an increased rate of cell death at 3.5 atm. The lower rate would lead to incomplete conversion of available glucose at the higher CO_2 levels if the fermentations were terminated at 45 h. Letting the fermentations continue to 65 h or more, there was little or no difference in the ultimate conversion of glucose except possibly at 3.5 atm. Two reservations accompany these conclusions. 1) Each condition was only tested once, so that no statistical test can be applied. The effects described here are greater, usually substantially greater, than the intrinsic uncertainties of the analytical methods. The reproducibility of the laboratory fermentation itself, however, was not tested due to limitations of time and resources. 2) Not all the carbon can be accounted for at 3.5

atm and in the air-supplemented run, though this may reflect a simple setup error. There was no indication of increased evaporation or acid production. The calculation of the percentage completion was based on the final ethanol concentration and was not influenced by any error in carbon recovery or amount of carbohydrate added. Similar slowing of the fermentation and decreases in the peak cell number have been observed at lower temperature by workers in the brewing industry *(6-9)*. Slowing of industrial alcohol fermentation of molasses by CO_2 was observed by Ukrainian workers *(28)*. However, if the concentration rather than the partial pressure of carbon dioxide is the determining factor *(6)*, it is somewhat surprising to observe appreciable slowing of the fermentation at pressures as low as 1.5 atmospheres, since the solubility of carbon dioxide is so much decreased at 32 ^{o}C. This slowing of the fermentation must arise in part from causes other than decreased cell count, since cell count was not decreased at 1.5 atm.

The laboratory experiments were run as fed-batch fermentations with pure glucose as the carbon source. This design was chosen to avoid any uncertainties associated with the use of complex substrates and enzymes *in situ.* The remainder of the medium was based on a well-characterized defined medium with ammonium chloride and yeast extract added as sources of nitrogen. Although we intended the lab runs to simulate the most important features of plant conditions, there were some noticeable differences. We were unable to get good fermentation rates or cell counts until we increased the level of N in the media. We now think it likely that the lower nitrogen media led to N-limitation of the inoculum; the runs employing the lower-N media are not reported here. Even with the higher N level, we could not match the cell count, the retention of cell vitality or the fermentation rate of the industrial runs. This is possibly due to the leaner semi-defined media used in the lab runs in comparison with the rich, complex ground-corn medium employed in the plant. Also, the base medium we used is not a particularly good match for the inorganic components of the corn mash. Supplementation with higher levels of N, lipids, or potassium did not improve the performance in shake flask experiments (not shown). The lab runs were a good match for the industrial runs in the critical variables of pH and temperature, particularly during the latter part of the fermentation when the CO_2 inhibition manifests itself.

Glycerol production represents an economically significant diversion of carbon, about 4-5%, from the production of ethanol (*18* and our results). Glycerol production under plant conditions is largely growth associated. Under lab conditions, glycerol production was largely non-growth-associated; this may be related to the fact that cell growth in the industrial fermentation was nitrogen limited, while growth in the main set of lab runs was not, or it may be related to the higher rate of cell death in the lab runs. The limiting factor in the lab fermentations was not identified. Glycerol formation was least in the 3.5 atmosphere lab run, and greatest at 0.5 atmospheres. This is contrary to Oura's hypothesis *(18)* but in accord with the results of Kuriyama *et al.* and Bur'yan & Volodrez *(19,20)*. These two papers employed semiaerobic continuous fermentations while our experiments were anaerobic batch fermentations. Oura based his suggestion of increased glycerol production at elevated CO_2 on the requirement of pyruvate carboxylase for CO_2 in

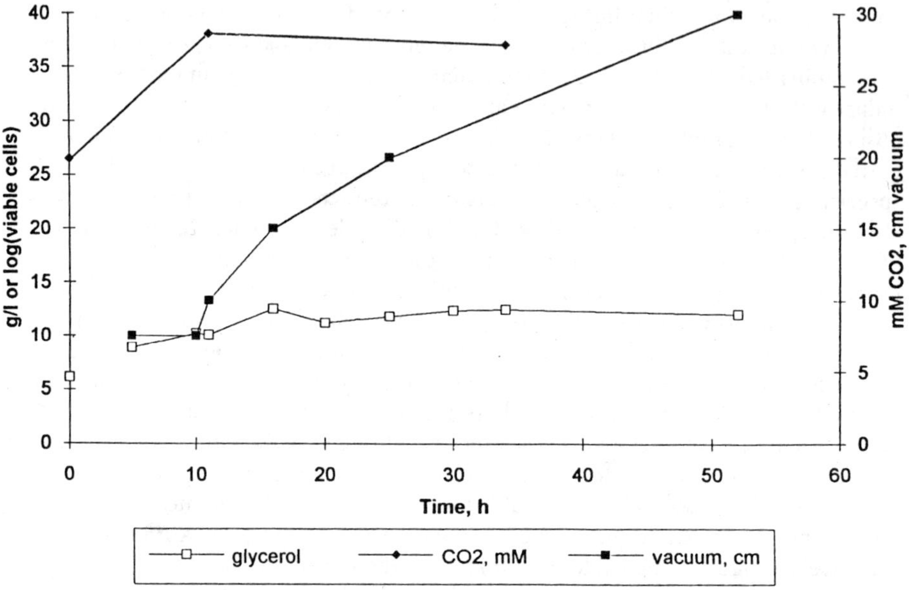

Figure 3a. Product and Vacuum Levels

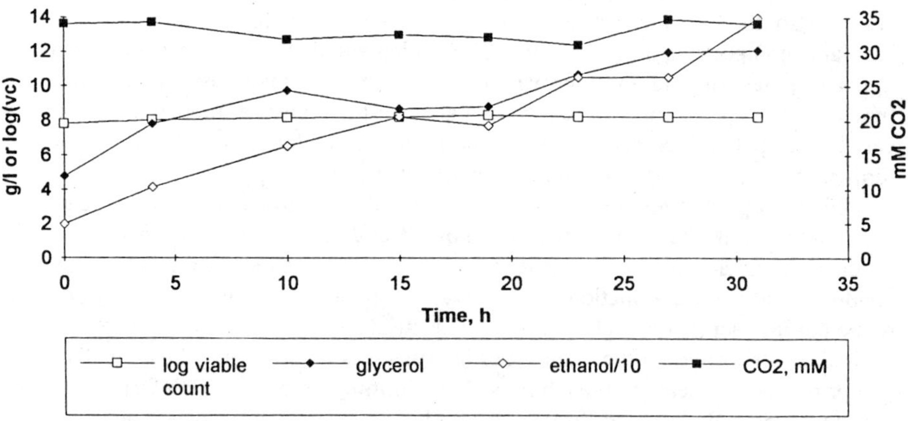

Figure 3b. Product Levels, Air Supplement 0–23 h

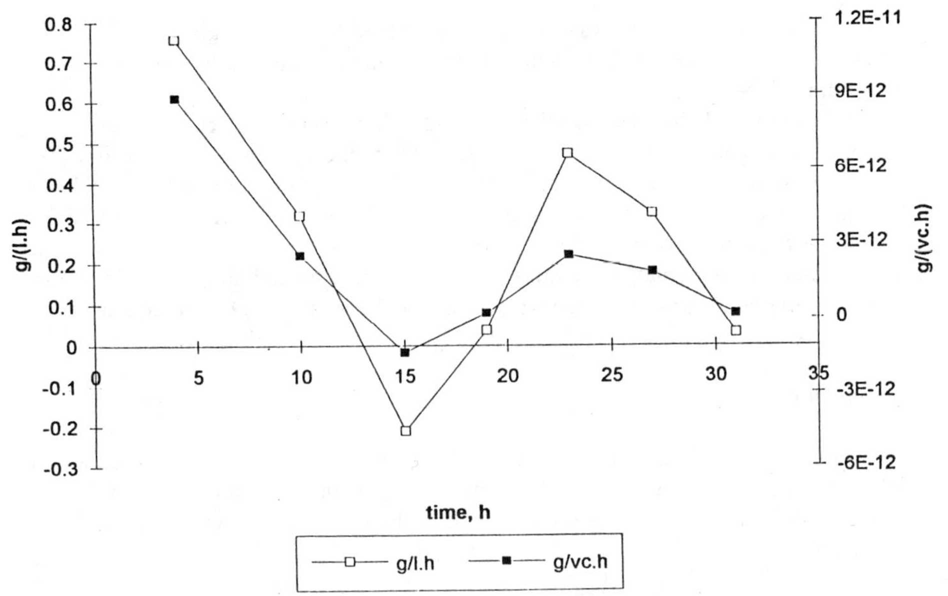

Figure 3c. Glycerol Productivity, Air Supplement 0–23 h

order to produce oxaloacetate, which is necessary for succinate production. However the K_m of yeast pyruvate carboxylase for $KHCO_3$ is in the range 2-8 mM *(29)*, so the enzyme should be easily saturated with CO_2. Kuriyama *et al.* suggest pyruvate dehydrogenase as the site sensitive to CO_2 but their results and ours may also be related to inhibition of pyruvate carboxylase by high CO_2 since Foster & Davis *(30)* found that high CO_2 inhibited *Rhizopus* pyruvate carboxylase. Although high levels of CO_2 suppressed some of the glycerol production the concomitant decrease in fermentation rate means that increased CO_2 levels would not be a good strategy for increasing yield of conventional batch fermentations under industrial conditions. Oura demonstrated that low levels of aeration in a continuous fermentation led to decreased glycerol production *(31)*. We were able to manipulate the rate of glycerol production with air supplementation, both in the lab and in the plant. The air-supplementation in each case was calculated to be just sufficient to permit the cells to retain redox balance without producing glycerol, assuming 50% oxygen uptake. In the lab, oxygen eliminated the initial spike of growth-associated glycerol productivity and decreased the non-growth-associated productivity on a per-cell basis but the accompanying higher cell count resulted no decrease in overall production. The oxygen uptake in the lab fermentation was substantially more than 50%, but decreases in the flow rate compensated for this so that the total oxygen uptake through the run was close to the intended amount. In the plant, oxygen decreased the growth-associated glycerol production on a per-cell basis but did not eliminate it. Additional glycerol was produced later in the run, resulting in no net improvement. In the lab, aeration was continued until 53 hours and then discontinued without resulting in an increase in glycerol production. It is plausible to think that an aeration schedule could be developed to decrease glycerol production, though more experimentation would be required to establish whether this could be achieved.

We were unable to influence the carbon dioxide level or the fermentation kinetics in the plant by operating the fermentor at reduced pressure. This might have been more effective with a larger pump which could keep up with the peak CO_2 production. The light aeration in the air-supplemented fermentation resulted in a small decrease in carbon dioxide level in the industrial fermentor, but no clear-cut improvement in fermentation kinetics. Other experiments including increased CO_2 back pressure and nitrogen sparge were considered and rejected due to cost or complexity.

Conclusions

Carbon dioxide levels as low as 1.5 atmospheres, the range prevailing in the shallow fermentors at Morris Ag-Energy, slows ethanol fermentation appreciably. In these experiments this effect became noticeable after the fermentation was about half complete.

Carbon dioxide at 2.5 and 3.5 atmospheres decreased the peak yeast cell count slightly and at 3.5 atmospheres was associated with more rapid cell death.

Higher carbon dioxide was accompanied by decreased glycerol production.

Air supplementation of the fermentation decreased glycerol productivity on a per-cell basis and led to higher cell counts. These effects combined to leave the overall glycerol production essentially unchanged.

Acknowledgments

This research was supported by a grant from the Minnesota Corn Research and Promotion Council, with additional support from AURI (Minnesota's Agricultural Utilization Research Institute), Morris Ag-Energy, and the Minnesota Department of Public Service. We thank Duaine Flanders and Richard W. Fulmer for encouragement and helpful advice, Gerald Bachmeier for arranging plant access, Nancy Goebel for help with assays and lab fermentations, and Sterling Keller and Craig Bremmon for help with instrumentation.

Literature Cited

1. Nour-El Dien, H. *Cheese Whey as a Substrate for Single-Cell Protein Production (Yeast Biomass)*. Ph.D. Thesis, 1980, Hungarian Academy of Sciences, cited by A. Halasz & R. Lasztity, *Use of Yeast Biomass in Food Production*, CRC Press, Boca Raton, 1991.
2. Parker, D., D.S. Baker, K.G. Brown, R.D. Tanner, & G.W. Malaney. *J. Biotechnol. 2*, 337-46, 1985.

3. Chen, S.L. & F. Gutmanis. *Biotech. Bioeng. 18*, 1455-62, 1976.
4. Norton, J.S. & R.W.E. Krauss. *Plant & Cell Physiol. 13*, 139-49, 1972.
5. Thibault, J. A. LeDuy, & F. Cote. *Biotech. Bioeng. 30*, 74-80, 1987.
6. Rice, J.F., J.R. Helbert, & J.C. Garver. *Am. Soc. Brewing Chem. Proc. 34*, 94-96, 1974.
7. Renger, R.S., S.H. van Hareren, and K.C.A.M. Luyben. *J. Inst. Brewing. 98*, 509-13, 1992.
8. Arcay-Ledezma, G.J., & J.C. Slaughter. *J. Inst. Brewing. 90*, 81-4, 1984.
9. Lumsden, W.B., J.H. Duffus, & J.C. Slaughter. *J. Gen. Micro. 133*, 877-81, 1987.
10. Cahill, J.T., P.A. Carroad, & R.E. Kunckee. *Am. J. Enol. Vitic. 31*, 46-52, 1980.
11. Dixon, N.M., & D.B. Kell. *J. Appl. Bact. 67*, 109-136, 1989.
12. Kaethner, T.M. & A.D. Bangham. *Biochim. et. Biophys. Acta 468*, 157-61, 1977.
13. Simon, S.A., & J. Gutknecht. *Biochem. Biophys. Acta 596*, 352-8, 1980.
14. Ryan, J.P. & H. Ryan. *Biochem. J. 128*, 139-46, 1972.
15. Jones, R.P., & P.F. Greenfield. *Enz. Micro. Technol. 4*, 210-23, 1982.
16. Pamphula, M.E. & V. Loureiro. *Biotech. Lett. 11*, 269-74, 1989.
17. Carlsen, H., H. Degn, & D. Lloyd. *J. Gen. Microbiol. 137*, 2879-83, 1991.
18. Oura, E. *Process Biochemistry*, April 1977, 19-36.
19. Bur'yan, N.I., & G.D. Vodorez. *Vinograd. Vinodel. SSSR* 31, 10-12, 1971.(*Chemical Abstracts*)
20. Kuriyama, H., W. Mahakarnchanakul, S. Maru, & H. Kobayashi. *Biotech. Lett. 15*, 189-94, 1993.
21. American Society of Brewing Chemists, Eagan MN. *Methods, 8th ed.* Yeast-3A
22. Lie, S. *J. Inst. Brewing 79*, 37-41, 1972.
23. Koops, J., H. Klomp, & R. Elgersma, *Neth. Milk Dairy J. 29*, 169, 1975.
24. Martin, PA. *J. Inst. Brewing 76*, 344-7, 1970.
25. Schumpe, A. in *Biotechnology, a comprehensive treatise in 8 volumes. v2*. H. Brauer, ed. VCH, Weinheim, 1985 pp 159-170
26. *International Critical Tables, v III*, McGraw-Hill, 1928 p279 (data of Muller, 1889, and Findley & Shen, 1911)
27. Oura, E. *Biotechnol. Bioeng. 16*, 1197-1212, 1974.
28. Malysh, L.V., V.K. Yanchevskii, A.D. Kovalenko, & A.A. Egorov. *Pishch. Prom-st. (Kiev)* 1, 46-7, 1986. (*Chemical Abstracts*)
29. Kreze, G.B. & H. Ronft. *Eur. J. Bioch. 119*, 573-9, 1981.
30. Foster, J.W., & J.B. Davis. *Archives of Biochemistry 21*, 135-142, 1949.
31. Oura, E. *Biotech. Bioeng. Symp. 4*, 117-27, 1973.

Chapter 6

Ethanol from Wood: Design and Operation of a Process Development Unit for Technoeconomic Process Evaluation

M. Galbe, M. Larsson, K. Stenberg, C. Tengborg, and G. Zacchi[1]

Department of Chemical Engineering I, Lund University, S–221 00 Lund, Sweden

Ethanol is an attractive alternative for replacing gasoline as a motor fuel and it can be produced from a number of cellulosic materials. However, processes in which biomass is involved are often very complex. By a combination of computer simulations and experiments in a bench-scale unit different process strategies can be evaluated, and the results from experiments can serve as feed-back to the process simulations. The results in this study show the impact of recycling process streams and the effect of different energy-integration options.

Ethanol, which is believed to be an interesting alternative for replacing gasoline as a motor fuel can be produced from a number of cellulosic materials including agricultural products (1-4). The main reason for turning towards processes utilising cellulosic materials is the abundance of various cellulose sources, such as forest waste (4-6). Ethanol production from sugar- or starch-containing crops is an industrially well-established technique which has been used for many years, mainly for the production of alcoholic beverages. However, the process technology for the conversion of cellulosic materials into ethanol has not yet been fully optimised. A number of different technologies have been proposed, the major difference being the way in which the material is hydrolysed and the fermentable sugars extracted.

Basically, the production of ethanol from, for example, wood can be performed by hydrolysing the material, thus releasing fermentable sugars. The sugars are fermented to ethanol using various micro-organisms, and the diluted ethanol is recovered in a distillation unit. Hydrolysis can be performed either by the use of dilute or concentrated acids (7-9), or by using cellulose-degrading enzymes (3,10-12). The advantages of acid hydrolysis are the well-established technology and short reactor residence times. The main drawbacks are the corrosive action of the acid on equipment, the production of large amounts of salts, such as gypsum, from the neutralisation of the acid and the non-selectivity of the acid. Enzymatic hydrolysis has

[1]Corresponding author

the potential to give higher yields, and it is performed at temperatures only slightly above room temperature, 40-50°C. However, enzymatic hydrolysis requires pretreatment in order to open up the compact structure of the cellulosic material (13-16).

When developing a new industrial process for ethanol production based on enzymatic hydrolysis, it is essential to have access to reliable experimental data. Most data found in the literature are based on laboratory-scale equipment, and determined under carefully controlled conditions, although some pilot facilities have been used to verify some of the laboratory data (17-19). Most of the data are also for individual steps, e.g. pretreatment, hydrolysis or fermentation, and are not always valid for a fully integrated process. This leads to many assumptions when estimating the ethanol production cost. This is reflected in the large discrepancy between estimated production costs. Several cost estimates for the production of ethanol can be found in the literature with prices ranging from 0.18 to 1.51 US$/L (20-22). Accurate design requires mid-scale process facilities with integrated process steps, e.g. bench-scale or pilot plants. A bench-scale unit is attractive, due to its flexibility and low equipment cost, for investigations of a variety of process parameters and process configurations. The more expensive pilot-plant facilities are, however, necessary, for final scale-up of the process.

Process development also requires technical and economic calculations in order to estimate equipment and energy requirements, to assess the overall production cost and to discriminate between various process configurations. Process simulations are also valuable for the identification of cost-critical steps and for the planning and configuration of experimental investigations in small-scale equipment. The data thus gathered can be used to refine the technical and economic models for more accurate calculations.

This chapter presents the combined use of a bench-scale unit and process simulation for the development and optimisation of an integrated bio-ethanol plant based on enzymatic hydrolysis of lignocellulosics.

Process Description

General Process. Figure 1 shows a schematic flowsheet of the enzymatic biomass-to-ethanol process. Lignocellulosic biomass consists mainly of cellulose, hemicellulose and lignin. In the production of ethanol the sugars in the hemicellulose and cellulose are converted to ethanol. As a by-product, lignin is formed which can be used as a solid fuel.

In the pretreatment stage the hemicellulose is solubilised into sugars. The cellulose structure is opened up and becomes more accessible to enzymes. Prior to pretreatment, the raw material must be chipped. There are many types of pretreatment methods and steaming has been recognised as one of the most effective for lignocellulosics (12, 23-27). Although companies such as Stake Technology (Stake Technology Ltd, Canada) currently sell equipment for the steam pretreatment of wood, the technology is still considered to be in the development stage (28). Furthermore, the use of catalysts to enhance the cellulose accessibility and the yield of hemicellulose sugars of a number of feedstocks has been recognised (14, 15, 29, 30). If catalysts such as sulphuric acid or sulphur dioxide are used, the corrosive effects of long-term usage on the equipment must be considered. The effect on the quality of the lignin by-product, due to possible sulphonation, must also be taken into account. If a catalyst is employed, the raw material must usually be pre-steamed to expel trapped air from the pores in the material. This can be performed using low-pressure steam.

In the hydrolysis step, the cellulose is hydrolysed by enzymes into glucose. To reduce the cost of concentrating the end-product, a high sugar concentration is desirable, which means a high dry matter content of the cellulosic material entering the hydrolysis stage. However, a high dry matter content in the reactor makes stirring difficult which reduces mass transfer. Fed-batch hydrolysis, i.e. gradual feeding of the material into the hydrolysis vessel, is an option that may be attractive.

Large quantities of enzymes are required for the hydrolysis to be fast due to the low specific activity of the cellulase system. The cellulose-degrading enzymes should preferably be produced on the same raw material as that used for ethanol production. A minor part, about 5%, of the pretreated material is used as a carbon source for some cellulase-producing organism, such as *Trichoderma reesei* (31-33). Unfortunately, this fungus does not excrete sufficient amounts of one of the enzyme components, β-glucosidase. As the consumption of β-glucosidase is much lower than that of the other cellulases, this enzyme may either be produced by another organism such as *Aspergillus phoenicis* (34, 35) or be purchased to avoid increased complexity in the process. Extensive research is necessary to find more effective enzymes which will reduce the hydrolysis time. It is also essential to find inexpensive substrates for enzyme production.

After hydrolysis, the solid residue is separated from the hydrolysate and washed. The residue consists mainly of lignin which can be dried and used as a high-quality solid fuel. Separation can be performed either through filtration in a filter press (36) or by using a decanting centrifuge (18). The sugars in the hydrolysate are then fermented to ethanol.

Hexoses, mainly glucose, mannose and galactose, are generally fermented by *Saccharomyces cerevisiae* in continuous fermenters with yeast recycling. The technology and equipment required for glucose fermentation in a biomass-to-ethanol facility are believed to be similar to those used in the sugar- and starch-to-ethanol processes or in the production of ethanol from spent sulphite liquors.

The pentose fermentation step is essentially still at the laboratory stage, although significant advances have been made in understanding the mechanism and in the modification of pentose-fermenting micro-organisms (37-39). None of the wild-type, adapted or engineered strains, have to our knowledge, been able to routinely ferment the pentose-rich, water-soluble stream obtained from steam-pretreated hardwoods or softwoods.

An alternative to separate hydrolysis and fermentation is simultaneous saccharification and fermentation (SSF), i.e. operation of the hydrolysis step in combination with the fermentation step. This operation reduces the end-product inhibition resulting from glucose and cellobiose build-up by continuously converting the glucose to ethanol. Although ethanol and other fermentation products may decrease the activity of individual cellulase enzymes, this inhibition is much weaker than the inhibition caused by equivalent glucose or cellobiose concentrations (40-42). The conditions for running SSF are a compromise between the optimal conditions for cellulose hydrolysis and those for fermentation. It is probably necessary to produce new yeast and enzymes for each SSF batch because of the difficulty in separating the cells and enzymes from the unhydrolysed solid residue (43).

The ethanol in the fermenter is rather diluted and must be recovered by distillation, which is considered to be a mature technology. The experiences from production of alcoholic beverages can be used to design and construct distillation units for recovery of fuel ethanol.

Four major by-product and liquid waste streams are expected from the enzymatic hydrolysis process: a waste stream from the pretreatment unit; spent fungal mycelia from the enzymatic production unit; unhydrolysed cellulose residue containing spent enzyme and lignin; and stillage from the distillation column which contains solubilised non-volatiles from the raw material, carbohydrate- and lignin-degradation products as well as by-products from the fermentation stage. The stillage waste stream is the largest in volume. It has high Biological and Chemical Oxygen Demands, (BOD_7 and COD, respectively), and a low pH. It often contains high levels of colouring compounds and has been noted for its high corrosiveness (44). Biological treatment of the stillage stream generally involves anaerobic treatment prior to aerobic treatment and a tertiary treatment to remove colouration. Recycling of liquid streams would minimise the fresh water requirement and lower the amount of waste water produced. However, this recycling leads to an increase in the concentration of various substances in the hydrolysis and fermentation steps (45). To avoid the accumulation of non-volatile inhibitors in the process, the stillage stream could be evaporated prior to recirculation (36). The volatile fraction from the evaporation step is then recycled and the non-volatile residue incinerated. This is described in more detail in the sections " Recycling of process streams" and "Process integration".

Bench-Scale Unit. In scaling up, the rate-limiting steps must be identified. The rates of all other process steps must be carefully assessed to determine whether they may decrease on scale-up and become new bottlenecks on a larger scale. Since microbial processes are complex, it is important to evaluate how various recycling scenarios will affect the microbial environment in continuous fermentation in a fully integrated process. The final investigations should preferably be performed on a pilot scale, but it is often more cost-effective to evaluate various options and operating conditions first on a bench scale.

At Lund University, a bench-scale unit was set up in 1995 for the development of a process for ethanol production from lignocellulosics based on enzymatic hydrolysis (36). Figure 2 shows a schematic flowsheet of the unit. The different unit operations are not physically connected so the material is passed manually from one step to the next. This makes the unit very suitable for studying various process configurations and also the recycling of process streams.

The pretreatment unit consists of a 10-litre pressure vessel (corresponding to approximately 1 kg of dry wood chips) and a flash tank to collect the pretreated material. An electric steam boiler supplies steam up to a maximum pressure of 30 bar, corresponding to a saturated steam temperature of 235°C. The temperature, pressure, and hold-up time in the reactor are monitored and controlled via a computer.

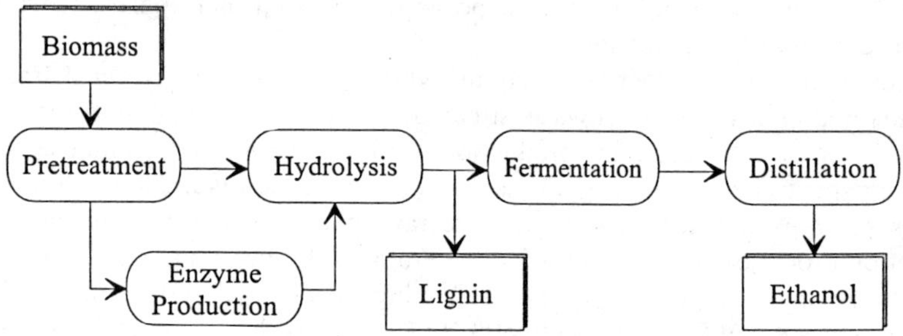

Figure 1. Production of ethanol using enzymatic hydrolysis.

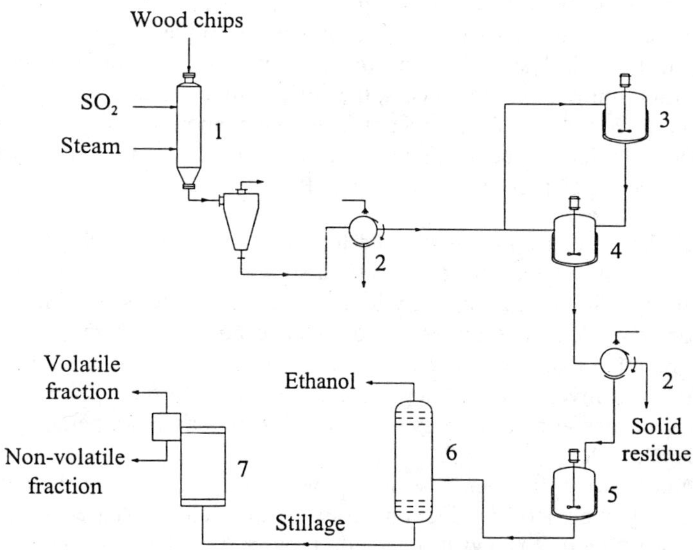

Figure 2. Schematic flowsheet of the bench-scale unit. 1: Pretreatment; 2: Filtration; 3: Enzyme production; 4: Hydrolysis; 5: Fermentation; 6: Distillation; 7: Evaporation.

A laboratory filter press is used to filter and wash the pretreated material. The material is stored in a 100-L stirred tank, slurried with water, then pumped through a filter cloth at a maximum pressure of 10 bar. After filtration, the filter cake is dewatered by applying a pressure of 15-17 bar and then washed under pressure with water. The unit can be used to determine scale-up design data, such as the filter-cake resistance.

The hydrolysis unit consists of two tanks with working volumes of 20 litres and 40 litres. The tanks are stirred by propeller-type agitators which have been proven to give efficient mixing of pretreated material up to 10% dry matter (DM). Both vessels are equipped with a water jacket to maintain a constant temperature in the range 30-90°C. The hydrolysis residue can be filtered in the filter press described above.

The enzyme production and fermentation units consist of fermenters with total volumes of 22 litre and 16 litre, respectively. The fermenters are equipped with sensors for control and data sampling of temperature, pH, dissolved oxygen and anti-foam addition. The outlet gas is analysed for oxygen and carbon dioxide.

The distillation unit consists of a 1.5 m packed column, corresponding to about 15-20 ideal stages, a jacket reboiler and a condenser. The distillation unit can be operated batchwise or continuously. The reflux ratio can be varied between 0.1 and 20.

The evaporator is designed for two different purposes: flash-sterilisation of fermentation broth, or concentration of the non-volatile components in liquids. Heat transfer is achieved via plate heat exchangers using live steam and cold water as heating and cooling media, respectively. The maximum capacity of the evaporator is 16 kg/h evaporated water. The evaporator can be used for the simulation of condensate withdrawal in a multiple-effect evaporator.

Experimental Runs. In the introductory experimental runs, willow, a fast growing energy crop, was used as substrate. The willow was pre-steamed with low-pressure steam (2 bar) prior to impregnation with gaseous sulphur dioxide in a plastic bag. The amount of sulphur dioxide added was approximately 4% of the DM. The impregnated material was then subjected to steam pretreatment at 205-210°C for 5 minutes. The pretreated material was slurried with water to approximately 5% DM, and filtered. After filtration, the filter cake was dewatered to approximately 40% DM.

Hydrolysis was performed using the fibrous material and the filtrate at a dry matter content of 5%, supplemented with Celluclast 2L and Novozym 188 (Novo AS, Denmark), corresponding to 15 FPU/g substrate. Commercial enzyme preparations were used, as the main purpose of the study was to investigate the influence of the recycling of waste streams. The temperature was maintained at 40°C and the pH at 4.8 during the entire hydrolysis time of 90 hours.

Fermentation of the willow hydrolysate was carried out in the 22-litre fermenter containing 16 litres medium at 30°C, pH 5.5, at a stirring speed of 300 rpm. The hydrolysate was supplemented with a rich medium, and inoculated with compressed baker's yeast, *S. cerevisiae*, to a final cell concentration (DM) of 6 g/L.

The fermentation broth was stripped in the evaporation unit, prior to distillation, to avoid problems associated with suspended particles in the distillation column. The evaporated fraction was then distilled. The COD, the BOD_7 and the fermentability of the stillage and the evaporation residue were determined.

The enzyme production was run in a parallel study. Enzyme production was performed in the 16-litre fermenter. A modified Mandels medium was used (46), in which the yeast extract and proteose peptone were replaced by dried yeast. A separate batch of willow, pretreated under the same conditions as described in the pretreatment section, was used as substrate for the enzyme production.

Some of the results from the experimental study in the bench-scale unit are described in the section "Recycling of process streams". A more detailed description of the equipment and the experimental procedure can be found in the original publication (36). Similar experiments have also been performed using a mixture of softwoods (47).

Process Simulations

Modern industrial processes are often very complex and it is almost impossible to evaluate every alternative by experiment alone. The process steps nearly always depend on each other, and a change in operating conditions in one step affects the performance of other units. Computer simulations can be used to identify problem areas which require further exploration. However, computer simulations can never replace experiments, but constitute an important tool when planning and evaluating laboratory or pilot-plant experiments.

Computer simulation of industrial processes requires mathematical models that are able to predict, for example, reaction rates, yields, vapour-liquid equilibria and energy requirements. The simulation can be performed either with a tailor-made program for the process of interest, or by using a general flowsheeting program, such as Aspen Plus (Aspen Tech, USA), ChemCad (Chemstations Inc., USA), or Process (Simulation Sciences Inc., USA). Another possibility is to use spreadsheet programs such as Excel or Lotus to create the simulation models. Most process simulators are not capable of handling complex solid and heterogeneous materials such as wood or yeast. These materials cannot be characterised in terms of well-defined physical properties, e.g. molecular weight or enthalpies. However, these problems are encountered when simulating the production of ethanol, and the components must be characterised in terms of average physical properties for the material.

We have chosen to use the commercial process simulator Aspen Plus to perform simulations of different process configurations. Aspen Plus is capable of handling heterogeneous and solid materials, and it is also possible to incorporate user-defined components and unit operation modules into the program. When defining user modules, results from experimental investigations, such as laboratory-scale or bench-scale trials, are fitted to empirical or mechanistic models and included in the process simulation.

Cost estimations are also important to determine the economic feasibility of a process. It is valuable to obtain the distribution of costs within a process to identify high-cost steps so that research and development efforts can be directed towards the most expensive process steps. By means of sensitivity analysis it is possible to predict how changes in a specific parameter influence the yield and the economy of the overall process. A simulation program, BioEconomics, developed by von Sivers and Zacchi (48) is used to estimate the production cost for ethanol from lignocellulosic materials.

In the following, three examples of how techno-economic simulations combined with experimental investigations in the bench-scale unit can be used for process development are presented.

Cost-Critical Steps

Economic calculations for the enzymatic process performed by von Sivers (48) show that the capital cost is the most dominating, amounting to 47% of the total cost, followed by the raw material cost (30 %). The cost estimate is based on experimental data from the literature using pine as raw material and a plant capacity of 100,000 ton dry matter/year. The calculations are based on separate hydrolysis and fermentation (SHF) and anaerobic fermentation of the waste water. Figure 3 shows the production cost for each process step. The most expensive step, constituting 15% of the total cost, is the steam production, including the lignin dryer, the steam boiler and the anaerobic treatment of the liquid waste, followed by pretreatment (13%) and enzymatic hydrolysis (12%). In the following, four examples of major contributors to the overall process economy are discussed.

Influence of the Raw Material. Since the raw material contributes a large part to the total ethanol production cost, the production cost is greatly affected by the cost of the wood and by the chemical and physical characteristics of the wood. These characteristics determine the difficulty in converting the cellulose and hemicellulose fractions in the wood to fermentable sugars at high yields, which in turn influences capital and operating costs. Softwood has proven to be more difficult to utilise than hardwood (30, 49). The chemical composition of the wood determines how much ethanol can theoretically be produced per tonne of raw material. The difference in composition between hardwood and softwood is that hardwood contains high levels of xylan, low levels of mannan and less lignin than softwood. The high level of xylan in hardwood makes it necessary to include a pentose fermentation step or the production of some co-product such as furfural or methane from the xylose, to make ethanol production from hardwood economically feasible. The pentose fermentation step requires further development before it can be used in a full-scale process, although much progress has been made in recent years. *Zymonas mobilis* (39), *S. cerevisiae* (37, 50), and *Escherichia coli* (38) have successfully been genetically transformed to ferment xylose to ethanol, but the organisms must also tolerate the inhibiting components in the hydrolysates. Otherwise, a detoxification step must be included, which has been shown to be quite expensive (51).

As the cost of the raw material is high, its maximum utilisation is important to lower the final cost of the ethanol. The overall ethanol yield has proven to be the most important factor for the ethanol production cost and it is necessary to develop the various steps in the process, i.e. pretreatment, hydrolysis and fermentation, to achieve as high a yield as possible. As a consequence of this, we have at Lund University a large R&D program focused on the development of these different process steps. However, it is also very important to examine how the various process steps perform in an integrated process, for example, how the accumulation of inhibitors due to recirculation will affect the ethanol yield. This has been investigated in the bench-scale process development unit where the whole process can be experimentally simulated. This is discussed in more detail in the section "Recycling of process streams".

Capital Cost. The enzyme production and the hydrolysis steps are the major contributors to the capital cost. These steps are rate limiting in the process and the high costs are due to the long residence time which requires numerous and large reactors. Decreased residence times in these two steps must be weighed against reduced enzyme and sugar yields. One way of reducing the capital cost for the enzymatic hydrolysis and fermentation steps is to use the SSF concept. There are several advantages with SSF. Only one reactor is needed for both hydrolysis and fermentation, no product inhibition of the enzymes in the hydrolysis arises when the glucose is converted directly to ethanol, and the risk of contamination decreases for the same reason. According to a study performed by Wright et al. (52) the SSF process leads to a reduction in the total production cost of about 30% compared with SHF. One drawback of SSF which remains to be overcome is recycling of the yeast. If the pretreated material is not delignified, it will be very difficult to separate the yeast from the lignin residue after SSF. The capital and productivity advantages attributed to the SSF configuration may decrease drastically due to the cost of producing new yeast in every fermentation batch or the cost of an additional delignification step. An alternative is to run the SSF continuously at low dilution rates (long residence times) so that the yeast has time to grow; but this will require extra-large fermenters. As the hydrolysis is rate limiting, SSF will, in any case, require large fermenters compared with separate hydrolysis and fermentation, where only the hydrolysis tanks are large. Fermenters are more complex and expensive than hydrolysis tanks and the gain in using one reactor instead of two will decrease. The bench-scale unit will, in the near future, be used to compare the SHF and the SSF methods in an integrated process.

Energy. Some operations, such as pretreatment of the raw material, distillation of the product, drying of the lignin, and, if included, evaporation of the stillage for recirculation of process water, are extremely energy demanding. In the steam pretreatment stage, high-quality steam of 20-30 bar is used, while in the drying section and in distillation steam of 3-6 bar is used. The steam consumption in the distillation step is very dependent on the ethanol concentration in the distillation unit. However, in a carefully designed and energy-integrated plant it is possible to reduce the energy costs by a considerable degree. This is described in more detail in the section "Process Integration".

By-products. The primary by-product from the large-scale production of ethanol will be lignin. Lignin can be utilised for many chemical applications, but due to the large amount of lignin which will be produced in a future transition from fossil fuels to fuel ethanol, the most realistic use of lignin is as a solid fuel. An alternative is to use the lignin for electricity production in a back-pressure power plant. The price obtained for the lignin will affect the cost of the ethanol (53, 54) but not as much as the cost of the raw material. Lignin and other by-products are produced in the bench-scale unit for further characterisation as the quality is of great importance for the price.

Recycling of Process Streams

As mentioned in the section "Influence of the raw material" it is important for the economic viability of the process to reach high yields and to utilise the raw material efficiently. In the following example, the effect of recycling liquid streams on the various unit operations was investigated through a combination of simulations and experiments in the bench-scale unit in order to reduce the use of fresh water in the process. In this case, it is also important to examine the build-up of inhibitory substances in the conversion steps, such as hydrolysis and fermentation, as they may affect the yield and productivity in these steps. One of the simulated process alternatives was then experimentally verified in the bench-scale process development unit. A base case was established (Figure 4) which serves as a reference case. The base case simulates a plant where fresh water is added if needed to modify the concentration of dry matter in a step, and also for washing steps. Obviously, this is not realistic for an industrial process, but it reflects the normal procedure used in most lab-scale experiments. A number of different process configurations were simulated to establish the concentration levels of soluble components at various locations in the process, but in the following, two examples are reviewed in more detail. The flowsheets shown have been simplified to speed up calculations and to make the results easier to interpret.

The simulations are based on a feed capacity of 20,000 kg/h wood chips with a moisture content of 50%. This gives the minimum amount of water that enters the system and which must be disposed of. The dry matter is assumed to consist of 36% cellulose, 24% hemicellulose, 21% lignin, and 19% solubles, which is the composition of willow (55). It is assumed that the conversion in the hydrolysis stage is 90%, in the fermentation step 95%, and the recovery in the distillation step is 99%.

The outputs from the process include ethanol as the major product, lignin as the major by-product, resulting from the filtering of the hydrolysis residue, and the stillage waste stream. In the process configuration for the base case, large volumes of fresh water are used. This will result in a very dilute distillation feed, containing around 2.5% (wt/wt) ethanol. Distillation of such a dilute feed makes the operation cost-sensitive for changes in the feed concentration for nearly all distillation technologies (56, 57). The base case yields a liquid waste stream of about 38 tonnes for every tonne of ethanol produced. This results in high costs for fresh water and for waste treatment. The main and probably the only advantage of this process alternative is the low concentration of inhibitors in all the reaction steps, due to the dilution with fresh water. One reason to recycle liquid process streams is to decrease the fresh water demand and to considerably reduce the amount of waste water. Other reasons for recycling are the opportunities of increasing the concentrations of glucose in the fermenter and ethanol in the distillation column.

Recycling of Distillation Stillage. Several recycling options are possible. The simulations were performed to investigate how various components were distributed

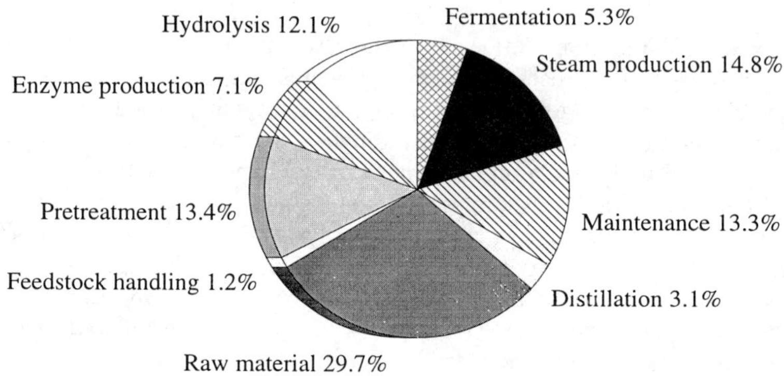

Figure 3. Cost divided into the various process steps.

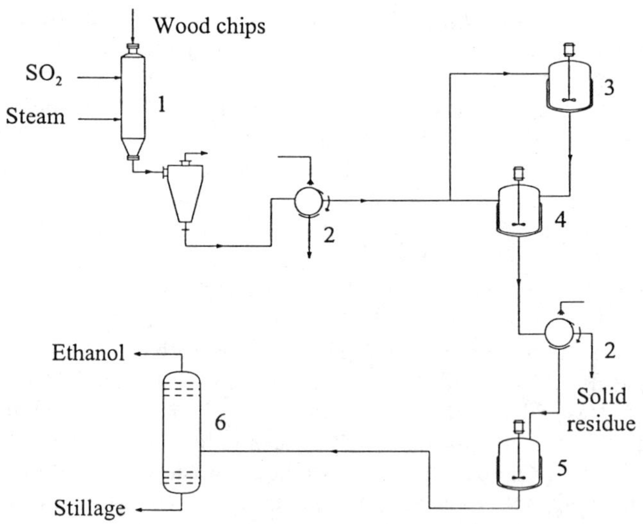

Figure 4. The base case flowsheet. 1: Pretreatment; 2: Filtration; 3: Enzyme production; 4: Hydrolysis; 5: Fermentation; 6: Distillation.

in the process assuming that they would not alter the yields in the various reaction steps. A number of components were chosen to represent very volatile (furfural), moderately volatile (acetic acid) and non-volatile ("soluble") components. Figure 5 shows a configuration where the stillage resulting from the distillation column is recycled to the washing steps. This reduces the amount of fresh water required in the washing steps to virtually zero. In this case, the stillage stream is reduced to about 7 tonnes/h but the concentration of by-products is much higher. The total amount of by-products is, of course, unaltered. Through this change in the original configuration, it is possible to increase the ethanol concentration to approximately 4.5% (wt/wt), which reduces the energy consumption in the distillation step with about 40%. However, the concentration of by-products is greatly increased and in the hydrolysis step is up to 18 times higher than that obtained in the base case for acetic acid, furfural, and non-volatile components (Figure 6). It can also be seen in the same figure that there is a 9-10 times higher concentration of acetic acid in the fermentation step compared with the base case, about 1-1.5% (wt/wt). This may cause a negative effect on the yield and on the productivity in the fermentation.

The results obtained from the simulation were examined experimentally in the bench-scale unit using willow as raw material (36). The experiments comprised pretreatment, enzymatic hydrolysis, fermentation, and distillation. The overall ethanol yield in this experiment was only 65% of the theoretical, which was lower than previously obtained in lab-scale investigations (14, 58). But enhanced yields are expected when, for example, hydrolysis is performed in fed-batch mode and fermentation is run continuously. To evaluate the effect of recycling, the stillage stream was fractionated into several volatile fractions and one non-volatile fraction by evaporation, thus simulating a multi-effect evaporation unit. The inhibitory effects of the various fractions were assessed by fermentation using *S. cerevisisae* after the addition of glucose. The non-volatile residue of the stillage was found to be inhibitory to fermentation already at a concentration five times higher than in the original stillage. The ethanol yield decreased from 0.37 g/g in a pure sugar reference to 0.31 g/g in the residue and the average fermentation rate decreased from 6.3 g/(L h) to 2.7 g/(L h). The acetic acid concentration in the residue was 9.2 g/L, a concentration previously found not to inhibit *S. cerevisiae* significantly (36), but it is more likely that lignin-degradation products are responsible for the inhibitory action. The evaporation condensates, containing the volatile components, showed no negative effects on fermentation.

The COD and the BOD_7 in the stillage stream, the volatile fractions, and the non-volatile residue were used to estimate the environmental impact of disposal. The most volatile part and the non-volatile residue exhibited a considerably higher COD and BOD_7 than the intermediate fraction (Figure 7). This indicates that the stream most suited for disposal is the intermediate part of the stillage stream and the parts most suited for recycling are the more volatile fractions. These results show that although the simulation indicates that a high degree of recirculation is an attractive option, in practice it is not possible to achieve an optimised process without the removal of non-volatile compounds. Incorporation of an evaporation unit is one way of removing the non-volatile residue from the process. Since this residue contains high amounts of organic compounds it is well suited to be used for steam generation.

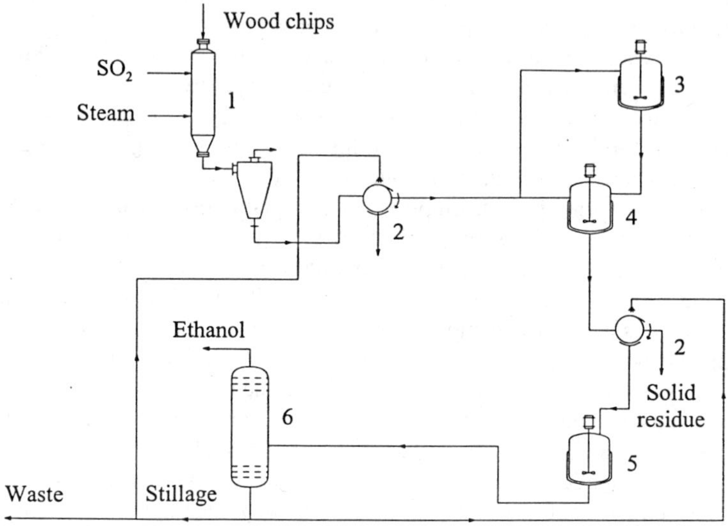

Figure 5. Recycling of the stillage. 1: Pretreatment; 2: Filtration; 3: Enzyme production; 4: Hydrolysis; 5: Fermentation; 6: Distillation.

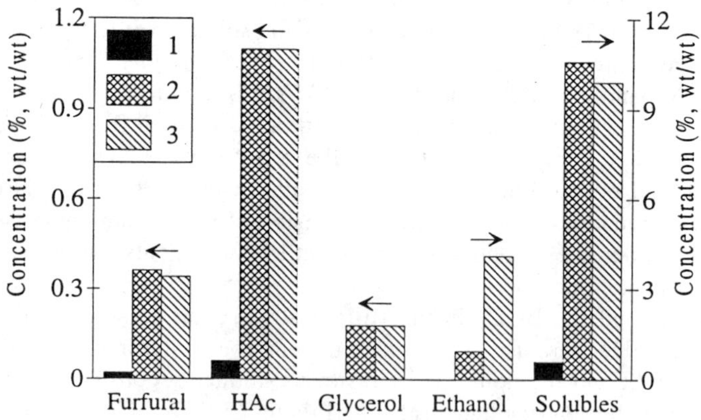

Figure 6. Concentration of ethanol and by-products in the hydrolysis. 1: Base case; 2: Recycling of the stillage; 3: Recycling of the distillation feed.

It is also possible to recycle part of the distillation feed back to the washing stage of the pretreated material. This configuration is shown in Figure 8. This alternative gives the same distribution of by-products as the former example (see Figure 6). But the ethanol concentration of the distillation feed could be increased to about 7% (wt/wt) (59), which is comparable to the concentration obtained when ethanol is produced from corn. This reduces the energy demand in the distillation by approximately 30%. However, in this particular case, the ethanol concentration in the hydrolysis reactor is increased to roughly 4% (wt/wt), which could yield a negative effect on the enzymatic hydrolysis. This configuration is presently being investigated using the bench-scale unit.

There are several other ways of increasing the ethanol concentration in the feed to the distillation unit (59). One of these is to recycle part of the stream from the hydrolysis tank. This will increase the ethanol concentration in the distillation feed, but it will also increase the risk of infection, since the recycling of sugar-containing liquids is involved.

Process Integration

The combination of simulation and bench-scale experiments may also be used to investigate the benefits of process integration regarding energy consumption. The integration can be performed internally within the ethanol plant or by integrating various parts of the ethanol plant with another type of plant, such as a pulp mill or a heat-generating plant. The latter requires, of course, detailed knowledge of the plant with which the ethanol plant is to be integrated. The internal integration will be exemplified by the incorporation of the distillation unit with a multiple-effect evaporation unit. Examples of integration of an ethanol plant and a pulp mill, as well as a power plant, will also be presented.

Internal Process Integration. The evaporation and distillation steps are large consumers of steam in the ethanol plant. The steam consumption in the evaporation and the distillation steps can be reduced by an increase in the dry matter content in the stillage or by increasing the ethanol concentration in the feed to the distillation step (60). This can be achieved by recirculation of various streams as described above. This might however, as shown, inhibit fermentation and/or hydrolysis depending on the degree of recirculation. Another way of increasing the ethanol content is to incorporate a stripper into the evaporation line, as shown in Figure 9. Instead of first distilling the fermentation broth and then evaporating the distillation stillage, the fermentation broth is fed to the stripper, evaporator effect 4, which is equipped with a reboiler, while the condensate from effect 5 is fed to the distillation unit.

In such case, the ethanol concentration in the feed to the distillation unit is increased from about 2-3% (wt/wt) to 18-20% (wt/wt), thus reducing the refining cost. With this configuration the steam consumption in the evaporation unit is not affected, while it is lowered by 60% in the distillation unit. For separate distillation and evaporation of a 100 tonnes/h liquid stream, the cost is 0.20 and 0.13 US$/kg product for dry matter contents in the feed of 4 and 6% (wt/wt), respectively (60). In both cases the liquid was concentrated to 65% (wt/wt) DM. With the incorporation of

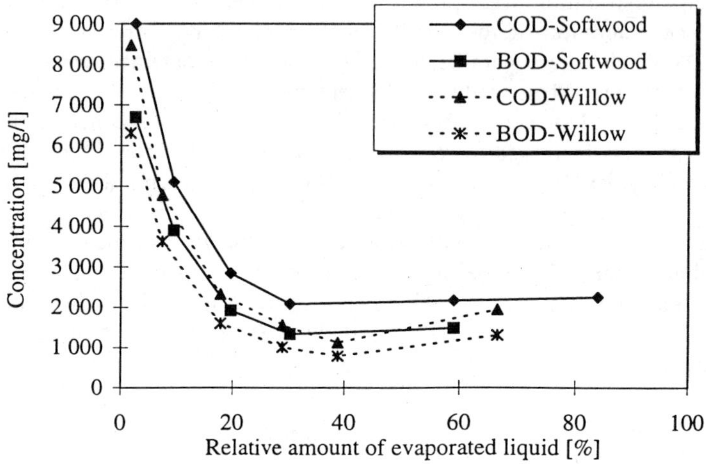

Figure 7. COD and BOD of various evaporated fractions.

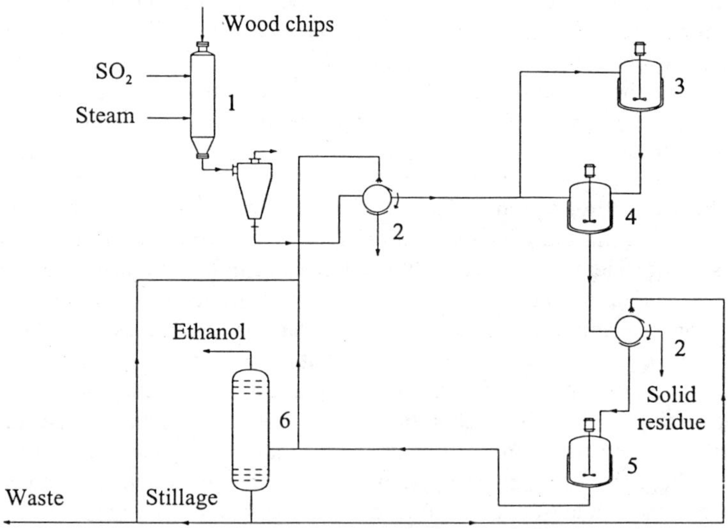

Figure 8. Recycling of the distillation feed. 1: Pretreatment; 2: Filtration; 3: Enzyme production; 4: Hydrolysis; 5: Fermentation; 6: Distillation.

a stripper into the evaporation unit, the cost decreases to 0.15 and 0.10 US$/kg product, as shown in Figure 10, illustrating the cost improvements when recycling is employed.

Further improvement in the economy can be achieved by integrating the energy-demanding steps in the entire process. One example is to release the high-pressure steam used in pretreatment to a back pressure of 3 bar and to use the secondary steam formed in other parts of the process, e.g. pre-steaming of the wood chips, in the distillation or in the evaporation unit (54).

External Process Integration. In an investigation by the Swedish engineering company ÅF-IPK (61), based on 150,000-250,000 tonnes/year raw material, the benefits of integrating an ethanol plant and existing processes handling biomass were evaluated. The plants investigated were a pulp mill, a power plant and a combined saw mill and peat-drying facility. The common infrastructure and the wood intake and storage are advantageous for all cases. The contribution of the common infrastructure is difficult to quantify in economic terms. The largest overall effects were obtained for a combination of an ethanol plant with either a pulp mill or a power plant. The saw mill has fewer possibilities for integration, since it is almost only the material handling that is common for the plants.

There are several opportunities for integrating redundant equipment or capacity in an ethanol plant and a pulp mill. The most interesting process step is that of steam generation as a pulp mill usually has an excess of low-pressure steam at 4 bar, which can be used in different process steps in the ethanol plant. The second most interesting process is effluent treatment. Due to the increased efforts in recent years to close the processes at pulp mills, the treatment plant will have surplus capacity which can be used by the ethanol plant. Furthermore, there is also the possibility of using a bark-fuelled boiler.

Assuming that the power plant is designed to receive large quantities of wood residue, the raw material handling, or part of it, could be shared with the ethanol plant. Furthermore, the boiler in the power plant can be used for steam generation in the ethanol plant. It is also possible to take advantage of the sorting of incoming wood to obtain a better raw material for ethanol production.

The potential for synergetic effects was determined to be in the range of 0.05-0.10 US$/kg ethanol, which should be compared with a calculated total production cost of 0.5-0.67 US$/kg ethanol for a plant with no integration. These benefits must be shared between the ethanol plant and the process with which it is integrated.

Conclusions

Processes for production of fuel-ethanol from lignocellulosic materials involving micro-organisms and biomass are very complex. Therefore, it can be difficult to design a full-scale or even a pilot-plant facility based on lab-scale data only. One of the most efficient ways of assessing the technological and economic feasibility of such processes is through the use of techno-economic computer modelling. Since a biomass-to-ethanol process consists of several process steps, all strongly interdependent, it is extremely difficult to identify the relative merits of a change in

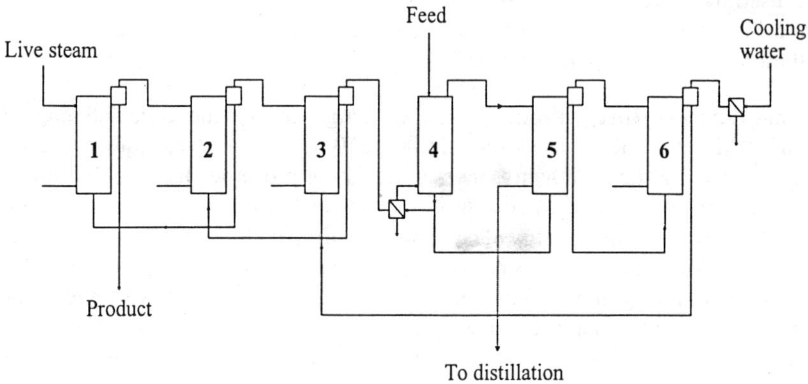

Figure 9. Schematic flowsheet of evaporation with stripper incorporated.

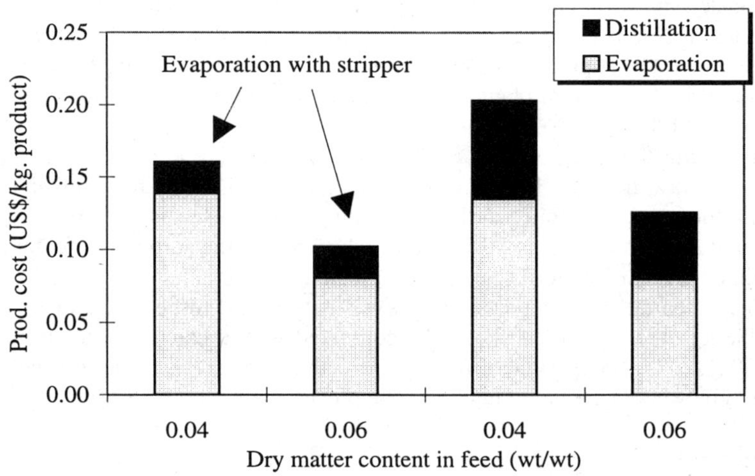

Figure 10. Cost for the evaporation and distillation units.

one subprocess and its influence on the final production cost of ethanol. By computer simulations different process strategies can be evaluated and experiments in a bench-scale unit can serve as feed-back to the process simulations.

A bench-scale unit is due to its flexibility very well suited for studying various process configurations and also the recycling of process streams. In this chapter we have shown how the use of a bench-scale unit in combination with process simulations can be used to study process integration. In an environmentally sustainable process, waste water must be minimised. Several process configurations with recycling of process streams were investigated which will result in reduced waste water streams as well as reduced energy consumption. The experimental investigation in the bench-scale unit showed that recycling of non-volatile compounds had an inhibitory effect on the fermentation process. This suggests an incorporation of a multi-effect evaporation line in the process. If internal energy integration is employed, the demand for extra energy can be reduced. The data thus gathered will now be included in Aspen Plus to refine the models which will lead to new simulation results. The aim is to reach an optimal process for production of ethanol which is economic and environmentally feasible. Final process optimisation should preferably be performed on a pilot-scale, since it is difficult to study the whole effect of energy integration options in a bench-scale unit.

Acknowledgments

The Swedish National Board for Industrial and Technical Development is gratefully acknowledged for its financial support.

Literature Cited

1. Matsui, S. Eleventh annual conference on alcohol and biomass energy technologies. 1991, 27-40 DE93716346; NTIS.
2. Katzen, R.; Fowler, D. *Appl. Biochem. Biotechnol.* **1994**, *45/46*, 697-707.
3. Vallander, L.; Eriksson, K. E. L. *Adv. Biochem. Eng./Biotechnol.* **1990**, *42*, 63-95.
4. Lynd, L. R.; Cushman, J. H.; Nichols, R. J.; Wyman, C. E. *Science* **1991**, *251*, 1318-1323.
5. Steiner, W.; Steinmueller, H.; Esterbauer, H.; Lafferty, R. M. *Biotekhnol. Biotekh.* **1987**, *6*, 3-9.
6. Polman, K. *Appl. Biochem. Biotechnol.* **1994**, *45/46*, 709-722.
7. Clausen, E. C.; Gaddy, J. L. *Biomass Energy Dev.* **1986**, *3*, 551-560.
8. Grethlein, H.; Converse, A.; Vick Roy, J. R.; Holland, J.; Grous, W.; Ward, J. 1987, Report: SERI/STR-231-3137, DE87001191; NTIS.
9. Parisi, F. *Adv. Biochem. Eng./Biotechnol.* **1989.** *38*, 53-87.
10. Walker, L. P.; Wilson, D. B. *Bioresour. Technol.* **1991**, *36*, 3-14.
11. Biermann, C. J.; McGinnis, G. D. *Holzforschung* **1990**, *44*, 229-233.
12. Eklund, R.; Galbe, M.; Zacchi, G.; *Enz. Microb. Technol.* **1990**, *12*, 225-228.

13. Excoffier, G.; Toussaint, B.; Vignon, M. R. *Biotechnol. Bioeng.* **1991**, *38*, 1308-1317.
14. Eklund, R.; Galbe, M.; Zacchi, G. *Bioresour. Bioeng.* **1995**, *52*, 225-229.
15. Clark, T. A.; Mackie, K. L. *J. Wood Chem. Technol.* **1987**, *7*, 373-403.
16. Holtzapple, M. T.; Lundeen, J. E.; Sturgis, R.; Lewis, J. E.; Dale, B. E. *13th Symp. Biotechnol. Fuels Chemicals: Proceedings*; The Humana Press: Totowa, NJ, 1992; Vol. 13, pp 5-21.
17. Cantarella, M.; Arcese, A.; Cantarella, L.; Gallifuoco, A.; Alfani, F. In *Biomass Energy, Ind. Environ.*; Grassi, G.; Collina, A.; Zibetta, H., Eds.; E.C. Conf., 6th 1991; Elsevier: London, UK, 1992; pp 444-449.
18. Nguyen, Q.; Duff, B.; Schell, D. *Bioenergy '94 using biofuels for a better environment. Conference: 6th national bioenergy conference, Reno, NV, USA.* **1994**, *1*, 19-26.
19. Hayn, M.; Steiner, W.; Klinger, R.; Steinmueller, H.; Sinner, M.; Esterbauer, H. In *Bioconversion of Forest and Agricultural Plant Residues.;* Saddler, J. N., Ed.; Biotechnology in agriculture 9; CAB International: Wallingford, UK, 1993; 33-72.
20. Wayman, M; Dzenis, A. *Can. J. Chem. Eng.* **1984**, *62*, 699-705.
21. von Sivers, M.; Zacchi, G. *Bioresour. Technol.* **1996**, *56*, 131-140.
22. Nystrom, J.; Greenwald, C. G.; Hagler, R. W.; Stahr, J. J.; Report: NYERDA 85-9, **1985**.
23. Foody, F. *Optimization of Steam Explosion Pretreatment*; Final Report for Contract No. DE-AX-02-79-ETZ3050; Submitted to U.S. Department of Energy Fuels from Biomass Program, 1980.
24. Mamers, H.; Menz, D. N. J. *Appita* **1984**, *37*, 644-649.
25. Dekker, R. F.; Karageorge, H.; Wallis, A. F. A. *Biocatalysis*, **1987**, *1*, 45-59.
26. Brownell, H. H.; Progress Report for the Project No. 04-53-12-402; Forintek Canada: Ottawa, Ontario, 1989.
27. Saddler, J. N.; Ramos, L. P.; Breuil, C. In *Bioconversion of Forest and Agricultural Plant Residues.;* Saddler, J. N., Ed.; Biotechnology in agriculture 9; CAB International: Wallingford, UK, 1993; 73-91.
28. Gregg, D.; Saddler, J. N. *Biomass Bioenergy* **1995**, *9(1-5)*, 287-302.
29. Mackie, K. L.; Brownell, H. H.; West, K. L.; Saddler, J. N. *J. Wood. Chem. Technol.* **1985**, *5*, 405-425.
30. Ramos, L. P.; Breuil, C.; Saddler, J. N. *Appl. Biochem. Biotechnol.* **1992**, *34/35*, 37-47.
31. Pourquie, J.; Warzywoda, M. In *Bioconversion of Forest and Agricultural Plant Residues.;* Saddler, J. N., Ed.; Biotechnology in agriculture 9; CAB International: Wallingford, UK, 1993; 107-116.
32. Esterbauer, H.; Steiner, W.; Labudova, I.; Hermann, A.; Hayn, M. *Bioresour. Technol.* **1991**, *36*, 51-65.
33. Chen, S.; Wayman, M. *Process Biochem.* **1991**, *26*, 93-100.
34. Reczey, K.; Stålbrand, H.; Persson, I.; Hahn-Hägerdal, B.; Tjerneld, F. *Appl. Biochem. Biotech.* **1990**, *24-25*, 637-649.

35. Duff, S. J. B.; Cooper, D. G-; Fuller, O. M.; *Enzyme Microb. Technol.* **1986**, *8(5)*, 305-308.
36. Palmqvist, E.; Hahn-Hägerdal, B.; Galbe, M.; Larsson, M.; Stenberg, K.; Szengyel, Z.; Tengborg, C.; Zacchi, G. *Bioresour. Technol.* In press.
37. Walfridsson, M. Ph. D. thesis, **1996**, Dept. Applied Microbiology, Lund University, Sweden. Report: LUTKDH(TKMB-1023)/115 (1996).
38. Alterthum, F.; Ingram, L. O. *Appl. Environ. Microbiol.* **1989**, *55*, 1943-1948.
39. Zhang, M.; Eddy, C.; Deand, K.; Finkelstein, M,; Picatagio, S. *Science*, **1995**, *267*, 240-243.
40. Meyers, S. G. *American Institute of Chemical Engineers Symposium Series* **1978**, *74*, 79-84.
41. Ghosh, P.; Pamment, N. B.; Martin, W. R. B. *Enzyme Microb. Technol.* **1982**, *4*, 425-430.
42. Ooshima, H.; Ishitani, Y.; Harano, Y. *Biotechnol. Bioeng.* **1985**, *27*, 389-397.
43. Singh, A.; Kumar, K. R.; Schugerl, K. J. *Biotechnol.* **1991**, *18*, 205-221.
44. Maiorella, B. L.; Blanch, H. W.; Wilke, C. R. *Process Biochemistry* **1983**, *12*, 5-8.
45. Galbe, M.; Zacchi, G. In *Bioconversion of Forest and Agricultural Plant Residues.;* Saddler, J. N., Ed.; Biotechnology in agriculture 9; CAB International: Wallingford, UK, 1993; 291-321.
46. Mandels, M.; Weber, J. *Adv. Chem. Ser.* **1969**, *95*, 391-414.
47. Larsson, M.; Galbe, M.; Zacchi, G. *Bioresour. Technol.* Accepted for publ.
48. von Sivers, M. Thesis, **1995**, Dept. Chemical Engineering I, Lund University, Sweden. Report: LUTKDH(TKKA-1004)/1-30/(1995).
49. Wayman, M.; Parekh, S.; Chornet, E.; Overend, R. P. *Biotechnol. Lett.* **1986**, *8*, 749-752.
50. Chen, Z. D.; Ho, N. W. Y. *Appl. Biochem. Biotechnol.* **1993**, *39*, 135-147.
51. von Sivers, M.; Zacchi, G.; Olsson, L; Hahn-Hägerdal, B. *Biotechnol. Prog.* **1994**, *10*, 555-560.
52. Wright, J. D.; Wyman, C. E.; Grohmann, K. *Appl. Biochem. Biotechnol.* **1988**, *18*, 75-90.
53. Nguyen, Q. A.; Saddler, J. N. *Bioresour. Technol.* **1991**, *35*, 275-282.
54. von Sivers, M.; Zacchi, G. *Bioresour. Technol.* **1995**, *51*, 43-52.
55. Eklund, R.; Galbe, M.; Zacchi, G. *J. Wood Chem. Technol.* **1988**, *8*, 379-392.
56. Zacchi, G.; Axelsson, A. *Biotech.Bioeng.* **1989**, *34*, 223-233.
57. Busche, R. M. *Biotech. Bioeng. Symp.* **1983**, *13*, 597-615.
58. Eklund, R.; Zacchi, G. *Enz. Microbiol. Technol.* **1995**, *17*, 255-259.
59. Galbe, M. Ph. D. thesis, **1994**, Dept. Chemical Engineering I, Lund University, Sweden. Report: LUTKDH(TKKA-1003)/1-98/(1994).
60. Larsson, M. Thesis, **1996**, Dept. Chemical Engineering I, Lund University, Sweden. Report: LUTKDH(TKKA-1006)/1-30/(1996).
61. Ångpanneföreningen-IPK, Report: NUTEK, Stockholm, Sweden P2332-1, **1994**.

Chapter 7

Biomass Conversion to Mixed Alcohol Fuels Using the MixAlco Process

M. T. Holtzapple, M. K. Ross, N.-S. Chang, V. S. Chang, S. K. Adelson, and C. Brazel

Department of Chemical Engineering, Texas A&M University, College Station, TX 77843–3122

The MixAlco Process converts biomass into mixed alcohol fuels. The biomass is first treated with lime to render it more digestible. Then, it is fed to a mixed culture of acid-forming microorganisms that produce salts of volatile fatty acids, such as calcium acetate, propionate, and butyrate. These salts are separated from the fermentation broth and thermally converted to ketones that are subsequently hydrogenated to alcohols, such as 2-propanol, 2-butanol, 2-pentanol, and 3-pentanol. Design data are presented related to the lime treatment, fermentation, thermal conversion, and hydrogenation. A preliminary economic evaluation indicates that alcohol fuels can be produced from negative-value biomass, such as municipal solid waste and sewage sludge, for about \$0.19/L (\$0.72/gal).

Figure 1 shows a schematic of the MixAlco Process which converts biomass (e.g. municipal solid waste, sewage sludge, agricultural residues, energy crops) into mixed alcohol fuels. To enhance digestibility, the biomass is treated with lime. Then, using a mixed population of acid-forming microorganisms such as those found in cattle rumen, the lime-treated biomass is converted to volatile fatty acids (VFA's) such as acetic, propionic, and butyric acids. To prevent the pH from decreasing as the acids are formed, a neutralizing agent is added to the fermentor; thus VFA salts – such as calcium acetate, propionate, and butyrate – exit the fermentor. These salts are concentrated to dryness and then are thermally converted to mixed ketones (e.g., 2-propanone, 2-butanone, 2-pentanone, 3-pentanone) that are subsequently hydrogenated to mixed alcohols (e.g., 2-propanol, 2-butanol, 2-pentanol, 3-pentanol).

In the fermentor, both lime and calcium carbonate are possible neutralizing agents. Lime is a much stronger alkali and therefore can attain the optimal rumen pH of 6.7 whereas calcium carbonate can achieve a pH of only 5.8 to 6.2. Although lower pH slows the fermentation rate, it also discourages methanogens and thus can

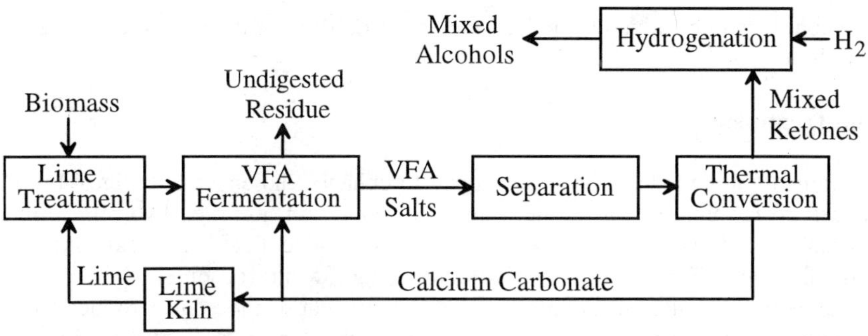

Figure 1. Schematic of the MixAlco Process.

increase the selectivity toward VFA's. This advantage, plus lower cost, favors calcium carbonate as the neutralization agent.

The fermentor temperature can be adjusted for mesophilic (~40°C) or thermophilic (~55°C) microorganisms. Of the total VFA's, mesophilic microorganisms produce 60-70% acetic acid whereas thermophilic microorganisms produce 80-90%; thus, temperature is an important parameter that affects the product distribution. Higher fermentor temperatures have an advantage of reducing the cooling water required to remove metabolic heat; however, for fuel production, the greater energy content in the higher VFA's favors lower fermentor temperatures.

A common approach to producing alcohol fuels from biomass is to enzymatically convert treated biomass with extracellular enzymes that hydrolyze polysaccharides to soluble sugars that are fermented to ethanol which is recovered by distillation. Compared to this approach, the MixAlco Process offers the following advantages:

- It is adaptable to a wide variety of feedstocks.
- Aseptic process conditions are not required.
- Inexpensive tanks can be employed.
- Expensive extracellular enzymes are not required.
- The fermenting organisms regenerate themselves.
- Cells and enzymes can be recycled without contamination risk.
- The fermenting organisms are stable.
- The process is robust.

Elements of the MixAlco Process have been investigated for many years. In 1914, Lefranc received a patent (*1*) on a process to convert waste biomass into butyric acid which was neutralized with calcium carbonate. The calcium butyrate was thermally converted to ketones for high-octane "Ketol" motor fuel (*2*). More recently, Playne (*3*) has further developed this technology by incorporating various pretreatments (*4*) and membrane separation techniques (*5*). In the U.S., only ethers and alcohols may be added to fuel, so the MixAlco Process converts the ketones into alcohols.

Lime Treatment

Numerous treatments have been developed to enhance the enzymatic digestibility of lignocellulosic biomass including: physical (e.g., ball milling, two-roll milling), chemical (e.g., dilute-acid hydrolysis, alkali), physico-chemical (e.g., steam explosion, Ammonia Fiber Explosion), and biological (e.g., white-rot fungi) (*6*). For the MixAlco Process, alkaline treatment is selected because the acids produced in the fermentor will neutralize the alkali, thus allowing recovery of the treatment agent. Of the various alkalis that are effective (e.g., sodium hydroxide, ammonia), lime was selected because of its low cost and compatibility with other process steps.

Compared to other alkalis, the literature on lime treatments is relatively sparse. Most of the studies have been performed by animal scientists seeking simple, room-temperature treatments to enhance ruminant digestibility. Because the treatment temperature was low, their results were poor; the general consensus is that lime is not as effective as other alkalis. However, by optimizing the reaction temperature and other conditions, lime is a very effective treatment agent.

Table I summarizes the results of some recent lime treatment studies using extracellular enzymes to hydrolyze the biomass. Compared to untreated biomass, lime-treated biomass has an enzymatic digestibility roughly ten fold larger. Because of its low lignin content, herbaceous biomass requires only lime treatment. However, because of its high lignin content, woody biomass requires the addition of oxygen to partially oxidize the lignin and remove it from the biomass. In addition, woody biomass requires more severe time and temperature.

Table II shows that lime treatment roughly doubles the ruminant digestibility of biomass. Comparing the digestibilities reported in Table I versus Table II, the digestibility within the ruminant is greater than that achieved with extracellular enzymes. This suggests that an industrial process based on a mixed culture of microorganisms may have advantages over one based upon extracellular enzymes.

Fermentation

Rapier (*10*) has determined that a mixture of 80% municipal solid waste (MSW) and 20% sewage sludge (SS) provides the optimal combination of energy and nutrients for a mixed culture of acid-forming microorganisms; therefore, this ratio was used in this study.

A series of semi-solid fermentations were operated using custom fermentors. The fermentors were horizontal, stainless-steel cylinders of 17.5-cm length and 10-cm diameter. A center shaft had finger-like projections that extended nearly to the cylinder wall. As the shaft rotated, it "kneaded" the fermentor contents through finger-like projections located on the cylinder wall. The 1.5-L fermentor was filled with about 0.5 L of fermenting MSW and SS.

Table I. Summary of Lime Pretreatment Results Using Extracellular Cellulase

Biomass	Temp (°C)	Time (h)	Lime Loading (g $Ca(OH)_2$/g dry biomass)	Water Loading (g H_2O/g dry biomass)	Oxygen Pressure (atm)	Untreated Sugar Yield[a] (g sugar/g dry biomass)	Lime-Treated Sugar Yield[a] (g sugar/g dry biomass)
Herbaceous							
Bagasse (*7*)	100	1	0.1	15	–	0.040	0.525
Wheat Straw (*7*)	85	3	0.1	10	–	0.065	0.579
Switchgrass[b]	120	3	0.1	13	–	0.102	0.574
Woody							
Aspenwood[c]	120	3	0.1	10	–	0.045	0.291
Aspenwood[c]	150	6	0.1	10	14.6	0.045	0.600

[a]Reducing sugar yield measured after three days using the dinitrosalicylic acid assay (*8*). Cellulase loading = 5 FPU/g dry biomass (*9*). Cellobiase loading = 28.4 CBU/g dry biomass as measured by Novo standard assay.

[b]Chang, V.S.; Burr, B.; Holtzapple, M.T. *Appl. Biochem. Biotechnol.*, in press.

[c]This work

Table II. Ruminant Digestibility of Untreated and Lime-Treated Agricultural Residues[a]

Agricultural Residue	48-h *In Situ* Digestion[b]	
	Untreated (g digested/g fed)	Lime Treated[c] (g digested/g fed)
Sugar-cane bagasse	0.308	0.627
African millet straw	0.451	0.899
Sorghum straw	0.541	0.829
Tobacco stalks	0.344	0.679

[a]Jagruti, J.; Holtzapple, M.T.; Ferrer, A.; Byers, F.M.; Turner, N.D.; Nagwani, M.; Chang, S.V. *Animal Feed Science and Technology*, in review.

[b]Determined as the weight loss from 2-g biomass samples placed in fine-mesh nylon bags located in the rumen of a fistulated steer.

[c]Treatment conditions: lime loading = 0.1 g $Ca(OH)_2$/g dry biomass, temperature = 100°C, water loading = 9 g H_2O/g dry biomass, time = 1 h (bagasse) or 2 h (African millet straw, sorghum straw, tobacco stalks).

Two or four fermentors were operated in series with solids flowing countercurrently to the liquid. Solid/liquid separation was achieved by centrifuging the fermentor contents and decanting the liquid in an anaerobic hood. This countercurrent operation allows high VFA concentrations to be generated in the fermentor receiving fresh, highly reactive solids. Because inhibition is low, it also allows high conversion in the fermentor receiving fresh liquid.

Table III shows the results from four countercurrent fermentations. Compared to rumen fermentations which typically require only a couple of days, the fermentor residence times are significantly longer due to the inhibition from the high VFA concentration (20 – 30 g/L versus 8 – 10 g/L). Fortunately, because the fermentors can be very inexpensive, the long residence time does not impose a severe economic penalty. The process time scales are similar to those for composting; thus, the process may be viewed as an anaerobic composting operation. The fermentor volume is proportional to the liquid residence time whereas the conversion is proportional to the solids residence time.

Fermentor A used two countercurrent stages and Fermentor B used four countercurrent stages. Similar VFA concentrations, conversions, yields, and selectivities were obtained; however, Fermentor A required 55% longer liquid residence time. This comparison shows the beneficial effect of increasing the number countercurrent stages.

Fermentor C also employed four countercurrent stages. Compared to Fermentor B, the ratio of solid:liquid residence times increased allowing the conversion to increase while holding the product concentration constant.

Fermentor D used essentially the same liquid and solid residence times as Fermentor B, but increased the nutrient content allowing both the VFA concentration and conversion to increase substantially. The conversion of Fermentor D represents 84% of the maximum possible. (Considering ash and lignin content, the maximum digestibility of the MSW/SS mixture is 77.5%.) Further research is required to determine the optimal nutrient package.

Separation

The VFA salt concentration exiting the fermentor is approximately 25 to 40 g/L, or approximately 25 to 40 parts of water per part of VFA salt. The pK_a of VFA's is 4.8 so at the fermentation pH (~5.8), only about 10% of the VFA is present as free unionized acid; the rest is ionic salt. Both the salt and free unionized acid are less volatile than water, so distillation is not a viable separation technique.

Playne (*5*) discusses many techniques for recovering VFA's from dilute aqueous solutions. Some proposed methods employ immiscible solvents (e.g., tributyl phosphate, trioctyl phosphine oxide, high-molecular-weight amines) that react with the free unionized acid and extract it from the broth. For solvent extraction to be effective, the fermentor pH must be acidic (4.8 to 5.2) which severely inhibits the microorganisms. Alternatively, if the fermentation is operated near neutrality, the fermentation broth can be acidified with mineral acids (which generates wastes) or carbon dioxide (which requires high pressures).

Table III. Countercurrent Fermentation[a]

Fermentation	A	B	C	D[i]
Number of Stages	2	4	4	4
Liquid[b] Residence Time (d)	17	11	22	11
Solid Residence Time (d)	45	30	72	31
MSW[c,d]:SS[e] (g:g)	80:20	80:20	80:20	80:20
Solids Concentration (g/L)	250	250	250	250
Temperature (°C)	40	40	40	40
Acetic Acid Concentration[f] (g/L)	16.0 ± 5.2	15.2 ± 2.1	15.6 ± 2.0	18.2 ± 2.4
Propionic Acid Concentration[f] (g/L)	4.0 ± 2.2	6.7 ± 1.3	3.3 ± 0.5	5.7 ± 1.5
Butryric Acid Concentration[f] (g/L)	1.8 ± 0.5	1.0 ± 0.1	2.9 ± 0.7	4.0 ± 2.0
Valeric Acid Concentration[f] (g/L)	1.0 ± 0.5	0.6 ± 0.2	1.0 ± 0.4	0.8 ± 0.3
Caproic Acid Concentration[f] (g/L)	0.5 ± 0.1	0.1 ± 0.03	0.8 ± 0.3	0.3 ± 0.2
Total Acid[f] (g/L)	23.1 ± 8.0	23.6 ± 3.7	23.6 ± 4.0	29.4 ± 4.0
pH Range[g]	5.6 – 5.9	5.8 – 6.1	5.8 – 6.2	6.0 – 6.2
Conversion (g digested[h]/g fed)	0.48	0.46	0.61	0.65
Yield (g VFA/g fed)	0.21	0.20	0.21	0.27
Selectivity (g VFA/g digested[h])	0.44	0.44	0.34	0.42

[a]Inoculum = rumen fluid and soil.

[b]Caldwell & Bryant (*11*) media modified by removing acetic, propionic, and butyric acids, but keeping the branched fatty acids. Methanogen inhibitor =1 mM 2-bromoethanesulfonic acid.

[c]MSW composition = 21.0% cardboard, 13.9% newspaper, 10.7% packaging, 10.1% printer paper, 7.8% leaves, 6.3% wood, 6.3% miscellaneous non-packaging, 5.0% books and magazines, 4.4% grass, 4.0% food waste, 3.1% tissue paper, 2.7% brush, 2.7% fats and oils, 1.9% greens. Pretreatment conditions for MSW: lime loading = 0.1 g $Ca(OH)_2$/g dry biomass, water loading = 10 g H_2O/g dry biomass, time = 1 h, temperature = 121°C.

[d]MSW gross composition (*12,13*) = 14.1% lignin, 4.2% ash, 81.7% carbohydrate + fat + protein.

[e]SS gross composition (*12,13*) = 39.5% ash, 60.5% carbohydrate + fat + protein.

[f]Error band represents ± two standard deviations.

[g]Calcium carbonate neturalization.

[h]Residue washed with acetic acid to remove $CaCO_3$. Cells remain in residue.

[i]Extra nutrients added. 0.073 g urea/g biomass. Four-fold increase in Caldwell & Bryant (*11*) nutrients (vitamins, heavy metals, phosphate) but not buffer, oxygen scavengers, or branched fatty acids.

Multi-effect evaporation can concentrate the salts, but it is too energy intensive. Membrane techniques (e.g. reverse osmosis, electrodialysis, water-splitting electrodialysis, carrier-mediated transport membranes) may also be considered (*5*), but the cost of membranes makes this prohibitive.

The MixAlco Process uses a proprietary separation procedure that is both capital and energy efficient; it overcomes a major obstacle to the economic use of mixed-acid fermentation.

Thermal Conversion

Until World War I, the major chemical route to acetone was the thermal conversion of "grey acetate of lime," i.e., calcium acetate derived from neutralizing pyroligneous acid (wood distillate) with lime (*14*). An excellent early kinetic study of this reaction was performed by Ardagh et al. (*15*) although the reaction is difficult (*16*).

The thermal conversion of calcium acetate, propionate, and butyrate to 2-propanone (acetone), 3-pentanone, and 4-heptanone, respectively, is a first-order reaction (*17*). The rate constants follow an Arrhenius dependence on temperature with the constants shown in Table IV. The Arrhenius constants for each species are similar except for calcium propionate below 398°C. The reactions are very rapid; at 440°C, it takes less than one minute to achieve 90% conversion.

The thermal decomposition is quite selective to liquid products with theoretical yields over 93% for calcium acetate and butyrate, and over 87% for calcium propionate. Gaseous products account for the remaining material. In the liquid product, over 90% is the primary product (2-propanone, 3-pentanone, 4-heptanone) with other ketones being the dominant by-products. These by-products will also contribute to the value of the final fuel. The composition of the liquid product was remarkably independent of the temperature, even at very high temperatures, indicating that the products are stable at the reaction conditions.

Hydrogenation

Figure 2 shows the hydrogenation rate of four ketones: 2-propanone (acetone), 2-butanone, 2-pentanone, and 3-pentanone (*18*). From typical VFA compositions in the fermentor, these four ketones comprise about 90% of the expected product. The hydrogenation was performed at 1 atm total pressure and near-ambient temperature. The reaction rates are expressed as moles of alcohol product per minute per gram of Raney nickel catalyst. The right-most data points of each figure indicate the hydrogenation rate of pure ketone. The data points to the left indicate the hydrogenation rate of ketone mixed with 2-propanol, the dominant product in the MixAlco Process.

Comparing the hydrogenation rates of the four ketones at 40°C (the only temperature that is common to all), they are all very similar and differ at most by a factor of two. At a catalyst concentration of 50 g Raney nickel/L, a temperature of 40°C, and a total pressure of 1 atm, the time required to hydrogenate 99% of

Table IV. Thermal Conversion of VFA Salts to Ketones

	Calcium Acetate[a]	Calcium Propionate[a]	Calcium Butyrate[a]
Activation energy (kJ/mol)	642.6 ± 28.0	[b]2327 ± 162 [c]691.7 ± 73.3	386.5 ± 15.8
Frequency factor (min^{-1})	109.4 ± 0.4	[b]414.7 ± 0.4 [c]121.1 ± 0.2	66.12 ± 0.22
Time required for 90% conversion at 440°C (min)	0.817	0.0283	0.915
Yield (g actual liquid/g theoretical liquid)	93.1 ± 0.9	87.5 ± 0.7	94.4 ± 0.9
Maximum temperature studied (°C)	455	475	508
2-Propanone (acetone) (%)	91.5 ± 4.3		
Isophorone (%)	2.78 ± 2.6		
2,4-Dimethyl phenol (%)	0.30 ± 0.63		
3-Pentanone (%)		96.9 ± 0.9	
2-Butanone (%)		1.06 ± 0.37	
5-Methyl-2-hexanone (%)		0.49 ± 0.13	
4-Heptanone (%)			94.9 ± 3.46
2-Pentanone (%)			2.84 ± 1.7
3-Heptanone (%)			0.44 ± 0.07
3-Hexanone (%)			0.39 ± 0.26
Other (%)	5.43 ± 1.71	1.57 ± 0.85	1.44 ± 1.79

[a] Error band represent ± two standard deviations. [b]398°C and below [c]above 398°C

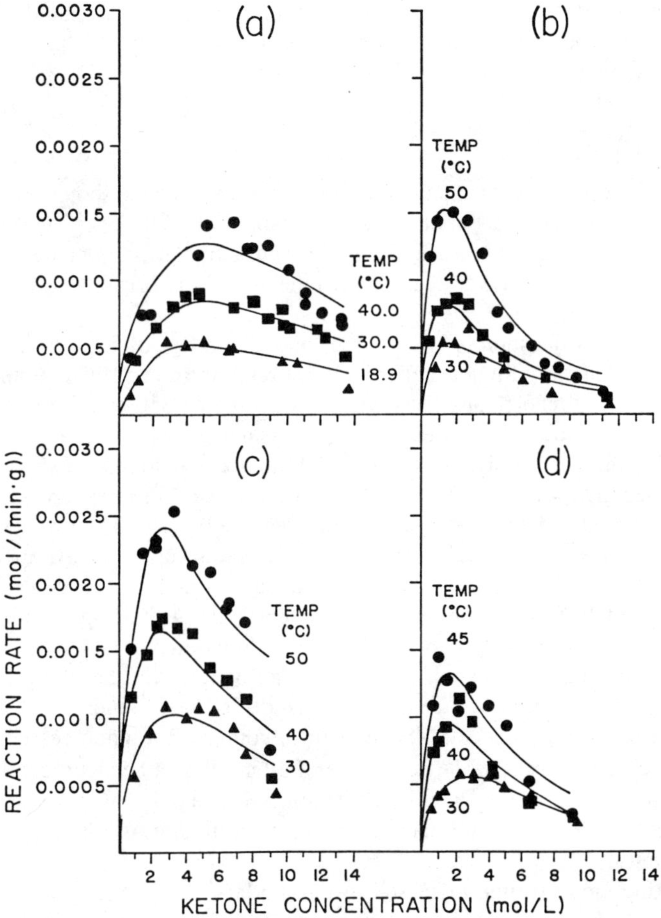

Figure 2. Hydrogenation of a) 2-propanone (acetone), b) 2-butanone, c) 2-pentanone, and d) 3-pentanone.

2-propanone is 4.4 h. The reaction rate increases linearly with hydrogen pressure (*18*), so the reaction time could be reduced to under one hour using a moderate pressure. Alternatively, the catalyst concentration could be increased.

The hydrogen would likely be derived from reformed natural gas, an abundant domestic energy source. In large-scale production, hydrogen costs about the same as gasoline per unit of energy (Appleby, J., Texas A&M University, personal communication, 1996), so there are no economic penalties associated with its use. The ketone may be viewed as a hydrogen carrier which avoids the need for high-pressure tanks to store gaseous hydrogen.

Economics

A preliminary economic evaluation of the MixAlco process shows that from free biomass, the mixed alcohols could be sold for about $0.19/L ($0.72/gal) with a 15% before-tax return on investment. This is a very attractive price compared to other fuel oxygenates such as methyl tertiary butyl ether ($0.24/L or $0.90/gal) and ethanol ($0.29/L or $1.10/gal).

An obvious question is "where does one obtain free biomass?" The answer is from MSW which currently has a disposal cost ranging from $10/wet tonne (Nevada) to over $110/wet tonne (Northeast) with an average of $50/wet tonne. Processing MSW to separate the refractories (e.g., metals and glass) from the organics (e.g., paper, cardboard, food scraps) costs about $21 to $62/wet tonne (*19*).

Rather than sending the organics to the MixAlco Process, an alternative use for the separated organics is "refuse derived fuel" (RDF); however, chlorine-containing plastics within RDF are very corrosive requiring expensive metallurgy for the boiler tubes and scrubbers to treat the exhaust gases. As a consequence, the energy value of RDF is negative; it actually costs about $55/wet tonne to combust MSW (*20*). Therefore, this competing use for the separated organics is not attractive.

To avoid the combustion problems described above, the undigested residue from the MixAlco process would likely be landfilled. The undigested residues are dense (specific gravity = 1.2 at 50% moisture) compared to unprocessed MSW in a landfill (specific gravity $\cong$ 0.71), so the life of a landfill can be significantly increased by employing the MixAlco Process. Assuming that metals and glass are recovered and that 63% of the remaining material is digested, the life of a landfill increases by about 3.5 times.

Taking credit for the increased life of the landfill and using $41/wet tonne as the MSW sorting cost, free biomass can be obtained from communities that currently pay $58/tonne to landfill unprocessed MSW.

Conclusions

The MixAlco Process converts a wide variety of biomass feedstocks (e.g., MSW, SS, agricultural residues, energy crops) into mixed-alcohol fuels. It has evolved from older technologies, some of which were practiced in the last century. Until now, a significant barrier to the commercialization of these older technologies has been the difficult separation of the VFA salts from the fermentation broth; however, a recently developed proprietary separation process has overcome this obstacle making the MixAlco process economically viable.

Although many of the steps in the MixAlco Process have been demonstrated in the past, important rate and yield data have not appeared in the open literature. Critical data for the lime treatment, fermentation, thermal conversion, and hydrogenation are presented which will allow further process development.

Preliminary economic evaluations indicate that the MixAlco Process can economically produce fuel oxygenates from negative-value biomass, such as MSW and SS.

Acknowledgments

The authors are grateful for the financial support provided by the National Renewable Energy Laboratory, the Energy Resources Program at Texas A&M University, and Terrabon Corporation.

Literature Cited

1. Lefranc, L. *British Patent* 17,776, **1914**.
2. Hurd, C.D. *The Pyrolysis of Carbon Compounds*; The Chemical Catalog Company: New York, NY, 1929; p. 486.
3. Playne, M.J. In *Advances in Biotechnology*; Moo-Young, M., Ed.; Pergamon Press: New York, NY, 1981, Vol. 2; pp. 85-90.
4. Playne, M.J. *Biotechnol. Bioeng.* **1984**, *26*, pp. 426-433.
5. Playne, M.J. In *Comprehensive Biotechnology*; Moo-Young, M.; Blanch, H.W.; Drew, S.; Wang, D.I.C., Eds.; Pergamon Press: New York, NY, 1985, Vol. 3; pp. 731-759.
6. Dale, B.E. *Annual Reports on Fermentation Processes*; Academic Press: New York, NY, Vol. 8, pp. 299-323.
7. Nagwani, M., MS thesis, Texas A&M University, College Station, TX, **1992**.
8. Miller, G.L. *Analytical Chemistry* **1959**, *31*, pp. 426-428.
9. Adney, B.; Baker, J. *Chemical Analysis and Testing Standard Procedure No. 6*; National Renewable Energy Laboratory, Golden, CO, 1992.
10. Rapier, C.R., MS thesis, Texas A&M University, College Station, TX, **1995**.
11. Caldwell, D.R.; Bryant, M.P. *Appl. Microb.*, **1966**, *14*, pp. 794-801.
12. Magill, K. *Chemical Analysis and Testing Standard Procedure No. 3 & 4*; National Renewable Energy Laboratory, Golden, CO, 1992.
13. Ehrman, T. *Chemical Analysis and Testing Standard Procedure No. 5*; National Renewable Energy Laboratory, Golden, CO, 1994.
14. Nelson, D.L.; Webb B.P. In *Kirk-Othmer Encylcopedia of Chemical Technology*; 3rd Ed., Wiley: New York, NY, 1978, Vol. 1; pp. 179-191.
15. Ardagh, E.G.R.; Barbour, A.D.; McClellan, G.E.; McBride, E.W. *Industrial and Engineering Chemistry* **1924**, *16*, pp. 1133-1139.

16. Schultz, H.P.; Sichels, J.P. *Journal of Chemical Education* **1961**, *38*, pp. 300-301.
17. Adelson, S.K., MS thesis, Texas A&M University, College Station, TX, **1994**.
18. Chang, N.-S., PhD dissertation, Texas A&M University, College Station, TX, **1994**.
19. Broder, J.D.; Eley, M.H.; Barrier, J.W. *Municipal Solid Waste and Waste Cellulosics Conversion to Fuels and Chemicals*, Tennessee Valley Authority, Muscle Shoals, AL, 1993, Vol. II.
20. Delong, J.V. *The Brookings Review* **1994**, *12*, pp. 34-39.

Chapter 8

Computer-Mediated Addition of Fresh Medium in Continuous Culture of *Zymomonas mobilis* by Monitoring Weight Changes

H. D. Zakpaa[1], Ayaaki Ishizaki[1], and K. Shimizu[2]

[1]Department of Food Science and Technology, Faculty of Agriculture, Kyushu University, Hakozaki, Higashi-ku, Fukuoka 812, Japan
[2]Department of Biochemical Engineering and Science, Faculty of Computer Science and Systems Engineering, Kyushu Institute of Technology, Iizuka 820, Japan

A major problem encountered in pH-stat continuous culture of *Zymomonas mobilis* which employed growth-dependent pH changes to control the rate of addition of fresh medium into the culture was that of low dilution rate. Glucose concentration in the culture broth also became very low after the initiation of continuous fermentation. A new control system for delivering fresh substrate into the fermentor is discussed that solves the problems of low dilution rate and low residual glucose. This new system involves linking the feeding of fresh substrate into the fermentor with the outflow of alkali (NaOH) from the alkali reservoir by monitoring weight changes of the two solutions. Dilution rates of about 0.33 h^{-1} and productivity values of about 11.00 $g.l^{-1}h^{-1}$ could be achieved with this method, while maintaining the concentration of ethanol in the fermentor at about 33.00 $g.l^{-1}$ when the ratio (α) of weight of glucose fed into the fermentor to the unit weight of alkali used in neutralizing acid produced in the broth was 25.00.

Zymomonas mobilis (*1*), a bacterium that occurs as motile short rods (*2*), has attracted considerable attention as a promising microorganism for large scale production of ethanol because of its unusual physiological and biochemical properties (*1*), and more recently because of its high efficiency in ethanol production.

Three main approaches namely; physiological (*3, 4-9*), genetic (*10-16*), and engineering (*17-23*) are being pursued with an aim towards improving the productivity of ethanol fermentation. With the engineering approach to fermentation process improvement, fermentors are operated in continuous mode instead of the more conventional batch mode, resulting in an increase in productivity. The genetic approach aims at increasing process productivity through the improvement of the metabolic characteristics of the organism employed, by attempting to correct a recognized weakness or deficiency, such as broadening the range of substrates which the organism can metabolize as carbon source. The physiological approach attempts to control parameters that affect process productivity, by varying environmental factors such as the chemical composition of the fermentation medium, concentration of essential nutrients or inhibitory substances, as well as pH and temperature (*4-9, 24*).

Computer technology has had significant impact on fermentation in recent times. The advent of inexpensive, computer-coupled fermentation systems has facilitated close monitoring and direct-digital control (DDC) of environmental variables such as temperature and dissolved oxygen concentration. These systems are now commonly used for on-line monitoring of several fermentation variables such as oxygen uptake rate, carbon dioxide evolution rate, and respiratory quotient. Such variables have been successfully used for identifying the physiological state of microorganisms in many fermentation processes (*25, 26*). The use of computer-coupled fermentation systems to control pH in DDC mode also allows for on-line measurements of the amount of acid produced and the acid-production rate during fermentation. These variables can prove significant in on-line characterization of an anaerobic fermentation process, like ethanol fermentation by *Z. mobilis*.

Low dilution rate and low residual glucose concentration in the fermentor broth were encountered in the previous set-up of computer-mediated, pH-dependent addition of fresh substrate in continuous culture of *Z. mobilis* (*27*). In this paper an alternative but similar method of supplying fresh substrate to the culture is designed and tested. This approach involved linking the feeding of fresh substrate into the fermentor with the outflow of alkali (NaOH) from the alkali reservoir into the fermentor, achieved through the monitoring of weight changes. The rationale behind this method is that by monitoring the NaOH addition resulting from decrease in the pH of the culture broth, better control of glucose concentration in the fermentor can be accomplished. This contrasts with previous methods (*27*), where fresh substrate was only fed when the pH in the fermenting culture rose above the set pH limit on the control unit as a result of excretion of NH_4^+ ions into the medium. In anaerobic culture employing *Z. mobilis*, this pH rise is delayed. By the time it occurred the cells had been starved of glucose and most had lost viability.

MATERIALS AND METHODS

Microorganism: *Zymomonas mobilis* NRRL-B14023 was used in this study. Stock culture was incubated in YM liquid medium (Difco laboratories, Michigan, USA) for 18 h at 30°C and stored for up to two weeks at 4°C in YM liquid medium.

Growth medium: The growth medium consisted of 100 g.l-1 glucose, 10 g.l^{-1} yeast extract, 1 g.l^{-1} KH_2PO_4, 1 g.l^{-1} $(NH_4)_2SO_4$, and 0.5 g.l^{-1} $MgSO_4.7H_2O$. Composition of fresh medium feed was the same as the medium employed as growth medium. The substrate medium had the same composition as the growth medium.

Inoculation and Preculture: *Z. mobilis* was first propagated at 30°C for 18 h without agitation by transferring 0.5 ml of culture from the stock culture to

10 ml of YM liquid medium. The starter culture was then transferred to 40 ml of seed culture medium at pH 5.5 and incubated for 18 hours at 30°C. This seed culture was transferred into 450 ml of growth medium in the fermentor to make a final volume of 500 ml.

Continuous culture: The schematic diagram of the experimental set-up is shown in Figure 1. Continuous culture was performed in a 1-*l* jar fermentor with a working volume of 500 ml. The composition of the medium and culture conditions were the same as used for the batch and continuous culture experiments. Fresh substrate of the same composition and same initial pH of 5.5 as that in the fermentor was fed intermittently into the fermentor. The amount of fresh substrate fed into the fermentor depended on the amount of alkali (0.5 M NaOH) pumped into the fermentor by the pH control unit (Biott Co. Ltd.,Tokyo Japan, model KVS-15-B) to neutralize the acid produced by the metabolizing cells, and to maintain the pH of the fermenting broth at the set level of 5.5. pH was monitored using a glass Pt/KCl electrode (Toko Chemical Laboratories Co., Ltd. Tokyo Japan). Fresh medium and 0.5 M NaOH solution were contained in 1 *l* and 100 ml graduated cylinders, respectively. The fresh substrate and alkali reservoirs were placed on sensitive electronic balances (Shinko Denshi Co. Ltd. Tokyo Japan, models SK-6000H and SK-600H) with weight limits of 6000 g and 600 g , respectively. The weight signals (voltage) from the feed substrate and alkali balances were inputted into a computer (NEC, PC 9801 VM) through an analog-to-digital (A/D) converter. The digital signal was converted to the voltage signal by means of a digital-to-analog (D/A) converter. Using software written in C^{++} computer language loaded on the computer hard disk, any decrease in weight detected by the alkali reservoir balance, resulting from the pumping of alkali into the fermentor to neutralize any acid produced, led to the switching on of a peristaltic pump (Tokyo Rikikai Co. Ltd., Tokyo Japan, EYELA Microtube pump MP-3) connected to the D/A converter. The pump then delivered specific weight of fresh substrate into the fermentor. This was achieved by monitoring the decrease in weight recorded by the fresh medium pump when the pump was activated. The pump was switched off when the required amount of fresh substrate was delivered. The same pump was employed to draw off exactly the same volume of culture broth equal to the volume of fresh substrate fed into the fermentor. This was to ensure that culture volume was maintained at constant level in the fermentor. Each experiment was carried out until steady state was achieved, after about 30 to 50 hours.

Computer-mediated addition of substrate through the monitoring of weight changes of the alkali and substrate reservoirs.

At the initiation of the experiment the amount of fresh substrate to be delivered per unit weight of alkali fed (α) was set on the computer. In the initial series of experiments, α was varied to determine its effect on the dilution rate and residual glucose concentration at steady state. The software also incorporated a short program for signal conditioning or smoothing, using a moving average calculation method, to filter out noise generated in the electrical circuits contained in the digital weighting balances. Continuous plot of the weight of alkali and fresh substrate reservoirs were displayed on the computer screen during the course of the experiment.

Analytical Methods: Ethanol concentration was estimated by gas chromatography (GC-8 APE; Shimadzu Corporation, Kyoto Japan) with PEG 80-100 mesh. The temperature of the column oven and the injector were 70°C and 90°C, respectively. Glucose concentration was determined with a glucose analyzer (Model 23 A, Yellow Spring Instrument Co. Ltd., Ohio USA). Biomass was estimated using correlation between optical density measurements at 562 nm using a spectrophotometer (Uvidec 320 AS Co. Ltd., Tokyo Japan) and dry cell weights.

RESULTS AND DISCUSSION

Despite its importance as an influence on microbial metabolism and growth, pH had not been employed much in control strategies, mainly due to the fact that coarse control of pH within broad ranges was adequate for most experimental and industrial purposes. Increasing knowledge of microbial physiology and availability of sophisticated computer-linked monitoring and control systems have led to more rapid advances in pH manipulation for industrial fermentations. Small changes in pH are used to monitor nutrient availability and supply as in the pH-stat (*28*). When ammonium salts are supplied to a culture as sole nitrogen source, uptake of ammonia leads to pH fall. Use of complex carbon and nitrogen sources, such as proteins and polypeptides, will generally lead to pH rise as excess nitrogen is released as ammonia.

Veeramallu and Agrawal (*25*), evaluated on-line the acid production rate (APR) of *Z. mobilis* ATCC 10988, using the double exponential smoothing of the acid production data. Both acid production and acid production profiles, as expected, increased exponentially with time during exponential batch culture. Upon further investigation, Agrawal and Lim (*29*) reported that under balanced-growth conditions, the APR could be related to the specific-growth rate (μ) in a constant volume batch fermentor as;

$$d\,[\,\ln(\mathrm{APR})\,]\,/\,dt \quad = \quad \mu \qquad \text{(constant)}$$

The slope of the APR profile on a semilogarithmic plot could, therefore, be used as check for the presence of balanced growth (constancy of slope), and to estimate the value of μ (magnitude of slope) in batch cultures.

The production of acid during growth of *Z. mobilis* is largely due to the cellular incorporation of nitrogen from ammonium ions, supplied as ammonium sulfate, in the nutrient medium. This process can be represented as:

$$NH_4^+ \;\text{------>}\; NH_3 \qquad + \qquad H^+ \quad \text{(acid)}$$

Thus, the rate of acid production is related to the rate of nitrogen assimilation which, in turn, is related to the rate of cell-mass production.

Figure 2 depicts a profile of pH change in batch culture of *Z. mobilis* in which pH was not controlled by the addition of alkali. The pH of the culture medium was first adjusted to 5.5 before the initiation of fermentation. When fermentation commenced pH gradually decreased from 5.5 to reach a minimum value of 3.9 after 11 hours of fermentation, when all glucose in the medium was exhausted. Under anaerobic conditions ATP is produced by substrate-level phosphorylation. The proton motive force needed to drive the transport process, was obtained by hydrolysis of ATP. ATP hydrolysis was involved in the generation and maintenance of the electrochemical potential ($\Delta\mu H^+$). The electrochemical potential is generated by the excretion of protons with the metabolic product into the fermenting culture by the action of a membrane bound, proton-translocating ATPase (*30*), this results in a decrease in the pH of the culture broth.

Upon the exhaustion of glucose in the medium pH began to rise to peak around 5.8. With the depletion of glucose the bacterium switched into metabolizing the amino acids present in the medium as carbon source. Nitrogen is released in the form of ammonia, with resulting rise in the pH of the culture broth.

Other factors such as the formation of acidic/basic products and the uptake of components such as phosphate also play a role in determining the final pH of the culture broth. The fact that there is pH decrease with the consumption of glucose suggest that it might be possible to feed fresh substrate into the fermentor using pH fall as the indicator.

At the beginning of the investigation of pH-mediated addition of substrate in continuous culture of *Zymomonas mobilis*, ordinary chemostat culture of the bacteria was carried out. The chemostat operates by fixing the dilution rate to a value lower than the maximum specific growth rate and it allows some condition in the fermentor, normally substrate concentration, to limit the growth of organism to equal that of the set value of the dilution rate of the medium being fed into the culture. In a chemostat culture of *Z. mobilis* (Table 1) at pH 5.5 a dilution rate of 0.15 h^{-1} was attained. Productivity and biomass yield of 4.62 g l^{-1} h and 0.03 g l^{-1}, respectively were obtained. However, with this anaerobic bacterium long-term economical cultivation proved difficult as there were a number of disadvantages inherent with the use of the chemostat culture with *Zymomonas mobilis*. In long term fermentation using a dilution rate of about 0.15 h^{-1} with such a setup, substrate limitation eventually resulted. When glucose became limiting, the cells were starved leading to lost of viability and cell death. The nonviable cells were then 'washed out'. Too high a dilution rate also led to waste of substrate as the glucose concentration in the spent stream was uneconomically high, eventually the cells 'washed out' as growth rate did not keep up with the rate at which cells were removed from the fermentor. The reciprocal of the dilution rate, the mean residence time, which denotes the average time a cell remained within the culture vessel became very small. Recycling led to ethanol inhibition in the fermentor. Hence the major disadvantage with the use of such a setup is that the dilution rate is fixed and does not vary with the metabolic activity of the microorganism. The system is inherently unstable with an anaerobic bacteria like *Zymomonas mobilis*.

Computer-mediated continuous culture using pH-dependent feeding was

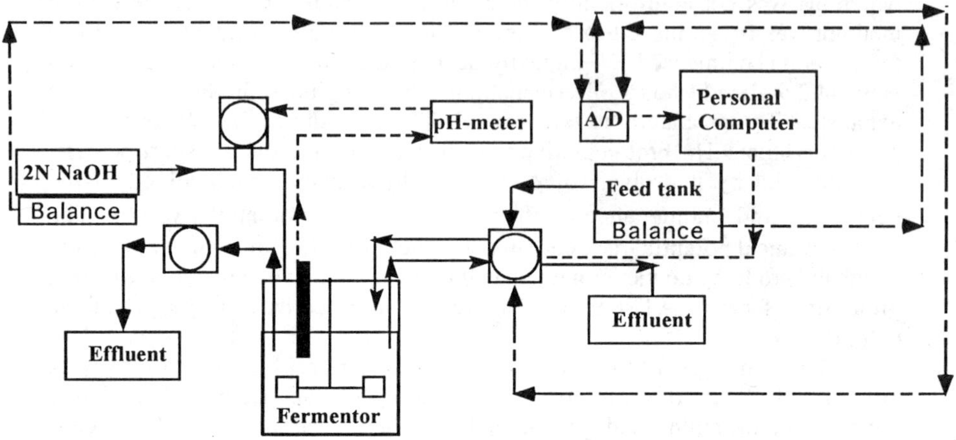

Figure 1. Schematic diagram of pH-dependent feeding of substrate through monitoring of weight changes.

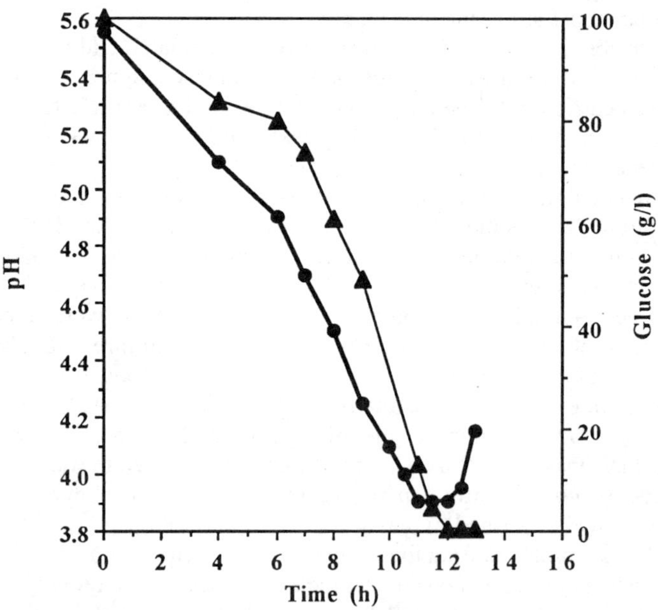

Figure 2. Profile of pH change in batch culture of *Z. mobilis* without pH control using NaOH solution.

investigated previously in our laboratory with an aim of overcoming the limitations of the chemostat and also to improve on the dilution rates and productivities which could be obtained in long term fermentations (*27*). However, 'stop flow' occurred when an upper limit and glucose concentration of 5.6 and 100 $g.l^{-1}$ respectively were investigated. The pH-stat employs pH changes, brought about by cell growth to control the rate of addition of fresh medium to the culture vessel. This is accomplished by allowing inflowing medium to return the pH value of the medium in the growth vessel to some preset value. Stop flow occurred because the pH set point was too high. pH did not rise very significantly until almost all the glucose in the culture broth was exhausted. When this occurred, the feeding of fresh glucose medium into the fermentor was delayed. By the time pH rose to the set limit to initiate the feeding of fresh medium, the cells had switched from the utilization of glucose as carbon source, to the deamination and use of the carbon skeletons of amino acids as carbon source, and had begun to loose viability. Once greater proportion of the cells lost viability, they ceased to metabolize any substrate supplied, hence resulting in no further pH changes. Flow of fresh substrate therefore stopped. Another problem was that after the initiation of continuous culture was that of very low residual glucose concentration in the culture broth. Again this suggests that rate of fresh medium addition did not meet demand during the active phase of the bacterial growth.

Table 2 shows some of the culture parameters of pH-dependent addition of fresh substrate through the monitoring of weight changes at various values of α using 100 $g.l^{-1}$ glucose fresh substrate media and 0.5 M NaOH to control pH. When α was 25.0 and 30.0 dilution rates of 0.33 h^{-1} and 0.38 h^{-1} respectively were obtained. The corresponding residual glucose concentrations were 19.70 $g.l^{-1}$ and 49.44 $g.l^{-1}$. As expected dilution rate increased with an increase in α. Ethanol concentration in the fermentor for α of 25.0 and 30.0 were 33.07 $g.l^{-1}$ and 18.02 $g.l^{-1}$ respectively. Higher dilution results in faster removal of the product from the fermentor. With small values of α less than 25, corresponding dilutions rates were not adequate to replenish the glucose consumed by the microorganisms hence the residual glucose concentration decreased rapidly to very low values, resulting in cell starvation (data not shown). On the other hand, when α was increased above 30 the corresponding dilution rates exceeded the μ_{max} value for the bacteria. The result was cell washout from the fermentor and uneconomically high values of residual glucose concentration. To investigate the effect of alkali concentration on the dilution rate, the concentration of NaOH solution used to control pH was doubled from 0.5M to 1.0M. Value of α was doubled as well. When the experiment was conducted dilution rate was not affected to any significant extent. Using 1.0M NaOH and α of 50.0 and 60.0 the corresponding dilution rate were 0.30 h^{-1} and 0.37 h^{-1} nearly the same as for the first set of experiments (Table 3). Ethanol concentration and residual glucose concentration were 37.74 $g.l^{-1}$ and 25.27 $g.l^{-1}$; 15.66 $g.l^{-1}$ and 36.88 $g.l^{-1}$ respectively.

Figure 3 shows the variation of the dilution rate, product concentration, biomass and residual glucose concentration for α equals 60.0. The plot for the dilution rate closely follows that of the residual glucose concentration, but with a

Table 1. Typical chemostat culture parameters at pH 5.5 and 6.0

	pH	
	5.5	6.0
Flow rate (ml h^{-1})	73.94	73.56
Dilution (h^{-1})	0.15	0.15
Productivity ($gl^{-1}h$)	4.62	3.00
Biomass yield (gl^{-1})	0.03	0.05
Specific glucose consumption rate ($gl^{-1}h^{-1}$)	13.51	11.45
Specific production rate ($gl^{-1}h^{-1}$)	1.97	0.87

Table 2. Culture parameters at various values of α using 100 $g.l^{-1}$ and 0.5M NaOH

	Experimental Conditions	
	$\alpha = 25.00$	$\alpha = 30.00$
Substrate (glucose) concentration ($g.l^{-1}$)	100.00	100.00
Concentration of alkali feed	0.5M NaOH	0.5M NaOH
Dilution rate (h^{-1})	0.33	0.38
Product concentration ($g.l^{-1}$)	33.07	18.02
Productivity ($g.l^{-1}h^{-1}$)	11.09	6.58
Biomass concentration ($g.l^{-1}$)	2.38	1.84
Residual glucose concentration ($g.l^{-1}$)	19.70	49.44

Table 3. Culture parameters at various values of α using 100 g.l^{-1} and 1.0M NaOH

	Experimental Conditions	
	α = 50.00	α = 60.00
Substrate (glucose) concentration (g.l^{-1})	100.00	100.00
Concentration of alkali feed	1.0M NaOH	1.0M NaOH
Dilution rate (h^{-1})	0.30	0.37
Product concentration (g.l^{-1})	37.74	25.27
Productivity (g.$l^{-1}h^{-1}$)	15.36	9.20
Biomass concentration (g.l^{-1})	2.53	2.02
Residual glucose concentration (g.l^{-1})	15.66	36.88

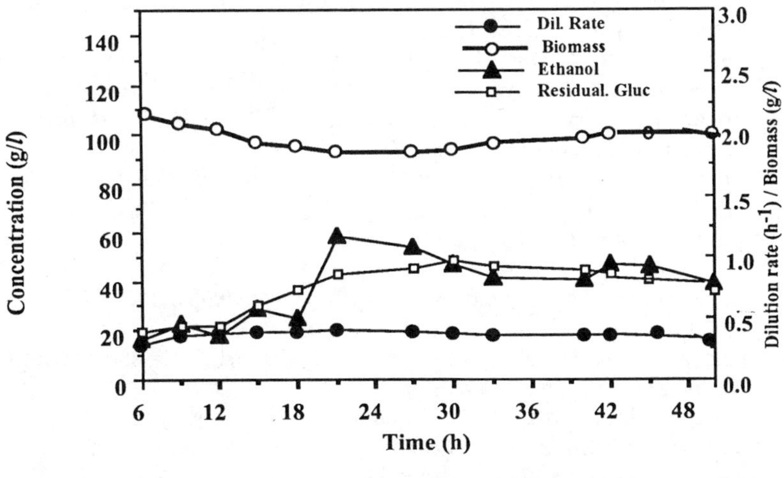

Figure 3. Variation of some culture parameters in pH-dependent feeding of fresh medium using α of 60 and 1.0M NaOH.

slight delay. Any increase in residual glucose in the fermentor results in an increase in acid production, which in turn leads to an increase in the volume of alkali fed into the fermentor to neutralize the extra acid. The final outcome is an increase in the dilution rate.

This method is an improvement over other pH-dependent control strategies employed with *Zymomonas mobilis* for controlling dilution rate and for preventing substrate limitation. Under certain experimental conditions, the residual glucose in the fermentor was slightly higher. Further investigations are being conducted to reduce residual glucose and to improve on the productivity by coupling the method with others to achieve in addition, fine control of the fermentation process.

Acknowledgment

The authors are grateful to Mr. Ishida and other members of Prof. K. Shimizu's laboratory, Faculty of Computer Science and Systems Engineering, Kyushu Institute of Technology, Iizuka for their valuable help with the software and hardware configuration.

Literature Cited

1. Swings, J. and DeLey, *J., Bacteriol. Rev.* **1977,** *41,* pp 1-44.
2. Buchann, R.E. and Gibbons, N. E. Ed.; *Bergey's mannual of determinative bacteriology;* 1974, p 353.
3. Morris, J. G., In *Comprehensive biotechnology : The principles of biotechnology: Scientific Fundamentals;* Murray Moo-Young Eds.; , Pergamon Press, Oxford, 1985, pp 358-359.
4. Lawford, G. R., Lavers, B. H., Good, D., Charley, R. C., Fein, J. E., and Lawford, H. G., *Intl. symp. on ethanol from biomas;* 1982 p 482.
5. Rogers, P. L., Lee, K. J., Skotnicki, M. L., and Tribe, D. E., *Advances in Biochemical Engineering.,* **1982,** *23,* pp 37-84.
6. Cromie, S. and Doelle, H. W., *Eur. J. Appl. Microbiol. Biotechnol.,* **1981,** *11,* pp 116-119.
7. Cromie, S. and Doelle, H. W., *Eur. J. Appl. Microbiol. Biotechnol.,* **1982,** *14,* pp 69-73.
8. Galani, I., Drainas, C., and Typas, M. A., *Biotechnol. Letts.,* **1985,** *7,* pp 673.
9. Fein, J. E., Charley, R. C., Hopkins, K. A., Lavers, B. H., and Lawford, H. G., *Biotechnol. Letts.,* **1983,** *5,* pp 1-6.
10. Skotnicki, M. L., Warr, R. G., Goodman, A. E., and Rogers, P. L., *Abstracts, 4th international symposium on genetics of industrial microorganisms;* Kyoto, Japan, 1982, pp. 32.
11. Lee, J. M., Skotnicki, M. L., and Rogers, P. L., *Biotechnol. Lett.,* **1982,** *4,* pp 615-620.
12. Ingram, L. O., Conway, T., Clark, D. P., Sewell, G. W., and Preston, J. F., *Appl. Environ. Microbiol.,* **1987,** *53,* pp 2420-2425.
13. Lawford, H. G. and Rousseau, J. D., *Biotechnol. Lett.,* **1991,** *13,* pp 191-196.

14. Sreekumar, O. and Basappa, S. C., *Biotechnol. Lett.*, **1991,** *13,* pp 365-370.
15. Park, S. C. and Baratti, J., *Biotechnol. Bioeng.*, **1991,** *38,* pp 304-313.
16. Su, P., Liu, C-Q., Lucas, R. J., Delaney, S. F., and Dunn, N. W., *Biotechnol. Lett.*, **1993,** *15,* pp 979-984.
17. Cysewski, G. H., and Wilke, C. R., *Biotechnol. Bioeng.*, **1977,** *19,* pp 1125.
18. Lee, S. W., Ebata, T., Liu, Y. C., and Tanaka, H., *J. Ferment. Bioeng.*, **1993,** *75,* pp 36-42.
19. Calibo, R. L., Matsumura, M., and Kataoka, H., *J. Ferment. Bioeng.*, **1989,** *67,* pp 40-45.
20. Ghose, T. K. and Tyagi, R. D., *Biotechnol. Bioeng.*, **1979,** *21,* pp 1387-1400.
21. Wang, H. Y., Robinson, F. M., and Lee, S. S., *Biotechnol. Bioeng. Symp.*, **1981,** *11,* pp 555-565.
22. Matsumura, M. and Markl, H., *Appl. Microbiol. Biotechnol.*, **1984,** *20,* pp 371-377.
23. Matsumura, M. and Markl, H., *Biotechnol. Bioeng.*, **1986,** *28,* pp 534-551.
24. Nipkow, A., Beyeler, W., and Fiechter, A., *Appl. Microbiol. Biotechnol.*, **1984,** *19,* pp 237-240.
25. Veeramallu, U. and Agrawal, P., *Biotechnol. Bioeng.*, **1988,** *31,* pp 770-782.
26. Wu, W., Chin, K., and Chiou, H., *Biotechnol. Bioeng.*, **1985,** *27,* pp 756.
27. Ishizaki, A., Tripetchkul, S., Tonokawa, M., Zhong-Ping, S., and Shimizu, K., *J. Ferment. Bioeng.*, **1994,** *77,* pp 541-547.
28. Forage, R. G., Harrison, D. E. F., and Pitt, D. E., In *Comprehensive biotechnology : The principles of biotechnology: Scientific Fundamentals;* Murray Moo-Young Ed.; Pergamon Press, Oxford; 1985 *1,* pp 251-280.
29. Agrawal, P. and Lim, H. C., *Biotechnol. Bioeng.*, **1984,** *26,* pp 1352.
30. Kalnemiek, U. Z., Pankova, L. M., and Shivinka, Y. E., *Mikrobiologiya,* **1987,** *52,* pp 721-723.

Chapter 9

Expression of *Microbispora bispora* Bgl B β-D-Glucosidase in *Streptomyces lividans*

Xiaoyan Xiong, William S. Adney, Todd B. Vinzant, Yat-Chen Chou, Michael E. Himmel[1], and Steven R. Thomas

Biotechnology Center for Fuels and Chemicals, National Renewable Energy Laboratory, 1617 Cole Boulevard, Golden, CO 80401

Active, thermostable *Microbispora bispora* Bgl B β-D-glucosidase was expressed in *Streptomyces lividans* TK24 cells transformed with plasmid pIJ702 carrying the *bglB* coding sequence under the control of *Streptomyces longisporus* STI-II trypsin inhibitor promoter. The recombinant enzyme, *Sl*Bgl B, has a molecular weight of 54 kDa, an isoelectric pH of 5.0, shows resistance to glucose inhibition, and is optimally active on *o*-nitrophenyl β-D-glucopyranoside at 57°C. This recombinant β-D-glucosidase was more active on aryl-glycosides than on cellobiose. We also report a successful mutagenesis strategy used to achieve increased levels of *Sl*Bgl B expression in this host organism. Screening mutants created by low fidelity PCR using Taq polymerase in the presence of manganese ion revealed a series of up-regulated clones, one yielding 235 mg/L of *Sl*Bgl B.

β-D-glucoside glucohydrolases (EC 3.2.1.21), or β-D-glucosidases, catalyze the hydrolysis of *O*-glycosyl bonds in aryl- and alkyl-glucosides, as well as in many β-linked disaccharides and some oligosaccharides. These enzymes are produced by plants, animals, and most known microbiota and have been extensively reviewed (*1,2*). Each of these enzymes displays a distinct pattern of relative activity on an array of β-glucosides. A subset of β-D-glucosidases are especially proficient at the hydrolysis of cellobiose, and are often referred to as cellobiases.

In general, thermotolerant enzymes have higher turnover rates and better tolerate the stresses of use in large-scale processes. Thus, they are of interest to industry (*3*). Thermophilic β-D-glucosidases have been found in *Clostridium stercorarium*, T_{opt} = 65°C (*4*); *Clostridium thermocellum* NCIB 10682, T_{opt} = 65°C (*5*); *Thermomonospora* sp. strain YX, T_{opt} = 55°C (*6*); and *Acidothermus cellulolyticus*, T_{opt} = 75°C (*7*). Caldophilic β-D-glucosidases have also been isolated from *Caldocellum saccharolyticum*, T_{opt} = 85°C (*8*);

[1]Corresponding author

Clostridium thermocopriae JT3-3, T_{opt} = 80°C (*9*); and *Thermus* strain Z-1, T_{opt} = 85°C (*10*).

Typically, native strains are not prolific producers of β-D-glucosidase activity, so researchers have used recombinant DNA expression strategies to obtain reasonable quantities of these enzymes. Genetically engineered expression from *E. coli* of various thermostable β-D-glucosidases including *C. saccharolyticum* (*8*), *C. thermocellum* NUB 10682 (*11*), *C. thermocellum* NCIB 10682 (*12*), and *Rhodothermus marinus* (*13*) has been demonstrated in several laboratories using lambda phage or plasmid vectors in *E. coli* and other host bacteria.

In 1986, Waldron and coworkers reported the isolation of a new thermotolerant actinomycete, *Microbispora bispora*, from warm compost (*14*). This microorganism produced thermostable β-D-glucosidase activity that was resistant to inhibition by glucose concentrations as high as 30% w/v. Upon screening a genomic library of *M. bispora* DNA, Wright (*15*) discovered two DNA fragments of 4.0 kb and 2.1 kb coding for two distinct β-D-glucosidases, Bgl A and Bgl B, respectively. The *bglB* gene encodes *M. bispora* β-D-glucosidase B, *M*Bgl B, which is highly resistant to feedback inhibition by glucose, but is expressed only very weakly from its native promoter in *E. coli*. We report the cloning of the coding sequence for *M*Bgl B under the control of the STI-II trypsin inhibitor promoter from *Streptomyces longisporus* (*16*) in plasmid pIJ702, as well as the expression and characterization of active, thermostable enzyme from transformed *Streptomyces lividans* TK24, *Sl*Bgl B. We also report on our efforts to mutagenize the STI-II promoter to achieve higher levels of *Sl*Bgl B expression in this host organism.

Materials and Methods

Bacterial Strains and Plasmids. The methylation-deficient *E.coli* strain JM110 was used to propagate plasmids pX29 and pGEM-BglB before DNA fragments were ligated into pIJ702. Plasmid pX29 (*17*) was kindly supplied by Dr. D.E. Eveleigh and carries the 2.1 kb *Xho*1 insert of *M. bispora* genomic DNA encoding the *bglB* gene. *S. lividans* TK24, kindly supplied by Dr. M.J. Bibb (University of Norwich, England), was employed as the host strain to express *Sl*Bgl B. The *Streptomyces* cloning vector, pIJ702, was also obtained from Dr. M.J. Bibb. Plasmid pIJ702 carries the *mel* gene (melanin production), which enables a black/white color selection when foreign DNA is inserted at the unique *Sph*1, *Bgl*II, or *Sac*1 sites (*18*) in the *mel* gene. *S. longisporus* ATCC 23931 was obtained from the American Type Culture Collection (Rockville, MD).

Growth Media. *E. coli* strains were grown in shake flasks in LB broth (*19*). Shake flask experiments and 10-L fermentations of *S. lividans* strains were conducted in tryptone soy broth (TSB) (*18*) or M media (*16*). *S. lividans* TK24 cells to be used for transformation were grown in YEME (*18*) before protoplasting. *S. lividans* transformants were regenerated on R2Y plates (*18*) that were overlayed with 300 mg thiostrepton diluted in 1 mL sterile water at 20 to 24 h after plating. Mutant screening plates were made by adding the colorogenic substrate analog, 5-bromo-4-chloro-3-indolyl β-D-glucopyranoside (0.5 mM; X-Glc; Sigma Chemical Co., St. Louis, MO), to minimal media plates (*18*).

Construction of *Sl*Bgl B Expression Plasmid. Chromosomal DNA of *S. longisporus* cells was prepared by the procedure of Hopwood et al. (*18*). A 0.5 kb DNA fragment that contained the promoter and signal sequences of the STI-II trypsin inhibitor gene (*16*) was amplified from *S. longisporus* chromosomal DNA using the following synthetic primers (see Figure 1). The upstream primer, "1pfw" (5'-AAGCTTGAAT TCGAGCTCGG CCAGCTCCTC GAAG-3'), primes synthesis of the coding strand. The downstream primer, "1pbk" (5'-CCGATTCGGT CATGGCGAGC GCGGCTCCGG TGAGG-3'), primes synthesis of the non-coding strand. A short DNA sequence coding the N-terminal peptide of *M. bispora* Bgl B was incorporated into the 5' end of the 1pbk primer to join the STI-II promoter with the *bglB* coding sequence using overlap extension PCR. Conditions for the first round PCR reactions were: one cycle at 96°C, 4 min: three cycles at 96°C, 1 min; 60°C, 1 min; and 75°C, 1 min: 27 cycles using the same sequence, except annealing was done at 70°C, 1 min: and finally, the sequence was terminated with one cycle of 96°C, 1 min; 72°C, 1 min; and 75°C, 10 min.

In a similar manner, a 0.36 kb fragment of *bglB* coding sequence from the N-terminus through the unique *Apa*1 site, was subjected to PCR using pX29 as a template with a second set of synthetic primers. The upstream primer, "2pfw" (5'-GGAGCCGCGC TCGCCATGAC CGAATCGGCC ATGAC-3'), primes synthesis of the coding strand, and the downstream primer, "2pbk" (5'-CACGAGCCGG TCGTAGAAGT CCAGG-3') primes synthesis of the non-coding strand. The C-terminal sequence of the STI-II signal peptide was incorporated into the 5'-end of 2pfw primer for the subsequent overlap extension reaction.

The overlap extension PCR reaction to fuse the STI-II promoter/signal fragment with the *M*Bgl B N-terminal peptide coding fragment was conducted using the two primary PCR products as templates and the primers, 1pfw and 2pbk. Conditions for the fragment fusion reaction were: one cycle at 96°C, 4 min: three cycles at 96°C, 1 min; 59°C, 1 min; and 75°C, 1 min: 27 cycles at 96°C, 1 min; 70°C, 1 min; and 75°C, 1 min: and finally, the sequence was terminated with one cycle at 96°C, 1 min; 72°C, 1 min; and 75°C, 10 min. The 0.86 kb PCR fusion product was gel purified and cleaved with the restriction enzymes *Sac*1 and *Apa*1 to yield a 0.86 kb fusion product before cloning.

The remainder of the *bglB* coding sequence fragment extending from the internal *Apa*1 site (coordinate 686 of the *bglB* published sequence; *17*) to a *Sal*1 site downstream from the termination codon (coordinate 2071 of published *bglB* sequence; *17*) was removed from pX29 by restriction digestion (*Apa*1/*Sal*1). After gel purification the 1.4 kb *Apa*1/*Sal*1 fragment was ligated to the 0.86 kb (*Sac*I/*Apa*I) promoter fragment under controlled conditions (i.e., 30 min at 16°C followed by inactivation of T4 ligase at 65°C for 10 min). The resulting 2.2 kb ligation product was gel purified and inserted into plasmid pGEM-7 (Promega, Madison, WI) at *Sac*1/*Xho*1 to produce pGEM-BglB, followed by transformation into *E. coli* JM110 competent cells.

The purpose of constructing the cloning intermediate, pGEM-BglB, was to permit easy handling of DNA transformation and plasmid DNA propagation in *E. coli*. Once sufficient pGEM-BglB DNA could be isolated from JM110 cells, the chimeric STI-II/*bglB* gene fragment was subcloned into plasmid pIJ702 by digestion of pGEM-BglB with *Sac*1. The upstream *Sac*1 site is identical to that used for cloning the chimeric gene into pGEM-BglB. The downstream *Sac*1 site resides in *M. bispora* DNA just upstream from the *Sal*1 site used to construct pGEM-BglB. Ligation of the gel-purified 2.2 kb chimeric gene fragment from pGEM-BglB into *Sac*1-cut pIJ702 produced plasmid pIJ-BglB.

Identification of Transformants Expressing Recombinant *Sl*Bgl B. pIJ-BglB was introduced into *S. lividans* TK24 protoplasts and regenerated on R2Y plates as described above. Individual white colonies appearing on R2Y plates were picked, transferred onto selection plates (see growth media and plate), or directly innoculated into 5 mL liquid broth (TSB or M media), shaken at 30°C for 4-5 d, and assayed for *Sl*Bgl B activity.

***Sl*Bgl B Activity Monitoring Assay**. Two assays were employed to monitor *Sl*Bgl B activity.

4-methyumbelliferyl-beta-D-glucoside (MUG) Assay. A semiquantitative assay for *Sl*Bgl B activity was conducted in microtiter plates. An aliqout of cell culture supernatant, or cell-free lysate, or column fraction was diluted in PC buffer (50 mM NaH_2PO_4, 12 mM citric acid, pH 6.5) buffer to a final volume of 50 mL, and placed in microtiter plate wells. An equal volume, 50 mL, of 2 mM MUG (Sigma Chemical Co.), prepared in PC buffer, was added to start the reaction. The reaction was allowed to proceed for 30 min at 60°C, then stopped by adding 100 mL of 0.15 M glycine/NaOH, pH 10.3. The release of 4-methylumbelliferone (MU) was measured wth a fluorescence microtiter plate reader (Cambridge Technology, model 7620, Waterton, MA) equipped with 360 nm excitation and 460 nm emission filters.

***p*NPG Assay**. An aliquot of sample in PC buffer was incubated with 1 mg/mL *p*NPG for 15 or 30 min at 60°C in a total reaction volume of 1 mL. The reaction was then quenched with 2 volumes of 1 M Na_2CO_3 and mixed well. Release of *p*-nitrophenolate anion was determined by measuring absorbance at 410 nm.

***Sl*Bgl B Purification**. Seventy-two hours after inoculation with a 1L mid-log phase culture, *S. lividans* cells from a 10-L fermentor were pelleted using a continuous centrifuge (CEPA, Model LE, New Brunswick Scientific, Brunswick, NJ) operated at 5,000 x *g*. Cells were resuspended in PC buffer at 1/10 the original culture volume. Intracellular *Sl*Bgl B was released by breaking cells ultrasonically (Braun-Sonic model L, Braun Biotech, Allentown, PA). Aliquots of 100 mL cell suspension were sonicated at full power (15 watt), on ice, for 15 x 30 s with 15 s resting periods between each sonication. Cell lysates were pooled and centrifuged at 10,000 x *g* for 30 min to remove cell debris and unbroken cells. Pellets were discarded, and the supernatant was prepared for hydrophobic interaction chromatography by adding enzyme-grade $(NH_4)_2SO_4$ (Sigma Chemical Co.) to bring the $(NH_4)_2SO_4$ concentration to 0.5 M.

After pre-equilibration of a Pharmacia Fine Chemicals (Piscataway, NJ) Phenyl-SE column (approximately 250 mL; 18.5 cm x 2.4 cm^2) in 20 mM Tris pH 8.0 buffer that contained 0.5 M $(NH_4)_2SO_4$, approximately 100 mL of the crude protein mixture in 0.5 M $(NH_4)_2SO_4$ was loaded onto the column using a peristaltic pump. The column was washed with 5 bed volumes of equilibration buffer at a flow rate of 2 mL/min to elute unbound proteins. Then, *Sl*BgB was eluted with a descending linear gradient of $(NH_4)_2SO_4$ in 20 mM Tris pH 8.0. The 10-mL fractions collected from Phenyl-SE chromatography were assayed for MUG activity and active fractions were pooled.

Pooled MUG-positive fractions were desalted and concentrated by ultrafiltration using a PM30 membrane (DC10, Amicon, Beverly, MA) before being loaded onto a 6-mL Pharmacia Resource-Q column that had been pre-equilibrated with 20 mM Tris pH 8.0 buffer. This column was operated at a flow rate of 1 mL/min and washed with 30 mL of equilibration buffer before eluting with a 0 to 1 M gradient of NaCl in 20 mM Tris pH 8.0

buffer. Fractions eluting from the Resource-Q column were assayed for MUG activity and active fractions were pooled, concentrated, and exchanged with PC buffer containing 100 mM NaCl.

The active fractions were finally loaded onto a prepacked Pharmacia Superdex-200 26/60 column that had been pre-equilibrated with 50 mM NaH_2PO_4, 12 mM citrate, pH 6.5 buffer. This column was operated at 1 mL/min and fractions were collected at 10-min intervals. Fractions were assayed and active fractions were pooled and stored for future characterization against MUG.

Temperature Activity Optimum Determination. Aliquots of purified *Sl*Bgl B that contained equivalent quantities of protein were incubated with 1 mg/mL *p*NPG for 30 min at various temperatures in PC buffer. *Sl*Bgl B activity was quantitated by absorbance at 410 nm and expressed as percent of maximal activity.

Zymograms. Cell extracts from *S. lividans* transformants were concentrated tenfold using an Amicon mini-filtration device and then electrophoresed on native precast acrylamide/bis-acrylamide 8-16% gels gradient gels (Novex, San Diego, CA). After electrophoresis, gels were soaked in PC buffer for 20 min at 4°C with one buffer change. To visualize *Sl*Bgl B activity in the gel, presoaked gels were transferred to a prewarmed solution that contained 2 mM MUG prepared in PC buffer, and incubated at 58°C for 10 min. After rinsing with distilled water, gels were photographed under UV illumination.

Substrate Specificity Measurements. *Sl*Bgl B reactions on various aryl-glycoside substrates were conducted at 60°C for 15 min, and quenched with 2 volumes of 1 M Na_2CO_3. Substrates for these reactions were all supplied at 1 mg/mL in PC buffer. Release of p-nitrophenolate anion was quantitated spectrophotometrically at 410 nm, as stated for the *p*NPG assay. Assays of *Sl*Bgl B on cellobiose were conducted for 60 min in the presence of 1 mg/mL cellobiose in PC buffer at 57°C. To determine cellobiose hydrolysis, hydrolysis was terminated by boiling the reaction mixture for 5 min. The glucose product from this reaction was determined by HPLC using HPX-87C columns (Bio-Rad, Richmond, CA) with water as mobile phase. K_{cat} values are defined as mmoles product/min/mmol pure enzyme.

Product Feedback Inhibition. The effect of glucose on *Sl*Bgl B activity was tested by assaying enzyme activity against pNPG (1 mg/mL) during 15 min assays at 60°C in PC buffer that contained 0, 50, or 100 mM glucose.

Isoelectric Point Determination. The isoelectric pH of *Sl*Bgl B was measured using a Pharmacia Phast-automated electrophoresis system with Pharmacia precast gel media, following the manufacturer's instructions.

Mutagenesis and Mutant Selection. The DNA fragment (0.86 kb *Sac*I/*Apa*I) intended for mutation was amplified from pGEM-BglB (template) and primers 1pfw and 2pbk with Amplitaq (Perkin Elmer, Norwalk, CT) in the presence of 0.5 mM $MnCl_2$ using a Perkin Elmer Thermocycler. The conditions used for the PCR were: one cycle of 96°C, 4 min: three cycles of 96°C, 1 min; 59°C, 1 min; and 75°C, 2 min: 27 cycles of 96°C, 1 min; 70°C, 1 min; and 75°C, 2 min: and finally, one cycle of 96°C, 1 min; 70°C, 1 min; and

75°C, 10 min. After PCR, the 0.86 kb fragment was purified from a 1.5% agarose TAE gel, digested with *Apa*I and *Sac*I, and ligated with unmutated *bglB* 1.38 kb *Apa*I/*Sal*I fragment. This population of randomly mutated fragments was then substituted for the corresponding fragment in pIJ-BglB and transformed into *S. lividans* TK24 as described previously. Regenerated colonies were individually transferred onto screening plates that contained the colorigenic indicator, X-Glc. Plates were incubated at 30°C for 4 to 48 h (i.e., until a blue halo developed around each colony).

Growth Curve Studies. Seed cultures of *S. lividans* were obtained by inoculating 5 mL of TSB media (*18*) with individual colonies and permitting growth for 2 to 3 d. Four-mL seed cultures were then used to inoculate 200 mL of M media in 1-L baffled shake flasks at 30°C for up to 10 d. A 10 mL sample was taken from each culture every 24 h. Samples were stored at 4°C until the end of the experiment, when all time points were analyzed simultaneously for volumetric dry cell weight and MUG activity in supernant and cytosolic fractions.

Dry Cell Weight (DCW) Determination. An aliquot of 5 mL cell culture was collected and filtered under vacuum through preweighed 0.45 mm cellulose-acetate membranes (Gelman Filtration Products, Ann Arbor, MI) and rinsed once with 10 mL distilled water. Membranes plus cells were then dried at 55°C overnight. After drying, membranes were reweighed. The difference in the two weights was taken as the dry cell weight per 5 mL (DCW/5 mL).

DNA Sequencing. DNA was sequenced using the manual dideoxy chain-termination method. Reactions were carried out as recommended for the Sequenase Quick-Denature Plasmid Sequencing Kit (USB, Cleveland, OH) using α-^{35}S dCTP (Amersham Life Science, Arlington Heights, IL).

Nomenclature. *M*Bgl B is the native enzyme expressed from *M. bispora*, *Ec*Bgl B is the recombinant enzyme expressed from *E. coli*, and *Sl*Bgl B is the recombinant enzyme expressed from *S. lividans*.

Results and Discussion

Design and Construction of An Expression Plasmid for *bglB*. *M*Bgl B is not actively secreted by *M. bispora*. However, in order to facilitate purification and characterization of the recombinant protein expressed by *S. lividans*, we decided to build a construct which would direct secretion of *Sl*Bgl B (Figure 1). In pIJ-BglB, synthesis of *Sl*Bgl B is driven by the strong *S. longisporus* STI-II serine protease inhibitor promoter and secretion signal peptide. Fornwald and coworkers (*16*) have reported the expression and efficient secretion of a biologically active human protein in *S. lividans* using the STI-II promoter and its signal peptide at levels as high as 300 mg/L. Only the coding sequence for the 28 amino acid portion of the STI-II leader peptide thought to direct secretion (e.g., signal peptide; `MRNTARWAAT LALTATAVCG PLTGAALA`; ref *16*) was incorporated into our construction. The 6 amino acid STI-II propeptide (`TPAAAP`), which lies immediately downstream of the signal peptide, was excluded from this construction.

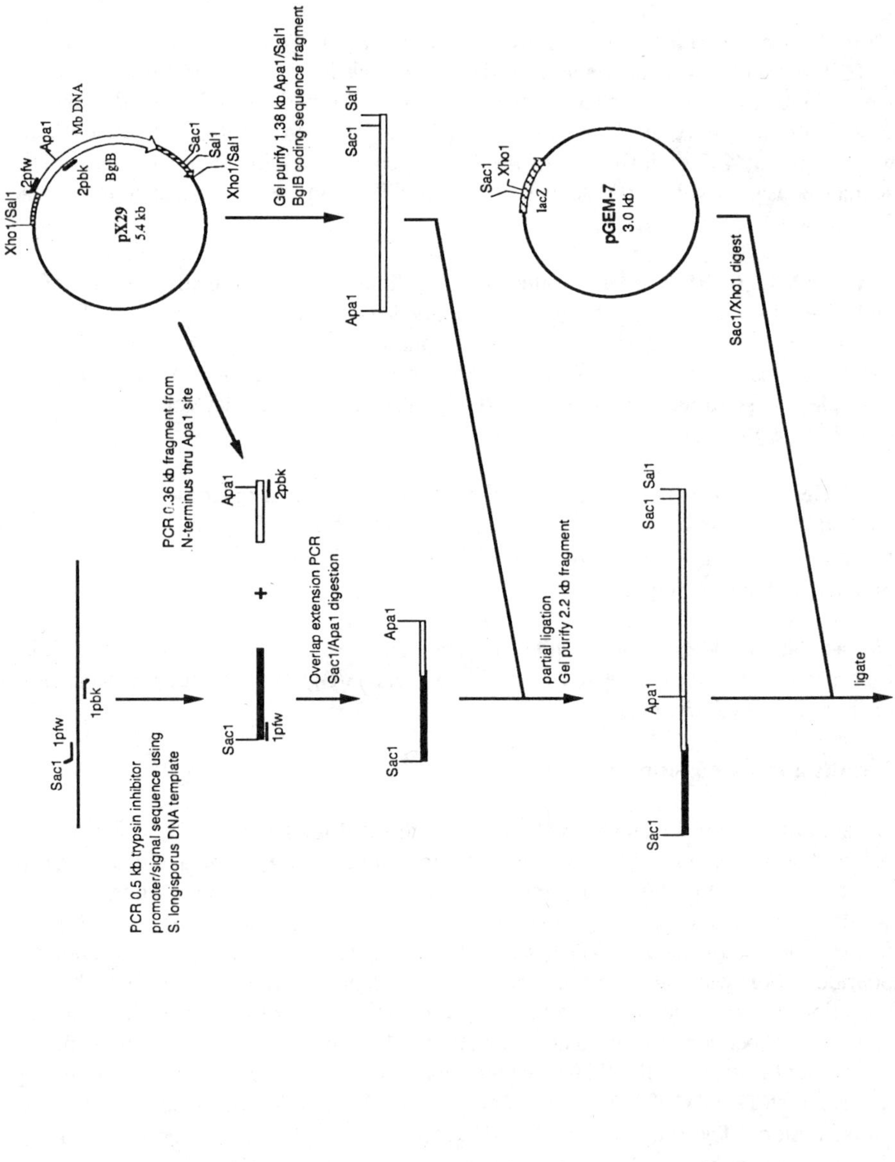

Xho1/Sal1
2pfw
Apa1
Mb DNA
2pbk
BglB
Sac1
Sal1
Xho1/Sal1
pX29
5.4 kb
Gel purify 1.38 kb Apa1/Sal1
BglB coding sequence fragment
Apa1
Sac1 Sal1
Sac1
Xho1
lacZ
pGEM-7
3.0 kb
Sac1/Xho1 digest
PCR 0.36 kb fragment from
N-terminus thru Apa1 site
Apa1
2pbk
Sac1 1pfw
1pbk
PCR 0.5 kb trypsin inhibitor
promoter/signal sequence using
S. longisporus DNA template
Sac1
1pfw
+
Overlap extension PCR
Sac1/Apa1 digestion
Apa1
Sac1
partial ligation
Gel purify 2.2 kb fragment
Sac1 Sal1
Apa1
Sac1
ligate

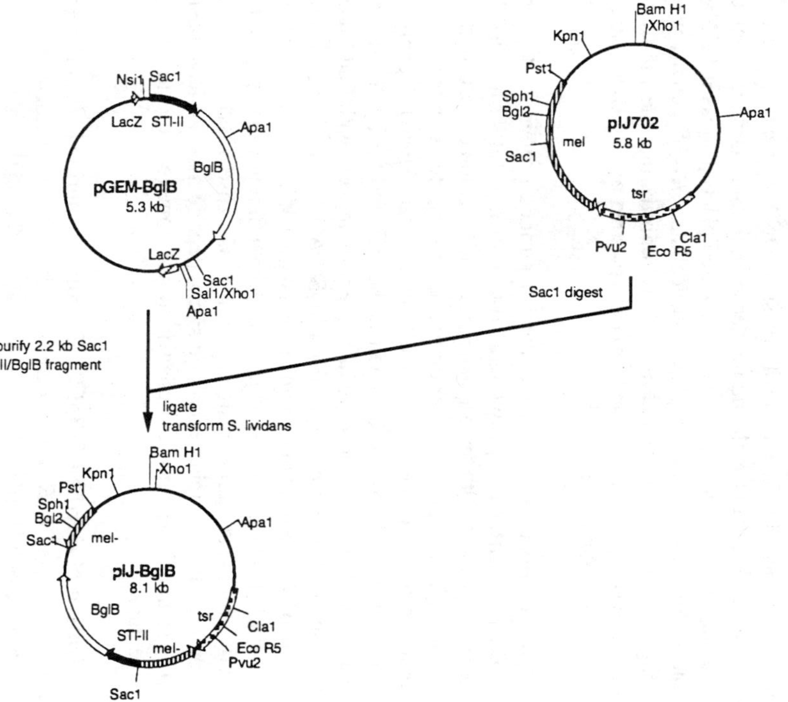

Figure 1. Schematic representation of construction of expression plasmid pIJ-BglB in *S. lividans*. An expression module for *Sl*Bgl B was constructed to be directed by the strong STI-II promoter and signal peptide. Abbreviations are as follows: STI-II, *S. longisporus* STI-II trypsin inhibitor promoter plus signal sequence; *Mbgl* B, *M. bispora* β-glucosidase B; *mel*, melanin gene.

Expression of *Sl*Bgl B. The host for *Sl*Bgl B expression, *S. lividans* TK24, had marginally detectable background MUG activity which was subtracted from all subsequent assays. Cell culture supernatant and cell extract from a pIJ-BglB transformant were assayed for activity against MUG. Unfortunately, most of the *Sl*Bgl B activity was retained in the intracellular fraction.

To determine whether expressed heterologous enzyme was distinct from the intrinsic enzyme, a zymogram was performed using concentrated cell extract. This gel (Figure 2) showed that the active protein band from pIJ-BglB transformants migrated differently than the intrinsic *S. lividans* TK24-expressed enzyme, as did the *Aspergillus niger* β-glucosidase, which was used as a positive control. Plasmid DNA was extracted from individual pIJ-BglB transformants and cut with endonuclease *Sac*I. All plasmids isolated contained the 2.2kb *Sac*I/*Sac*I chimeric *bglB* gene. When isolated pIJ-BglB plasmid DNA was re-transformed into *S. lividans* TK24 protoplasts, intracellular *Sl*Bgl B activity was apparent in the resulting thiostrepton-resistant clones. This confirms that the chimeric, cloned *Sl*Bgl B gene sequence is expressed in active form from plasmid pIJ-BglB, although the enzyme is not secreted as originally intended.

Purification of *Sl*Bgl B. To obtain sufficient amounts of purified protein for kinetic studies, a single *Sl*Bgl B transformant was grown in a 10 L fermentor for 72 h at 30°C. Cells were harvested, resuspended in 1/10th the original fermentation volume in PC buffer, and disrupted by sonication. The cytoplasmic fraction was then subjected to a three-step column chromatography procedure beginning with Pharmacia Phenyl-Sepharose (hydrophobic interaction), Pharmacia Resource-Q (anion exchange), and finally Pharmacia Superdex-200 (size exclusion). After this procedure, a single protein band with molecular weight of 54 kDa was observed on Commassie stained SDS-PAGE gels. The apparent molecular weight of purified *Sl*Bgl B is slightly higher than native *M*Bgl B (i.e., 52 kDa; D. Eveleigh, personal communication). We speculate that the difference may be due to unprocessed or incorrectly processed secretion signal peptide. Because *Sl*Bgl B was not expressed as a secreted protein, the 28 amino acid-long signal peptide may not have been cleaved. However, an N-terminal sequence analysis which was done to prove this point returned no interpretable data.

Functional Characterization of Purified *Sl*Bgl B. To compare purified *Sl*Bgl B with the recombinant enzyme expressed in *E.coli*, (*Ec*Bgl B), *Sl*Bgl B was characterized in several aspects. (1) The isoelectric pH (pI) of *Sl*Bgl B was about 5.0. This value is close to the pI of *Ec*Bgl B, 4.6, determined by Wright and coworkers (*17*). Unfortunately, no pI value for *M*Bgl B has been reported. (2) A temperature-activity profile for purified *Sl*Bgl B was constructed. The recombinant enzyme closely resembled the native enzyme in that it retains thermostability, displaying an optimium temperature of about 57°C (Figure 3). *Ec*Bgl B has temperature optima of 60°C. However, in comparison with *Ec*Bgl B, the *Sl*Bgl B temperature/activity profile was sharper (*17*). The *Ec*Bgl B protein retains 70% of maximal activity at 70°C, but the *Sl*Bgl B protein retains only 40% of maximal activity at this temperature. (3) An important characteristic of *M*Bgl B is its extreme resistance to glucose feedback inhibition. Wright et al. (*17*) reported that crude *Ec*Bgl B preparations retain 80% of maximal activity in the presence of 40% (2.2 M) glucose. We monitored the sensitivity of *Sl*Bgl B rates of reaction on *p*NPG in the presence of 50 and 100 mM glucose (Figure 4) and discovered that activity was only slightly affected. (4) The

Figure 2. Expression of recombinant β-glucosidase B in *S. lividans* (*Sl*Bgl B) demonstrated by zymogram. Cell extracts from *S. lividans* were electrophoresed on a native PAGE gradient gel (8-16%). After electrophoresis, the gel was washed twice briefly with cold PC buffer and incubated with a solution that contained 2 mM MUG at 60°C for 10 min. Resulting fluorescent bands were viewed under UV illumination and photographed immediately. Lane 1, pIJ-BglB #7; lane 2, pIJ-BglB #9; lane 3, *S. lividans* TK24.

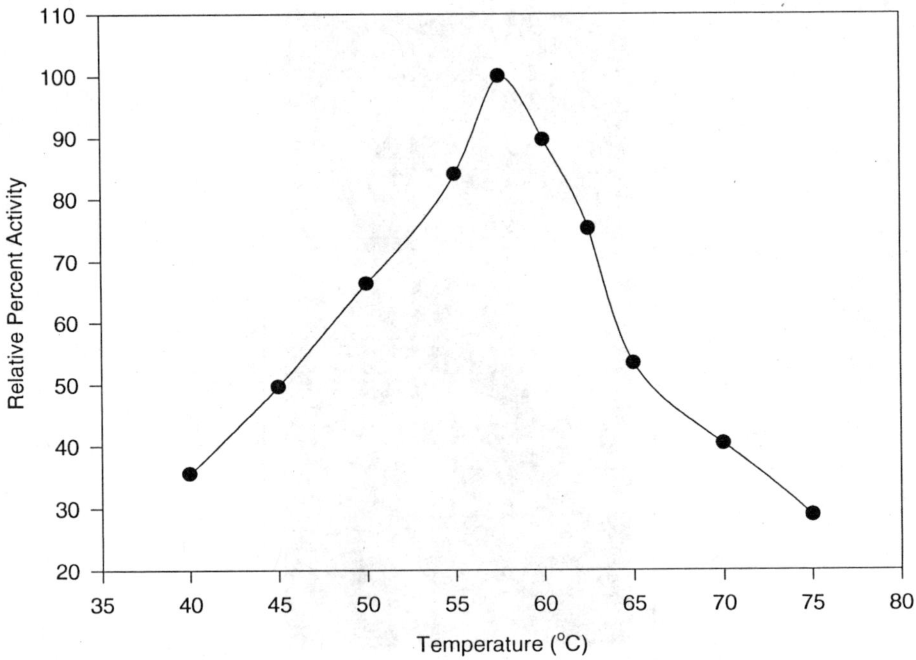

Figure 3. Temperature optima curve of purified *Sl*Bgl B. *Sl*Bgl B activity was assayed against 1 mg/mL *p*NPG at temperatures ranging from 40 to 75°C in PC buffer for 30 min.

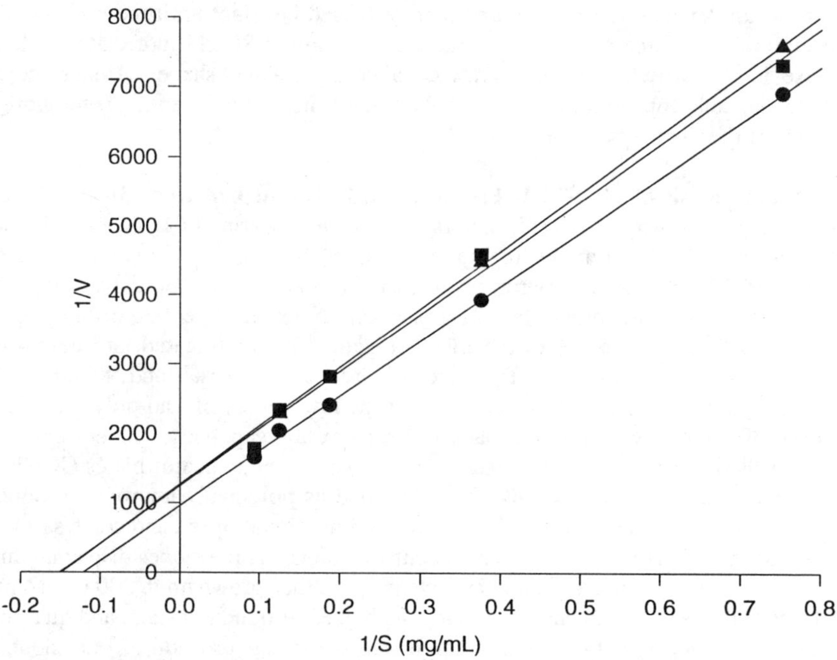

Figure 4. Double reciprocal Lineweaver-Burk plot of glucose feedback inhibition on *Sl*Bgl B activity assayed using *p*NPG as substrate. Reactions were carried out at 60°C in 50 mM NaH_2PO_4, 12 mM citrate buffer pH 6.5, at various concentrations of *p*NPG in combination with 0 (circles), 50 mM glucose (squares), or 100 mM glucose (triangles). UV light absorption at 410 nm was recorded continously for a period of 10 min using a Cary 3E spectrophotometer.

relative activity of highly purified *Sl*Bgl B on a variety of nitrophenyl-glycosides and cellobiose was also assessed. Table 1 shows that *Sl*Bgl B hydrolyzes *o*NPG at 20 times the rate of *p*NPG. Other substrates were hydrolyzed at much slower rates. Surprisingly, *Sl*Bgl B did not have comparable activity to *Ec*Bgl B on cellobiose (*17*).

***Sl*Bgl B Production.** A 10-day shake flask experiment for *Sl*Bgl B strain #7 was carried out in M media. Samples of the culture were taken every 24 h to track accumulating *Sl*Bgl B activity. As shown in Figure 5, productivity increased dramatically after 48 h, reached the maximum level at about 90 h, and then remained constant until the last day of the fermentation. It is interesting to note that accumulation of *Sl*Bgl B correlates with a pH increase in the growth medium and with aberrant cellular shape. Under the light microscope, cells appeared as shorter, kinked filaments with uneven diameter along the long axis of the filaments.

PCR Mutagenesis of the STI-II Promoter and Mutant Isolation. In an attempt to improve the productivity of *Sl*Bgl B, a library of mutants bearing PCR-generated random mutations in STI-II promoter region and signal peptide was created by cassette mutagenesis. The (*Sac*I/*Apa*I) fragment containing the STI-II promoter and signal peptide, as well as the fragment coding the 121 amino acid N-terminal peptide of *M*Bgl B, was amplified by PCR in presence of 0.5 mM manganese ion and ligated into gel purified, previously *Sac*I/*Apa*I digested pIJ-BglB (see Figure 1 and Materials and Methods). This PCR mutagenesis strategy takes advantage of the low fidelity of Taq polymerase in the presence of manganese ion and it's lack of proofreading capability. Clones expressing higher levels of *Sl*Bgl B were selected by first screening on indicator plates (X-Glc). Of 500 independent clones plated, 13 were identified as potentially improved production strains. For further testing, each was grown up in 5 mL liquid media and cell lysates were assayed for specific MUG activity versus control strains. Three clones displaying higher MUG activity were identified. These three clones were then grown up in 200 mL M media in 1-L shake flasks for 5 d to ensure reaching the highest production stage and quantitated for specific MUG activity. After a three-step screening procedure, one clone that produced *Sl*Bgl B at 203 mg/L was discovered and named pIJ-BglB (see Table 2). This is three times the volumetric productivity of the parent chimeric gene construct.

To confirm that the improved performance of this strain was the result of a change in the plasmid DNA and not of a mutation at some other location, the plasmid from this *S. lividans* clone was extracted and re-transformed into TK 24 protoplasts. A single transformant was grown in 1-L shake flask cultures and examined for MUG activity. The productivity of this strain, 235 mg/L, was essentially identical to that of the parent mutant strain (Table 2). This experiment demonstrates that the increased *Sl*Bgl B yield is probably due to one or more changes carried by the plasmid.

Mutated Promoter DNA Sequence. To identify the changes that PCR mutagenesis generated in pIJ-BglB*, the entire STI-II promoter region before and after mutagenesis was sequenced (Figure 6). Two interesting results were observed. The "wild-type" STI-II sequence that this work employed in the initial chimeric constructions, and which was obtained via PCR from *S. longisporus* genomic DNA, has several regions not in agreement with the previously published sequence (*16*). Apparently, either the STI-II promoter employed by the SmithKline Beecham group has undergone several mutagenic events from

Table 1. Substrate Specificity of Purified *Sl*Bgl B

Substrate	k_{cat}[1]
o-nitrophenyl β-D-glucopyranoside	142.2
o-nitrophenyl β-D-galactopyranoside	27
p-nitrophenyl β-D-glucopyranoside	7.2
p-nitrophenyl α-D-glucopyranoside	1.2
p-nitrophenyl β-D-galactopyranoside	0.65
p-nitrophenyl N-acetyl β-D-glucosamide	0.35
cellobiose	0.11
p-nitrophenyl β-D-lactopyranoside	0.09

[1]k_{cat} = μmole product /min/μmol pure enzyme

p-nitrophenyl β-D-fucoside, *p*-nitrophenyl 1-thio β-D-glucopyranoside, *p*-nitrophenyl β-D-galactopyranoside, *p*-nitrophenyl β-D-glucuronide, *p*-nitrophenyl α-D-mannopyranoside, *p*-nitrophenyl β-D-mannopyranoside, and *p*-nitrophenyl β-D-xylopyranoside yielded no detectable evidence of hydrolysis. All hydrolytic reactions were performed in the presence of 1 mg/ml substrate at 60°C in 50 mM NaH_2PO_4-citrate pH 6.5 buffer, for 15 min, except for reactions containing cellobiose which were conducted at 57°C for 60 min. *Sl*Bgl B activities were determined as described in Materials and Methods.

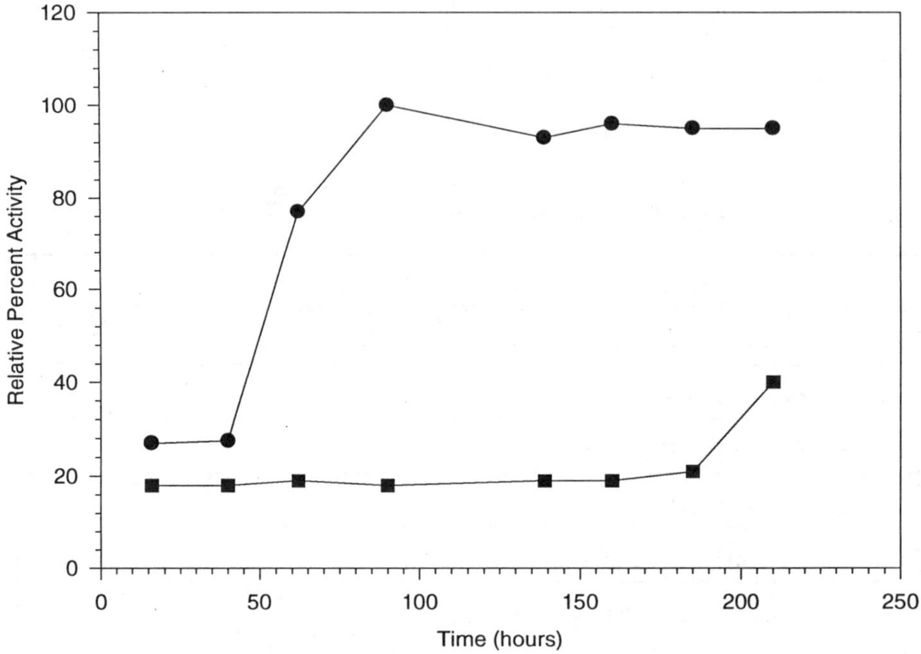

Figure 5. Accumulating *Sl*Bgl B activity in 10-d batch cultures. Squares represent *Sl*Bgl B activity in culture supernatant. Circles represent *Sl*Bgl B activity in cytosolic fraction.

Table 2. Comparason of *Sl*Bgl B Production in *S. lividans* TK24 Cells Carrying Unmutated or Mutated Expression Plasmid pIJ-BglB

*Sl*Bgl B Productivity (mg/L)

Transformation No.	*S. lividans* TK24 control	*S. lividans* TK24:: pIJ-BglB[1]	*S. lividans* TK24:: pIJ-BglB*[2]
1	20	60	205
2	15	120	235

[1]*Sl*Bgl B activity from cells with unmutated pIJ-BglB plasmid; [2]*Sl*Bgl B production in cells carrying mutated pIJ-BglB* plasmid. Data from both first transformation and retransformation experiments are shown.

```
         ↓EcoRI      ↓SacI                              *         ↓NaeI
SKB      GAATTCGAGC TCGGCCAGCT CCTCGAAGAG ATCCTCGGGC AGCCGGCCGG        50
CHR                                             C
MUT                                             A

SKB      -GAACCAGCC CGCAGCTTCT CTCGTTGCTC TGTCGTAATC ATGTAATGAC       100
CHR      C   A   CG
MUT      C   C   CG

SKB      GATTACTCGA CGGGACCATG AACGCAAGGG GGTGCGGGGG AGTCCCGCGA       150
CHR                                                       GC
MUT                                                       GC

                                 ↓StuI            *
SKB      CAGCTCAACC GGAATGTTTC AGGCCTATTA ACTAAGCGCA GGAAATCGGC       200
CHR                                         G      A
MUT                                         G      G

SKB      CACTTGGCTG CTTCGGGCGA TCAAGAACCG CTCAGTTCCA GGGGTCATGC       250
CHR
MUT

SKB      GGTCGAACTC TGTGACTTCG CGCCACTGAT TCAACACGCA AGGTTACTGA       300
CHR
MUT

         *                         ↓SacII
SKB      AACACATGGG GTCGAGGTGT TTTTCCGCGG CGGTACATGC GTGC -GACTC      350
CHR      A                                  -  - -    -   A
MUT      G                                  -  - -    -   A

SKB      GCGCTCGCCG GTCCGGCACC AAACCGGAAC GGGTCGGCAC ACCCTCGAAT       400
CHR          ↑
MUT
                                ▽
SKB      CCTGCGGAAG GATGCACACA ATGCGGAACA CCGCGCGCTG GGCAGCCACC       450
CHR
MUT

SKB      CTCGCCCTCA CGGCCACCGC CGTCTGCGGA CCCCTCACCG GAGCCGCGCT       500
CHR
MUT

SKB      CGCC                                                         504
CHR
MUT
```

Figure 6. Sequence comparison of the "wild-type" and mutant promoter region of the serine protease inhibitor gene STI-II from *Streptomyces longisporus*. SKB: previously published sequence data from the SmithKline group (*16*); CHR: sequence data from cloned chromosomal STI-II (this work); mSTI-II: PCR-mutagenized STI-II promoter region. Only differences among the sequences are displayed. Restriction enzyme recognition sites are labeled on top of the sequence with arrows pointing down. Stars above the sequence indicate the location of point mutations caused by PCR mutagenesis. Upward directed arrow indicates transcription start site. Triangle points to methionine initiator codon. Underlined sequences are locations where PCR primers bind. Dash line indicates single nucleotide deletion.

the wild-type sequence, or the PCR reaction used to clone this promoter for this work has introduced several mutations into the promoter region.

Especially critical is the region around -10 relative to the transcription start site, where little sequence similarity can be distinquished. This may help to explain why we were unable to achieve *Sl*Bgl B production levels in the range of those reported by Fornwald and coworkers (*16*). DNA sequencing also reveals three other point mutations that are apparently the result of the PCR-based mutagenesis reaction. At position -318 relative to the transcription start site, a C has changed to A. At positions -164 and -53, an A has changed to G (Figure 6). The mutation at -53 may be a very interesting nucleotide change with respect to elevated expression levels, because it lies just upstream of the important -35 (relative to the transcription start site) consensus promoter sequence. The significance of the other two point mutations is not understood.

In *E. coli,* up-regulation loci, new promoter elements which enhance promoter activity by 30-fold, situated between -40 and -60 relative to the transcription start site (i.e., immediately upstream of the -35 region) have been indentified in many promoters (*20*). Our work suggests that up-regulatory mutations may also occur immediately upstream from the consensus *E. coli* promoter sequences in *S. lividans* genes and may also be a general phenomenon in other bacteria.

Conclusions

Although detailed information concerning the kinetic and physicochemical characteristics of the native *M*Bgl B are not available, Wright (*15*) successfully cloned this enzyme in *E. coli* and found the molecular weight of the recombinant *Ec*Bgl B to be 52,000 Da by SDS-PAGE. The *E. coli* expressed *Ec*Bgl B was found largely in the cytoplasmic fraction and demonstrated pH and temperature optima on cellobiose, *p*NPG, and MUG of 6.2 and 60°C, respectively. The levels of *Ec*Blg B expression from *E. coli* lysates, although adequate for biochemical evaluation, are inappropriate for industrial applications (*15*). Expression of *M. bispora* Bgl B in a more closely related *Actinomycete* host, such as *S. lividans*, might be expected to yield higher levels of expression and minimally modified recombinant enzyme.

The present study reports the expression of active *bglB* gene product in *S. lividans* TK24 at reasonably high levels (i.e., 235 mg/L). Also, PCR mutagenesis increased approximately 2-3-fold the expression of this recombinant enzyme, which displays similar physicochemical properties to those found for the *E. coli* enzyme. However, for unexplained reasons, the *Sl*Bgl B is considerably more active on the aryl-glucosides, *o*NPG or *p*NPG, than on cellobiose. This apparent contradiction with the earlier work of Wright (*15*) may reflect uncharacterized mutations or host-specific differences in post-translational modifications. Unfortunately, these results diminish the potential role that *Sl*Bgl B could play in a biomass saccharification process.

Acknowledgments

This work was funded by the Biochemical Conversion Element of the U.S. Department of Energy Biofuels Energy Systems Program.

Literature Cited

1. Woodward. J.; Wiseman, A. *Enzyme Microb. Technol.* **1982,** *4*, 73-79.
2. β-*Glucosidases: Biochemistry and Molecular Biology*; Esen, A., Ed., American Chemical Society:Washington, DC, 1993, Vol. 533.
3. Wolnak, B.; Scher, M. *Industrial Use of Enzymes*; Bernard Wolnak and Associates: Chicago, IL, 1990.
4. Bronnenmeier, K.; Staudenbauer, W.L. *Appl. Microbiol. Biotechnol.* **1988**, *28*, 380-386.
5. Aït, N.; Creuzet, N.; Cattaneo, J. *J. Gen. Microbiol.* **1982**, *128*, 569-577.
6. Hagerdal, B.; Ferchak, J.D.; Pye, E.K. *Biotechnol. Bioeng.* **1980**, *22*, 1515-1526.
7. Adney, W.S.; Tucker, M.P.; Nieves, R.A.; Thomas, S.R.; Himmel, M.E. *Biotechnol. Lett.* **1995**, *17*, 49-54.
8. Love, D.R.; Streiff, M.B. *Bio/Technology* **1987**, *5*, 384-387.
9. Jin, F.; Toda, K. *J. Ferment. Bioeng.* **1989**, *67*, 8-13.
10. Takase, M.; Horikoshi, K. *Appl. Biochem. Biotechnol.* **1988**, *29*, 55-60.
11. Schwarz, W.; Bronnenmeier, K.; Staudenbauer, W.L. *Biotechnol. Lett.* **1985**, *7*, 859-864.
12. Romaniec, M.P.M.; Huskisson, N.; Barker, P.; Demain, A.L. *Enzyme Microb. Technol.* **1993**, *15*, 393-400.
13. Spilliaert, R.; Hreggvidsson, G.O.; Kristjansson, J.K.; Eggertsson, G.; Palsdottir, A. *Eur. J. Biochem.* **1994**, *224*, 923-930.
14. Waldron Jr., C.R.; Becker-Yallone, C.A.; Eveleigh, D.E. *Appl. Microbiol. Biotech.* **1986**, *24*, 477-486.
15. Wright, R.M. *Cloning and Characterization of the Gene Encoding Microbispora bispora Bgl B: A Thermostable Beta-Glucosidase*; PhD Thesis, Graduate School of Biomedical Sciences, Rutgers University, New Brunswock, NJ, 1993.
16. Fornwald, J.A.; Donovan, M.J.; Gerber, R.; Keller, J.; Taylor, D.P.; Acuri, E.J.; Brawner, M.E. *Bio/Technology* **1993**, *11*, 1031-1036.
17. Wright, R.M.; Yablonsky, M.D.; Shalita, Z.P.; Goyal, A.K.; Eveleigh, D.E. *Appl. Environ. Microbiol.* **1992**, *58*, 3455-3465.
18. Hopwood, D.A.; Bibb, M.J.; Chater, K.F.; Kieser, T.; Bruton, C.J.; Kieser, H.M.; Lydiate, D.J.; Smith, C.P.; Ward, J.M.; Schrempf, H. *Genetic Manipulation of Streptomyces - A Laboratory Manual*; The John Innes Foundation: Norwich, England, 1985.
19. *Molecular Cloning: A Laboratory Manual;* Sambrook, J.; Fritsch, E.F.; Maniatis, T., Eds.; 2nd Edition, Cold Spring Harbor: New York, NY, 1989.
20. Ross, W.; Gosink, K.K.; Salomon, J.; Igarashi, K.; Zou, C.; Ishihama, A.; Steverinov, K.; Gourse, R.L. *Science* **1993**, *262*, 1407-1413.

Chapter 10

Biodiesel: The Use of Vegetable Oils and Their Derivatives as Alternative Diesel Fuels

Gerhard Knothe, Robert O. Dunn, and Marvin O. Bagby

Oil Chemical Research, National Center for Agricultural Utilization Research, Agricultural Research Service, U.S. Department of Agriculture, 1815 North University Street, Peoria, IL 61604

Vegetable oils and their derivatives (especially methyl esters), commonly referred to as "biodiesel," are prominent candidates as alternative diesel fuels. They have advanced from being purely experimental fuels to initial stages of commercialization. They are technically competitive with or offer technical advantages compared to conventional diesel fuel. Besides being a renewable and domestic resource, biodiesel reduces most emissions while engine performance and fuel economy are nearly identical compared to conventional fuels. Several problems, however, remain, which include economics, combustion, some emissions, lube oil contamination, and low-temperature properties. An overview on all aspects of biodiesel is presented.

The use of vegetable oils in diesel engines is nearly as old as the diesel engine itself. The inventor of the diesel engine, Rudolf Diesel, reportedly used groundnut (peanut) oil as a fuel for demonstration purposes in 1900 (*1*). Some other work was carried out on the use of vegetable oils in diesel engines in the 1930's and 1940's. The fuel and energy crises of the late 1970's and early 1980's as well as accompanying concerns about the depletion of the world's non-renewable resources provided the incentives to seek alternatives to conventional, petroleum-based fuels. In this context, vegetable oils as fuel for diesel engines were remembered. They now occupy a prominent position in the development of alternative fuels. Hundreds of scientific articles and various other reports from around the world dealing with vegetable oil-based alternative diesel fuels ("biodiesel") have appeared in print. They have advanced from being purely experimental fuels to initial stages of commercialization. Nevertheless, various technical and economic aspects require further improvement of these fuels.

Numerous different vegetable oils have been tested as biodiesel. Often the vegetable oils investigated for their suitability as biodiesel are those which occur abundantly in the country of testing. Therefore, soybean oil is of primary interest as biodiesel source in the United States while many European countries are concerned with rapeseed oil, and countries with tropical climate prefer to utilize coconut oil or palm oil. Other vegetable oils, including sunflower, safflower, etc., have also been investigated. Furthermore, other sources of biodiesel studied include animal fats and used or waste cooking oils. Sources of biodiesel with some emphasis on developing countries have been discussed (*2*).

Several problems, however, have impaired the widespread use of biodiesel. They are related to the economics and properties of biodiesel. For example, neat vegetable oils reported to cause engine deposits. Attempting to solve these problems by using methyl esters causes operational problems at low temperatures. Furthermore, problems related to combustion and emissions remain to be solved. The problems associated with the use of biodiesel are thus very complex and no satisfactory solution has yet been achieved despite the efforts of many researchers around the world. This article will briefly discuss economics and regulatory issues as well as conventional diesel fuel (petrodiesel) and then focus on research on the use of biodiesel in a diesel engine.

Economics and Regulatory Issues

Economic reasons have been one of the major obstacles in the use of biodiesel. Diesel fuel (DF) derived from vegetable oils is more expensive than petroleum-based DF. The feedstock for biodiesel is already more expensive than conventional DF. For example, in the United States, a gallon of soybean oil costs approximately two to three times as much as a gallon of conventional DF. However, in the case of conversion of vegetable oils or fats to their esters, the resulting glycerol co-product, which has a potential market of its own, may offset some of the costs.

In most European countries, however, transportation fuels are so heavily taxed that tax incentives can be applied to encourage the use of biodiesel in the form of lower or no taxes on the biofuel and higher taxes on the petroleum-based fuel (*3,4*). This subsidy artificially cheapens the biodiesel to make it competitive. In many developing countries, the overriding concern is to become independent of the imported commodity petroleum. In the United States, the tax mechanism is inapplicable because of the comparatively low taxes on transportation fuels. Artificially regulating the demand for fuels from specific sources by means of taxation is currently politically not feasible.

Nevertheless, biodiesel is attractive for other reasons. Besides being a renewable resource and therefore creating independence from the imported commodity petroleum and not depleting natural resources, health and environmental concerns are the driving forces overriding the economic aspects in some cases. These concerns are manifested in various regulatory mandates of pollutants, particularly CAAA (Clean Air Act Amendments of 1990) and EPACT (Energy Policy Act of 1992) in the United States, which present opportunities for alternative fuels such as biodiesel. A life-cyle analysis of biodiesel (*5*) has shown that it is competitive with other alternative fuels such as compressed natural gas (CNG) and methanol in the urban transit bus market.

It is generally recognized that biodiesel has lower emissions, with the exception of nitrogen oxides (NO_x), than conventional petroleum-based DF. For example, due to its lack of sulfur, biodiesel does not cause SO_2 emissions. The lower emissions have caused biodiesel to be used in urban bus fleets and to make it especially suitable for other niche markets such as mining and marine engines. Besides environmental and health reasons with accompanying Government regulations, focusing on the use of biodiesel in niche markets is rendered additionally attractive because not enough vegetable oil is produced to supply the whole diesel market with biodiesel.

Numerous reports exist showing that fuel economies of certain biodiesel blends and conventional DF are virtually identical. In numerous on-the-road tests, primarily with urban bus fleets, vehicles running on blends of biodiesel with conventional DF (usually 80% conventional DF and 20% biodiesel; for a list of most biodiesel demonstration programs in the United States, see Ref. 6) required only about 2-5% more of the blended fuel than of the conventional fuel. No significant engine problems were reported as discussed later.

Conventional Diesel Fuel. Diesel Engines.

In contrast to gasoline which is spark-ignited, DF after injection is ignited by the heat of compression in a diesel engine. The diesel engine is therefore also termed a compression-ignition (CI) engine. The differences in the ignition processes entail significant differences in chemical composition and physical properties of the fuels.

Conventional DF is, like gasoline, obtained from cracking of petroleum. It is a fraction boiling at an initial distillation temperature of 160° (90% range of 290-360°C) (*7*), also termed middle distillates because of its boiling range in the mid-range of cracking products.

The ignition quality of DF is commonly measured by ASTM D613 and reported as the cetane number (CN). Ignition quality is defined by the ignition delay time of the fuel in the engine. The shorter the ignition delay time, the higher the CN. To rank different compounds on the cetane scale, hexadecane ($C_{16}H_{34}$; also called cetane), which has a very short ignition delay, has been assigned a CN of 100. At the other end of the scale, 2,2,4,4,6,8,8-heptamethylnonane (HMN; also $C_{16}H_{34}$), which has poor ignition qualities, has been assigned a CN of 15. It should be noted that the cetane scale is arbitrary and that compounds with CN > 100 (although the cetane scale does not provide for compounds with CN > 100) or CN < 15 have been identified. The ASTM specification for conventional DF (ASTM D975) requires a minimum CN of 40.

The CN scale clarifies an important aspect of the composition of, or, on a more fundamental level, the molecular structure of the compounds comprising DF. Long-chain, unbranched, saturated hydrocarbons (alkanes) have high CNs and good ignition quality while branched hydrocarbons (and other materials such as aromatics) have low CNs and poor ignition quality.

Since both too high and too low CN can cause operational problems (in case of too high CN, combustion can occur before the fuel and air are properly mixed, resulting in incomplete combustion and smoke; in case of too low CN, engine roughness, misfiring, higher air temperatures, slower engine warm-up and also incomplete combustion occur), most engine manufacturers designate a range of required CN for their engines. In most cases, this range is around CN 40-50.

Conventional DF is classified into different grades by ASTM D 975. This classification is the following: No. 1 diesel fuel (DF1) comprises volatile fuels oils from kerosene to intermediate distillates. They are applicable for high-speed engines whose operation involves frequent and relatively wide variations in engine load and speed. Such fuel is required for use at abnormally low temperatures. No. 2 diesel fuel (DF2) includes distillate gas oils of lower volatility. This grade is suitable for use in high-speed engines under relatively high loads and uniform speeds. DF2 can be used in engines not requiring fuels having the greater volatility and other properties specified for No. 1 diesel fuels. DF2 is the transportation diesel fuel to which biodiesel is usually compared. No. 4 diesel fuel (DF4) covers the more viscous distillates and their blends with residual fuel oils. It is usually satisfactory only for low-speed and medium-speed engines operated under sustained load at nearly constant speed.

Besides the just discussed characteristics of conventional DF, other properties such as heat of combustion, pour point, cloud point, and viscosity are of great significance. These properties also play very important roles in the use of biodiesel.

The two general types of diesel engines are the direct injection (DI) engine and the indirect injection (IDI) engine (*8*). In DI engines, the fuel is directly injected into the combustion chamber in the cylinder. In IDI engines, the fuel is injected into a prechamber which is connected with the cylinder through a narrow passage. Rapid air transfer from the main cylinder into the prechamber promotes a very high degree of air motion in the prechamber which is particularly conducive to rapid fuel air mixing (*8*). Combustion beginning in the prechamber produces high pressure and the fuels are subjected to high shear forces. The IDI engine is no longer used for heavy bus and truck engines due to somewhat lower efficiency and higher fuel consumption than the DI system (*8*). However, for special purposes, such as underground work, IDI engines are still made in the heavier class due to low exhaust emissions. For smaller vehicles such as cars and light trucks, the IDI system is used because of its ability to cover a wider speed range. The low exhaust emissions in combination with the wider speed range may lead to a continued use of IDI engines in urban areas, where the demand for low emissions can be more important than a somewhat higher fuel consumption combined with low annual mileage. The IDI engine is also less sensitive to fuel quality (*8*). Tests of biodiesel as a fuel have been performed on both DI and IDI engines.

Biodiesel. Definition of Biodiesel

The term biodiesel has no unambiguous definition. It stands for neat vegetable oils used as DF as well as neat methyl esters prepared from vegetable oils or animal fats and blends of conventional diesel fuel with vegetable oils or methyl esters. With increasing emphasis on the use of esters as DF, however, the term "biodiesel" increasingly refers to alkyl esters of vegetable oils and animal fats and not the oils or fats themselves. In an article on proposed ASTM standards, biodiesel was defined (*9*) as "the mono alkyl esters of long chain fatty acids derived from renewable lipid feedstock, such as vegetable oils or animal fats, for use in compression ignition (diesel) engines." Nevertheless, clear distinction between these different vegetable oil-based or -derived alternative diesel fuels is necessary.

For use in the United States, the U.S. Department of Energy has stated (*10*), "that biodiesel is already covered in the statutory and proposed regulatory definitions of "alternative fuel" which refer to any "fuel, other than alcohol, that is derived from biological materials." The Department, therefore, is considering amending the proposed definition of "alternative fuel" specifically to include neat biodiesel." The definition of biodiesel was not extended to include biodiesel blends, with the Department of Energy stating that "the issue of including biodiesel mixtures or blends comprised of more than 20 percent biodiesel is currently under study. However, this subject is complex and will require significantly more data and information, and a separate, future rulemaking, before DOE can make a determination as to whether to include them in the definition of "alternative fuel."

Vegetable oils.

Most vegetable oils are triglycerides (TGs; triglyceride = TG). Chemically, TGs are the triacylglyceryl esters of various fatty acids with glycerol (Figure 1).

Some physical properties of the most common fatty acids occurring in vegetable oils and animal fats as well as their methyl esters are listed in Table I. Besides these fatty acids, numerous other fatty acids occur in vegetable oils and animal fats, but their abundance usually is considerably lower. Table II lists the fatty acid composition of some vegetable oils and animal fats that have been studied as sources of biodiesel.

CH_2OOR						CH_2OH
|						|
$CHOOR$	+	$3\ CH_3OH$	→	$3\ CH_3OOCR$	+	$CHOH$
|						|
CH_2OOR						CH_2OH
Triglyceride		Methanol		Methyl ester		Glycerol

Figure 1. Structure of triglycerides and principle of the transesterification reaction (shown for methyl esters; $R = (CH_2)_xCH_3$ or unsaturated rests according to the fatty acids listed in Table I).

The most common derivatives of TGs (or fatty acids) for fuels are methyl esters. These are formed by transesterification of the TG with methanol in presence of usually a basic catalyst to give the methyl ester and glycerol (see Figure 1). Other alcohols have been used to generate esters, for example, the ethyl, propyl, and butyl esters.

Selected physical properties of vegetable oils and fats as they relate to their use as DF are listed in Table III. For esters these properties are given in Table IV. Also listed in Table III are the ranges of iodine values (centigrams iodine absorbed per gram of sample) of these oils and fats. The higher the iodine value, the more unsaturation is present in the fat or oil.

That vegetable oils and their derivatives are suited as DF is shown by their CNs (Table III) which generally are in the range suitable for or close to that of DF. The heat

Table I. Selected properties of some common fatty acids and esters

Trivial (Systematic) name[a]; Acronym[b)]	*Mol. wt.*	*m.p.[c] (°C)*	*b.p.[c,d] (°C)*	*Cetane No.*	*Heat of Combustion[e] (kg-cal/mole)*
Caprylic acid (Octanoic acid); 8:0	144.22	16.5	239.3		
Capric acid (Decanoic acid); 10:0	172.27	31.5	270	47.6 (98.0)[f]	1453.07 (25°),
Lauric acid (Dodecanoic acid); 12:0	200.32	44	131[1]		1763.25 (25°),
Myristic acid (Tetradecanoic acid); 14:0	228.38	58	250.5[100]		2073.91 (25°),
Palmitic acid (Hexadecanoic acid); 16:0	256.43	63	350		2384.76 (25°),
Stearic acid (Octadecanoic acid); 18:0	284.48	71	360d		2696.12 (25°),
Oleic acid (9Z-Octadecenoic acid); 18:1	282.47	16	286[100]		2657.4 (25°),
Linoleic acid (9Z,12Z-Octadecadienoic acid); 18:2	280.45	-5	229-30[16]		
Linolenic acid (9Z,12Z,15Z-Octadecatrienoic acid); 18:3	278.44	-11	230-2[17]		
Erucic acid (13Z-Docosenoic acid); 22:1	338.58	33-4	265[15]		
Methyl caprylate (Methyl octanoate); 8:0	158.24	...	193	33.6 (98.6)[f]	1313
Methyl caprate (Methyl decanoate); 10:0	186.30	...	224	47.7 (98.0)[f]	1625
Methyl laurate (Methyl dodecanoate); 12:0	214.35	5	266[766]	61.4 (99.1)[f]	1940
Methyl myristate (Methyl tetradecanoate); 14:0	242.41	18.5	295[751]	66.2 (96.5)[f]	2254
Methyl palmitate (methyl hexadecanoate); 16:0	270.46	30.5	415-8[747]	74.5 (93.6)[f]	2550
Methyl stearate (Methyl octadecanoate); 18:0	298.51	39.1	442-3[747]	86.9 (92.1)[f]	2859
Methyl oleate (Methyl 9Z-octadecenoate); 18:1	296.49	-20	218.5[20]	47.2[g]	2828

Continued on next page

Table I. Continued

Trivial (Systematic)name[a]; Acronym[b)]	*Mol. wt.*	*m.p.[c] (°C)*	*b.p.[c,d] (°C)*	*Cetane No.*	*Heat of Combustion[e] (kg-cal/mole)*
Methyl linoleate (Methyl 9*Z*, 12*Z*-octadecadienoate); 18:2	294.48	-35	215^{20}	28.5[g]	2794
Methyl linolenate (Methyl 9*Z*, 12*Z*,15*Z*-octadecatrienoate); 18:3	292.46	-57 -52	$109^{0.018}$	20.6[g]	2750
Methyl erucate (Methyl 13*Z*-docosenoate); 22:1	352.60	...	$221\text{-}222^{5}$	76.0	3454

a) *Z* denotes *cis* configuration.
b) The numbers denote the number of carbons and double bonds. For example, in oleic acid, 18:1 stands for eighteen carbons and one double bond.
c) Melting points and boiling points given in Ref. 28, pp. C-42 to C-553. Melting points and boiling points of 12:0 - 18:0 and 18:3 esters given in Ref. 181.
d) Superscripts in boiling point column denote pressure (mm Hg) at which the boiling point was determined.
e) See Ref. 27.
f) Cetane number from Ref. 21. Number in parentheses indicates purity (%) of the material used for CN determinations as given by the author. Other CNs given in Ref. 21 not tabulated here (purities in parentheses): ethyl caprate (10:0) 51.2 (99.4); ethyl myristate (14:0) 66.9 (99.3); propyl caprate (10:0) 52.9 (98.0); isopropyl caprate (10:0) 46.6 (97.7); butyl caprylate (8:0) 39.6 (98.7); butyl caprate (10:0) 54.6 (98.6); butyl myristate (14:0) 69.4 (99.0).
g) CN from Ref. 17. CNs (lipid combustion quality numbers) deviating from Ref. 21 as given in Ref. 17: Methyl laurate 54, methyl myristate 72, methyl palmitate 91, methyl stearate 159.

contents of various vegetable oils (Table III) are also nearly 90% that of DF2 (*11-13*). The heats of combustion of fatty esters and triglycerides (*14*) as well as fatty alcohols (*15*) have been determined and shown to be within the same range.

The suitability of fats and oils as DF results from their molecular structure and high energy content. Long-chain, saturated, unbranched hydrocarbons are especially suitable for conventional DF as shown by the CN scale. The long, unbranched hydrocarbon chains in fatty acids meet this requirement. Saturated fatty compounds have higher CNs. Other observations (*16*) are (i) that (a) double bond(s) decrease(s) quality (therefore, the number of double bonds should be small rather than large, (ii) that a double bond, if present, should be positioned near the end of the molecule, and (iii) no aromatic compounds should be present. A correlation to the statement on double bond position is the comparison of the CNs of methyl oleate (Table I), methyl petroselinate (methyl 6(*Z*)-octadecenoate and methyl *cis*-vaccenate (methyl 11(*Z*)-octadecenoate). The CN of methyl petroselinate (petroselinic acid occurs in less common oils such as parsley and celery seed oils) is 55.4 and that of methyl *cis*-vaccenate (vaccenic acid occurs in fats such as butter and tallow) is 49.5 (*17*). In that study the CN of methyl

Table II. Major fatty acids (in wt.-%) of some oils and fats used or tested as alternative diesel fuels.[a] All values combined from Refs. 176 and 181.

Oil or Fat	*Fatty Acid Composition (Wt.-%)*							
	12:0	*14:0*	*16:0*	*18:0*	*18:1*	*18:2*	*18:3*	*22:1*
Babassu	44-45	15-17	5.8-9	2.5-5.5	12-16	1.4-3		
Canola			4-5	1-2	55-63	20-31	9-10	1-2
Coconut	44-51	13-18.5	7.5-10.5	1-3	5-8.2	1.0-2.6		
Corn			7-13	2.5-3	30.5-43	39-52	1	
Cottonseed		0.8-1.5	22-24	2.6-5	19	50-52.5		
Linseed			6	3.2-4	13-37	5-23	26-60	
Olive		1.3	7-18.3	1.4-3.3	55.5-84.5	4-19		
Palm		0.6-2.4	32-46.3	4-6.3	37-53	6-12		
Peanut		0.5	6-12.5	2.5-6	37-61	13-41		1
Rapeseed		1.5	1-4.7	1-3.5	13-38	9.5-22	1-10	40-64
Safflower			6.4-7.0	2.4-29	9.7-13.8	75.3-80.5		
Safflower, high-oleic			4-8	2.3-8	73.6-79	11-19		
Sesame			7.2-9.2	5.8-7.7	35-46	35-48		
Soybean			2.3-11	2.4-6	22-30.8	49-53	2-10.5	
Sunflower			3.5-6.5	1.3-5.6	14-43	44-68.7		
Tallow (beef)		3-6	25-37	14-29	26-50	1-2.5		

a) These oils and fats may contain small amounts of other fatty acids not listed here. For example, peanut oil contains 1.2% 20:0, 2.5 22:0, and 1.3% 24:0 fatty acids (*181*).

oleate was 47.2, the lowest of these 18:1 methyl esters. The double bond of methyl petroselinate is closer to one end of the molecule. It also has the longest uninterrupted alkyl chain of these compounds, which may play a role because alkanes have higher CNs as discussed above. This complements the observations in Ref. 16. Another possibility is benzene formation by a disproportionation reaction from cyclohexane, which in turn would arise from cleavage of methyl oleate (*17*). The low CN of benzene would account for the lower CN of methyl oleate. The other 18:1 compounds would not form cyclohexane due to the different positions of the double bond.

Table III. Fuel-related properties and iodine values of various fats and oils [a]

Oil or Fat	*Iodine Value*	*CN*	*HG (kJ/kg)*	*Viscosity (mm^2/s)*	*CP (°C)*	*PP (°C)*	*FP (°C)*
Babassu	10-18	38					
Castor	82-88	?	39500	297 (38°)	---	-31.7	260
Coconut	6-12						
Corn	103-140	37.6	39500	34.9 (38°)	-1.1	-40.0	277
Cottonseed	90-119	41.8	39468	33.5 (38°)	1.7	-15.0	234
Crambe	93	44.6	40482	53.6 (38°)	10.0	-12.2	274
Linseed	168-204	34.6	39307	27.2 (38°)	1.7	-15.0	241
Olive	75-94						
Palm	35-61	42					
Peanut	80-106	41.8	39782	39.6 (38°)	12.8	-6.7	271
Rapeseed	94-120	37.6	39709	37.0 (38°)	-3.9	-31.7	246
Safflower	126-152	41.3	39519	31.3 (38°)	18.3	-6.7	260
High-oleic safflower	90-100	49.1	39516	41.2 (38°)	-12.2	-20.6	293
Sesame	104-120	40.2	39349	35.5 (38°)	-3.9	-9.4	260
Soybean	117-143	37.9	39623	32.6 (38°)	-3.9	-12.2	254
Sunflower	110-143	37.1	39575	37.1 (38°)	7.2	-15.0	274
Tallow	35-48	-	40054	51.15 (40°)	-	-	201
No. 2 DF		47	45343	2.7 (38°)	-15.0	-33.0	52

a) Iodine values combined from Refs. 176 and 181. Fuel properties from Ref. 11. All tallow values from Ref. 177 (No CN given in Ref. 177, calcd. cetane index 40.15).

The combustion of the glyceryl moiety of the TGs could lead to formation of acrolein and this in turn to the formation of aromatics (*16*), although no acrolein was found in precombustion of TGs (*18*). This may be one reason why fatty esters of vegetable oils perform better in a diesel engine than the oils containing the TGs (*16*). On the other hand, as discussed above, benzene may arise from the oleic moiety also.

Table IV. Fuel-related physical properties of esters of oils and fats [a]

Ester	*CN*	*HG (kJ/kg)*	*Viscosity (mm^2/s)*	*CP (°C)*	*PP (°C)*	*FP[b] (°C)*
Methyl						
Cottonseed[c]	51.2	-	6.8 (21°)	-	-4	110
Rapeseed[d]	54.4	40449	6.7 (40°)	-2	-9	84
Safflower[e]	49.8	40060	-	-	-6	180
Soybean[f]	46.2	39800	4.08 (40°)	2	-1	171
Sunflower[g]	46.6	39800	4.22 (40°)	0	-4	-
Tallow[h]	-	39949	4.11 (40°)	12	9	96
Ethyl						
Palm[i]	56.2	39070	4.5 (37.8°)	8	6	19
Soybean[f]	48.2	40000	4.41 (40°)	1	-4	174
Tallow[j]				15	12	
Propyl						
Tallow[j]				17	12	
Isopropyl						
Soybean	52.6[k]			-9[l]	-12[l]	
Tallow[j]				8	0	
n-Butyl						
Soybean[f]	51.7	40700	5.24 (40°)	-3	-7	185
Tallow[j]				13	9	
2-Butyl						
Soybean[l]				-12	-15	
Tallow[j]				9	0	

a) CN = cetane number; CP = cloud point, PP = pour point, FP = flash point. b) Some flash points are very low. These may be typographical errors in the references or the materials may have contained residual alcohols. c) Ref. 42. d) Ref. 55. e) Ref. 178. f) Ref. 17. g) Ref. 179. h) Ref. 177. i) Ref. 180. j) Ref. 95. k) Ref. 127. l) Ref. 123.

However, the high viscosity of the TGs is a major contributing factor to the onset and severity of durability problems when using vegetable oils (*19-20*).

The above statements on CNs correlate with the values given in Tables I, III and IV. For example, corresponding to components of conventional DF, saturated fatty compounds show higher CNs than the unsaturated compounds. CNs generally increase with increasing chain length (*21*). The CNs of mixtures are influenced by the nature of their components. Correlation of data from Tables II, III and IV shows that major high-CN components lead to relatively high CNs of vegetable oils or their esters.

In some literature it is emphasized that biodiesel is an oxygenated fuel, thus implying that their oxygen content plays a role in making fatty compounds suitable as DF by "cleaner" burning. However, the responsibility for this suitability rests mainly with the hydrocarbon portion which is similar to conventional DF. Furthermore, the oxygen in fatty compounds may be removed from the combustion process by decarboxylation, which yields incombustible CO_2, as precombustion (*18*), pyrolysis and thermal decomposition studies discussed below imply. Also, pure unoxygenated hydrocarbons, like cetane, have CNs higher than biodiesel. Fatty alcohols, whose oxygen content is lower than that of the corresponding esters, also have CNs higher than the corresponding methyl esters as determined with ASTM D613. For example, the CN of 1-tetradecanol is 80.8 (*22*). The CNs of fatty alcohols also increase with chain length with 1-pentanol having a CN of 18.2 (*22*). The CNs of 1-hexadecanol and 1-octadecanol were not determined in this work due to their high melting points (*22*), but ignition delay with the constant volume combustion apparatus (CVCA) vessel discussed below was measured. The CNs of some fatty alcohols were lower when employing the CVCA. Fatty ethers (*23*) were also shown to have CNs higher than the corresponding fatty esters and were suggested as DF extenders. Their main disadvantage compared to esters is their less straightforward synthesis.

The CNs of esters correlate well with boiling points (*21*). Quantitative correlations and comparison to numerous other physical properties of fatty esters confirmed that the boiling point gives the best approximation of CN (*22*).

ASTM D613 is used in determining CNs. For vegetable oil-derived materials, an alternative utilizes a CVCA (*24*). The amount of material needed for CN determination was reduced significantly with this bomb and it also allows studying materials with high melting points that cannot be measured by ASTM D613. Estimated cetane numbers (ECN) were determined on a revised scale permitting values greater than 100. In this case, the ECN of methyl stearate is 159 and that of methyl arachidate (20:0) is 196 (*24*). The ECNs of other esters were methyl laurate 54, methyl myristate 72, methyl palmitate 91, and methyl oleate 80. ECNs of fatty alcohols were 1-tetradecanol 51, 1-hexadecanol 68, 1-octadecanol 81, oleyl alcohol 51, linoleyl alcohol 44, linolenyl alcohol 41, and palmitoleyl alcohol 46. The ECNs of the TGs trilaurin and trimyristin exceeded 100, while the ECN of tripalmitin was 89, tristearin 95, triolein 45, trilinolein 32, and trilinolenin 23. The term "Lipid Combustion Quality Number" with an accompanying scale was suggested instead of CN to provide for values in excess of CN 100.

Often the "cetane index" of a fuel is published and should not be confused with CN. This is an ASTM-approved alternative method for a "non-engine" predictive equation of CN for petroleum distillates (*25* and references therein). Equations for predicting CNs are usually not applicable to non-conventional DFs such as biodiesel or other lipid materials (*26*) . Cetane indices are not given here. A method for estimating the cetane indices of vegetable oil methyl esters has been presented (*27*).

Besides CN, heat of combustion (HG) is another property of fatty compounds that is essential in proving the suitability of these materials as DF (*14*). Heats of combustion of fatty compounds, oils and fats as well as their methyl esters are listed in Tables I, III, and IV. For purposes of comparison, the literature values (*28*) for the heat of combustion of hexadecane (cetane), the high CN standard for conventional DF, is 2559.1 kg-cal (at 20°C). The data in Table I show that the heats of combustion of fatty compounds are similar to those of the compounds of similar CH content (long-chain, unbranched alkanes such as hexadecane) ideally comprising conventional DF. For example, the heat of combustion of methyl palmitate is 2550 kg-cal, that of methyl stearate is 2859 kg-cal, and that of unsaturated methyl oleate is 2828 kg-cal.

Even the combined CN and heat data do not suffice to determine the suitability of a material as DF. This is shown by the data in Tables III, which list the viscosities as well as cloud and pour points of numerous vegetable oils and fats. The viscosity of vegetable oils is approximately one order of magnitude greater than that of conventional DF. The high viscosity with resulting poor atomization in the combustion chamber was identified early as a major cause of engine problems such as nozzle coking, deposits, etc. (*14*, *29-31*). Therefore, neat oils have been largely abandoned as alternative DFs.

Four possible solutions to the viscosity problem have been evaluated (*32*). The most common applied solution to this problem is the preparation of the methyl esters by transesterification. The three other solutions to the problem of high vegetable oil viscosity are dilution (blending) with conventional DF or other suitable hydrocarbons, microemulsification or (co-solvency), and pyrolysis. These processes are also discussed below. As shown in Table IV, the methyl esters of oils and fats have viscosities approaching that of DF2.

The methyl esters, however, have higher cloud and pour points than their parent oils and fats and conventional DF (Tables III and IV). This is important for engine operation in cold or cooler environments. The cloud point is defined as the temperature at which the fuel becomes cloudy due to formation of crystals which can clog fuel filters and supply lines. The pour point is the lowest temperature at which the fuel will flow. It is recommended by engine manufacturers that the cloud point be below the temperature of use and not more than 6°C above the pour point.

Biodiesel Standards.

Besides favorable economics and environmental and health benefits, the development of reliable standards, which will instill confidence in biodiesel users, engine manufacturers, and other parties, is a milestone in facilitating commercialization (*6*). Austria (ÖNORM C 1190) and Germany (DIN V 51606) have established similar standards for neat biodiesel. In the United States, an ASTM standard was suggested (*9*). Table V gives the German standard and Table VI lists the proposed ASTM standard. The standards contain specifications particular to biodiesel (for example, glycerol quantitation) which are not given for conventional DF.

Table V. German biodiesel standard DIN V 51606

Fuel Property	*Unit*	*Test Method*	*Limit (min.)*	*Limit (max.)*
Density at 15°C	g / ml	ISO 3675	0.875	0.900
Kinematic Viscosity at 15°C	mm^2 / s	ISO 3104	3.5	5.0
Flash Point (Pensky-Martens)	°C	ISO 2719	100	
CFPP[a]	°C	DIN EN 116		
April 15- September 30				0
October 1- November 15				-10
November 16 - February 28				-20
March 1 - April 14				-10
Sulfur Content	wt.-%	ISO 4260		0.01
Carbon Residue - Conradson (10% distillation residue)	wt.-%	ISO 10370		0.30
Cetane Number		ISO 5165	49	
Ash	wt.-%	ISO 6245		0.01
Water	mg / kg	ASTM D 1744		300
Total Contamination	mg / kg	DIN 51419		20
Copper Strip Corrosion (3 h at 50°C)		ISO 2160		1
Acid Number	mg KOH / g	DIN 51558 Part 1		0.5
Methanol	wt.-%	tbs[b)]		0.3
Monoglycerides	wt.-%	tbs		0.8
Diglycerides	wt.-%	tbs		0.1
Triglycerides	wt.-%	tbs		0.1
Free Glycerine	wt.-%	tbs		0.02
Total Glycerine	wt.-%	tbs		0.23
Iodine Value	g Iodine / 100g	DIN 53241 Part 1		115
Phosphorus	mg / kg	tbs		10

a) CFPP = Cold-filter plugging point. b) tbs = to be standardized.

The iodine value (IV; see Table III) has been included in the European standards and is based on rapeseed oil as biodiesel feedstock. It is set at IV = 115, which would exclude soybean oil (neat vegetable oils and their methyl esters have nearly identical IVs) as biodiesel feedstock. The discussion in the previous section, however, shows that

Table VI. Suggested ASTM standard for pure (100%) biodiesel [a]

Property	*ASTM Method*	*Limits*	*Units*
Flash Point	93	100.0 min	°C
Water & Sediment	1796	0.050 max.	vol.-%
Carbon Residue, 100% sample	4530[b]	0.050 max.	wt.-%
Sulfated Ash	874	0.020 max.	wt.-%
Kinematic Viscosity, 40°C	445	1.9-6.0	mm^2 / s
Sulfur	2622	0.05 max.	wt.-%
Cetane	613	40 min.	
Cloud Point	2500	By customer	°C
Copper Strip Corrosion	130	No. 3b max.	
Acid Number	664	0.80 max.	mg KOH / g
Free Glycerol	GC[c]	0.20 max.	wt.-%
Total Glycerol	GC[c]	0.40 max.	wt.-%

a) This specification is the process of being evaluated by ASTM. A considerable amount of experience exists in the U.S. with a 20 percent blend of biodiesel with 80 percent petroleum-based diesel. Although biodiesel can be used in the pure form, use of blends of over 20 percent biodiesel should be evaluated on a case-by-case basis until further experience is available.
b) Or equivalent ASTM testing method.
c) Austrian (Christina Plank) update of USDA test method (author's note: refers to Refs. 97 and 104).

this is not without problems (*33*). Biodiesel from vegetable oils with high amounts of saturates (low IVs) will have a higher CN while the low-temperature properties are poor. Biodiesel from vegetable oils with high amounts of unsaturates (high IVs) will have low CN while the low-temperature properties are better. Thus, CN and low-temperature properties run counter to each other and this must affect IVs for biodiesel standards. Another argument against inclusion of the IV in biodiesel standards is the observation that different fatty acid compositions give identical IVs (e.g., neat methyl oleate has the same IV as a 1:1 mixture of methyl stearate and methyl linoleate). The IV also does not take into consideration structural factors of fatty compounds as discussed above where the CNs depend on double bond position, etc. Furthermore, once in place, the IV will hinder further research and development. It is possible that plants with desirable high-cetane fatty acid profile can be genetically engineered and bred (for example, substituting Δ6 unsaturated C18:1 acids for Δ9 unsaturated ones) or that combustion-improving additives are developed which are highly effective even for high degrees of unsaturation. It was suggested that it appears better to limit the amount of higher unsaturated fatty acids (e.g. linolenic acid) than to limit the degree of unsaturation by means of the IV (*34*). Note that soybean oil, rapeseed oil, and canola oil (low-erucic rapeseed oil) have very similar 18:3 fatty acid content (Table II), which is the most problematic in the formation of engine deposits through polymerization. However, linseed oil methyl ester (high 18:3 content and IV) satisfactorily completed 1000 hours of testing in a DI engine while neat linseed oil caused the engine to fail (*35* and references therein). These observations make the IV even more debatable.

Since most esters have higher CNs than neat vegetable oils and conventional DF, the esters could accommodate higher CNs than the minimum of 40 given in the ASTM standard for conventional DF. For example, the lowest reported CN for methyl soyate is 46.2 (see Table IV).

The German biodiesel standard includes the so-called Cold-Filter Plugging Point (CFPP) that pertains to the low-temperature flow properties of biodiesel. This low-temperature property test is used in Europe, South America, and the Pacific rim. In North America, a more stringent test, the Low-Temperature Flow Test (LTFT), is used and specified by ASTM D4539. Although the LTFT is more useful in evaluating low-temperature flow properties, ASTM requires only specification of cloud point for certification.

Combustion Chemistry. Emissions. Engine problems and deposits.

Besides the properties discussed above and accompanying operational problems, the question of combustion, emissions, and engine deposits of biodiesel fuels is of extreme significance and will be discussed here.

Generally, similar types of compounds are observed in the exhaust emissions of conventional DF and vegetable oil-derived fuels. This is additional proof of the suitability of fatty compounds as DF because there presumably exist similarities in their combustion behavior.

Emissions from any kind of engine are the result of the preceding combustion within in the engine. The combustion process, in relation to the properties of the fuel, and its completeness are responsible for any problems associated with the use of biodiesel, such as formation of deposits, etc. To understand the formation of emissions and deposits, and possibly direct the combustion to suppress undesirable emissions and deposits, it is essential to study the combustion of the fuel.

Ideally, the products of complete combustion of hydrocarbons are carbon dioxide (CO_2) and water according to the equation (shown for alkanes (saturated hydrocarbons)):

$$C_nH_{2n+2} + (1.5n + 0.5)O_2 \rightarrow nCO_2 + (n + 1)H_2O$$

Combustion in a diesel engine occurs mainly through a diffusion flame and is therefore incomplete (*8*). This causes the formation of partially oxidized materials such as carbon monoxide (CO), other oxygenated species (aldehydes, etc.), and hydrocarbons.

In the case of biodiesel, liberation of CO_2 (decarboxylation), as indicated above, from the ester moiety of the triglyceride or methyl ester occurs besides combustion formation of CO_2 from the hydrocarbon portions of biodiesel. The formation of CO_2, an incombustible compound despite its high oxygen content (although mistakenly assumed by some that it can serve as a combustion enhancer because of its high oxygen content), shows that one has to be judicious in choosing oxygenated compounds as combustion enhancers because the combustion-enhancing properties will depend on the nature of the oxygen (bonding, etc.) in those compounds. Therefore, the higher oxygen content of biodiesel does not necessarily imply improved combustion compared to conventional DF because of removal of this oxygen from the combustion process by decarboxylation, but CO_2 may contribute to combustion in other ways.

Exhaust emissions observed in the combustion of conventional DF and biodiesel are smoke, particulates (particulate matter), polyaromatic hydrocarbons (PAHs), hydrocarbons, CO, and oxides of nitrogen (NO_x; also referred to as nitrous oxides, or nitrogen oxides). An important difference are sulfur-containing emissions which are not formed from biodiesel due to its lack of sulfur. Note that rapeseed contains low amounts of sulfur but variations such as canola have not only lower erucic acid content but also reduced sulfur (*36*).

The composition of particulate matter has been studied for conventional diesel fuels (*37*). Particulates from conventional DF have a high carbon to hydrogen ratio of approximately 10:1 (*38*). Thus, particulates are mainly carbon in forms of crystallites. As temperatures decrease below 500°C, the particles are coated with adsorbed and condensed species, which include unburned hydrocarbons, various oxygenated hydrocarbons, PAHs and nitrogen dioxide (in case of conventional DF, also sulfur-containing species). With rapeseed methyl ester as fuel in DI engines, particulate matter showed large amounts of volatile and extractable compounds adsorbed on the soot, which caused the particulate emissions to be higher than with conventional DF (*39*).

PAHs are compounds composed of fused aromatic rings that may carry alkyl substituents such as a methyl group. They are of concern because many of them are known carcinogens.

Hydrocarbons represent a broad category of compounds including hydrocarbons and oxygenated species such as aldehydes, ketones, ethers, etc.

Nitrogen oxides (NO_x) arise by the reaction of nitrogen and oxygen from air at an early stage in the combustion process (*40*). NO_x emissions are difficult to control because such techniques may increase other emissions or fuel consumption (*8*).

Emissions of Neat Vegetable Oil Fuel. While neat vegetable oils are competitive with conventional DF in some emission categories, problems were identified for other kinds of emissions. For example, it was shown that PAH emissions were lower for neat vegetable oils, especially very little amounts of alkylated PAHs, which are common in the emissions of conventional DF (*41*). Besides higher NO_x levels (*42*), aldehydes are reported to present problems with neat vegetable oils. Total aldehydes increased dramatically with vegetable oils (*42*). Formaldehyde formation was also consistently higher than with DF2. It was reported that component TGs in vegetable oils can lead to formation of aromatics via acrolein (CH_2=CH-CHO) from the glycerol moiety (*16*).

Another author observed significantly lower emissions of C3 aldehydes (for example, acrolein) for methyl esters of rapeseed oil than for the oil itself (*43*). Another study (*44*) attributes increased emissions of aldehydes and ketones when using vegetable oils as fuels to the formation of acidic water during decomposition of the oils. This acidic water could be an indication for the formation of short-chain oxygenates which likely ignite poorly compared to the long-chain carbon-rich fatty compounds.

Engine Problems with Neat Vegetable Oil Fuel. Most references in this section report that, at least in short-term trials, neat oils gave satisfactory engine performance and power output, often equal to or even slightly better than conventional DF. However, vegetable oils cause engine problems. This was recognized in the early stages of renewed interest in vegetable oil-based alternative DFs. Studies on sunflower oil as fuel noted coking of injector nozzles, sticking piston rings, crankcase oil dilution, lubricating oil contamination, and other problems (*29-31*). These problems were confirmed and studied by other authors (*45-52*). A test for external detection of coking tendencies of vegetable oils was reported (*53*). The causes of these problems were attributed to the polymerization of TGs via their double bonds which leads to formation of engine deposits as well as the low volatility and high viscosity with resulting poor atomization patterns. An oxidative free-radical mechanism was suggested as governing TG polymerization in lubricating oil contamination when using sunflower oil as fuel (*54*). Fumigation with propane was studied as a means to reduce injector coking (*55*). The engine problems have caused neat vegetable oils to be largely abandoned as alternative DF and lead to the research on the aforementioned four solutions (*32*).

Emissions of esters. Generally, most emissions observed for conventional DF are reduced when using esters. NO_x emissions are the exception. In an early paper reporting emissions with methyl and ethyl soyate as fuel (*20*), it was found that CO and hydrocarbons were reduced but NO_x were produced consistently at a higher level than with the conventional reference DF. The differences in exhaust gas temperatures corresponded with the differences in NO_x levels. Similar results were obtained from a study on the emissions of rapeseed oil methyl ester (*43*). NO_x emissions were slightly increased, while hydrocarbon, CO, particulate and PAH emissions were in ranges similar to the DF reference. As mentioned above, the esters emitted less aldehydes than the corresponding neat rapeseed oil. Unrefined rapeseed methyl ester emitted slightly more aldehydes than the refined ester, while the opposite case held for PAH emissions. A 31% increase in aldehyde and ketone emissions was reported when using rapeseed methyl ester as fuel, mainly due to increased acrolein and formaldehyde, while hydrocarbons and PAHs were significantly reduced, NO_x increased slightly, and CO was nearly unchanged (*56*). The study on PAH emissions (*41*), where also the influence of various engine parameters was explored, found that the PAH emissions of sunflower ethyl ester were situated between DF and the corresponding neat vegetable oil. Reduced PAH emissions may correlate with the reduced carcinogenity of particulates when using rapeseed methyl ester as fuel (*57*). The general trend on reduced emissions except NO_x was confirmed by later studies (*58*), although some studies report little changes in NO_x (*59-60*). In a DI engine, sunflower methyl ester produced equal hydrocarbon emissions but less smoke than a 75:25 blend of sunflower oil with DF (*61*). Using a diesel oxidation catalyst (DOC) in conjunction with soy methyl ester was reported to be a possible emissions reduction technology for underground mines (*62*). Soy methyl esters were reported to be more sensitive towards changes in engine parameters than conventional DF (*63*).

Precombustion of Triglycerides. As discussed, every DF, conventional or vegetable oil-based, experiences an ignition delay, which is the basis of CN measurements. The fuel passes through a temperature and pressure gradient directly after injection but before combustion begins. Chemical reactions already occur in this precombustion phase. In an initial study (*64*), the unsaturated TGs triolein, trilinolein, and trilinolenin were studied at temperatures up to 400 °C in air or N_2 in a reactor simulating conditions in a diesel engine. The compounds arising in this phase were fatty acids of different chain lengths (some even longer than those in the parent fatty acids), various aliphatic hydrocarbons, and smaller amounts of other compounds such as aldehydes. The parent acids were the most prominent compounds in the precombustion mixture. Component patterns were largely independent of the starting material and reaction conditions. In a second study (*65*), tristearin and tripalmitin were studied besides the three unsaturated TGs at temperatures of 450 °C in air and N_2. Presumably due to the higher temperature, different component patterns were observed. Besides mainly unsaturated aliphatic hydrocarbons and unsaturated aldehydes, various aromatics, including benzene, toluene, compounds with unsaturated side chains, and polyaromatic hydrocarbons were detected. The atmosphere (air or N_2) had considerable influence on product formation. The number of components was less for samples of tripalmitin, tristearin and triolein for reactions under N_2 than under air while this finding was reversed for trilinolein and trilinolenin. No fatty acids, glycerol or acrolein (as decomposition product of glycerol) were detected. Extensive decarboxylation occurred, showing that the oxygen in biodiesel does not necessarily contribute to its combustion as an oxidizer. The compounds identified are also found in the exhaust emissions of engines running on conventional DF. It is therefore necessary to influence not only combustion but also precombustion to improve the combustion properties and emissions of biodiesel.

Cetane Improvers. Various compounds such as alkyl nitrates are used as cetane-enhancing additives in conventional DF (*66*). Few studies on such compounds in biodiesel exist. One paper reports (*67*) that in a turbulence combustion chamber and at an intake air temperature of 105 °C, 8% hexyl nitrate in vegetable oils (cottonseed, rape, palm) was necessary to exhibit the same ignition delay as conventional DF. The use of nitrate esters of fatty acids as cetane improvers in DF was reported in a patent (*68*).

Dilution of vegetable oils with conventional diesel fuel.

Dilution is an additional possible solution to the viscosity problem of vegetable oils as discussed above. Results with this technology have been mixed and engine problems similar to those found with neat vegetable oils as fuels were observed here also. A model on vegetable oil atomization showed that blends of DF2 with vegetable oil should contain from 0 to 34% vegetable oil if proper atomization was to be achieved (*69*).

A 75:25(vol-%) petrodiesel / sunflower oil blend had a viscosity of 4.88 mm^2/s at 40 °C, exceeding the ASTM maximum value of 4.0. The blend was not recommended for long-term use in the DI diesel engine (*64*). A 75:25 (vol-%) petrodiesel / high-oleic safflower oil blend with a viscosity of 4.92 mm^2/s passed the 200 hr EMA (Engine Manufacturers Association) test. The different results were attributed to the degree of unsaturation of the respective vegetable oil (*32*). The more unsaturated oil (sunflower) that accumulates in the crankcase and hot engine parts tends to oxidize and polymerize due to its reactivity. Accumulation of such products in the lube oil could lead to lubricant thickening. A lube oil change is called for by the EMA test after 100 hr and at that time the viscosity of the lube oils had not varied greatly in either test.

Other reports include successfully using a 70:30 winter rapeseed oil / DF1 mixture (*47*) or blends of ≤ 15% rapeseed oil with DF2 (*71*), and an 80:20 DF2 / safflower oil blend with reduced CO and hydrocarbon emissions (*72*). A 75:25 DF / crude sunflower oil blend produced greatest solids contamination in the lubricating oil (*49*) similar to the results mentioned above, while another report mentions satisfactory performance of a 75:25 DF / sunflower oil blend (*61*). In early studies on sunflower oil, 80:20 DF / sunflower oil blends (*31*) were run for prolonged periods of time before exhaust smoke increased due to carbon build-up or power loss ensued. Another engine, due to inadequate atomization, showed more of the engine problems associated with neat vegetable oils.

The CP of a 50:50 DF2/ high-oleic safflower oil was -13 °C and the PP was -15 °C, and similar blends with high-linoleic safflower oil had CP -13 °C and PP -15 °C or winter rapeseed oil had CP -11 °C and PP -18 °C (*55*).

A 50:50 blend of Stoddard solvent (a dry-cleaning fluid, viscosity 0.95 mm^2/s, estimated CN 50, heat of combustion 46,800 kJ/kg, CP < -16 °C, PP < -35 °C, flash point 42.2 °C) with soybean oil gave low CP (-18.9 °C) and PP (-31.7 °C) but performed less well in a diesel engine than DF2 (*73*).

Transesterification.

The conversion of component TGs to simple alkyl esters (transesterification) with various alcohols reduces the high viscosity of oils and fats (see also Figure 1). Base catalysis of the transesterification with reagents such as sodium hydroxide is preferred over acid catalysis because the former is more rapid (*74*). Transesterification is a reversible reaction. The transesterification of soybean oil with methanol or 1-butanol proceeded with pseudo-first order or second order kinetics, depending on the molar ratio of alcohol to soybean oil (30:1 pseudo-first order, 6:1 second order; NaOBu catalyst) while the reverse reaction was second order (*75*).

Methyl esters are the most "popular" esters for several reasons. One reason is the low price of methanol compared to other alcohols. Generally, esters have significantly lower viscosities than the parent oils and fats (Tables III and IV). Accordingly, they improve the injection process and ensure better atomization of the fuel in the combustion chamber. The effect of the possible polymerization reaction is also decreased. The advantages of alkyl esters were noted early in studies on the use of sunflower oil and its esters as DF (*29-31*). Another advantage of the esters is possibly more benign emissions, for example, with the removal of glycerol (which is separated from the esters) the formation of undesirable acrolein may be avoided, as discussed above. These reasons as well as ease and rapidity of the process are responsible for the popularity of the transesterification method for reducing the viscosity-related problems of vegetable oils. The popularity of methyl esters has contributed to the term "biodiesel" now usually referring to vegetable oil esters and not neat vegetable oils.

In the early studies on sunflower esters, no transesterification method was reported (*29-31*). Another early study used H_2SO_4 as the transesterification catalyst (*76*). It was then shown, however, that in homogeneous catalysis, alkali catalysis is a much more rapid process than acid catalysis in the transesterification reaction (*74*, *77*). At 32°C, transesterification was 99% complete in 4 h when using an alkaline catalyst (NaOH or NaOMe). At 60°C and a molar ratio alcohol:oil of at least 6:1 and with fully refined oils, the reaction was complete in 1 h to give methyl, ethyl, or butyl esters. The reaction parameters investigated were molar ratio of alcohol to vegetable oil, type of catalyst (alkaline vs. acidic), temperature, reaction time, degree of refinement of the vegetable oil, and effect of the presence of moisture and free fatty acid. Although the crude oils could be transesterified, ester yields were reduced because of gums and extraneous material present in the crude oils.

Besides sodium hydroxide and sodium methoxide, potassium hydroxide is another common transesterification catalyst. Both NaOH and KOH were used in early work on the transesterification of rapeseed oil (*78*). Recent work on producing biodiesel (suitable for waste frying oils) employed KOH. With the reaction conducted at ambient pressure and temperature, conversion rates of 80 to 90% were achieved within 5 minutes, even when stoichiometric amounts of methanol were employed (*79*). In two steps, the ester yields are 99%. It was concluded that even a free fatty acid content of up to 3% in the feedstock did not affect the process negatively and phosphatides up to 300 ppm phosphorus were acceptable. The resulting methyl ester met the quality requirements for Austrian and European biodiesel without further treatment. In a study similar to previous work on the transesterification of soybean oil (*74*, *77*), it was concluded that KOH is preferable to NaOH in the transesterification of safflower oil of Turkish origin (*80*). The optimal conditions were given as 1 wt-% KOH at 69±1°C with a 7:1 alcohol : vegetable oil molar ratio to give 97.7% methyl ester yield in 18 minutes.

Patents. Most patents dealing with transesterification emphasize the engineering improvement of the process. Using patented procedures, a transesterification process permitting the recovery of all byproducts such as glycerol and fatty acids has been described (*81*). The use of alkaline catalysts is also preferred on the technical scale, as is documented by patents using sodium hydroxide, sodium methoxide, and potassium hydroxide (*82-85*). Different esters of C_{9-24} fatty acids were prepared with Al_2O_3- or Fe_2O_3- containing catalysts (*86*). A sulfonated ion exchange catalyst was preferred as catalyst in the esterification of free fatty acids (*87*).

Other procedures. Besides the methods discussed here, other catalysts have been applied in transesterification reactions (*88*). Some recently studied variations of the above methods as applied to biodiesel preparation are briefly discussed here.

Methyl and ethyl esters of palm and coconut oils were produced by alcoholysis of raw or refined oils using boiler ashes, H_2SO_4 and KOH as catalysts (*89*). Fuel yields > 90% were obtained using alcohols with low moisture content and EtOH-H_2O azeotrope.

Instead of using the extracted oil as starting material for transesterification, sunflower seed oils were transesterified in situ using macerated seeds with methanol in the presence of H_2SO_4 (*90*). Higher yields were obtained than from transesterification of the extracted oils. Moisture in the seeds reduced the yield of methyl esters. The cloud points of the *in situ* prepared esters appear slightly lower than those prepared by conventional methods.

Another study (*91*) reported the synthesis of methyl or ethyl esters with 90% yield by reacting palm and coconut oil from the press cake and oil mill and refinery waste with MeOH or EtOH in the presence of easily available catalysts such as ashes of the waste of these two oilseeds (fibers, shell, husk), lime, zeolites, etc. Similarly, it was reported that the methanolysis of vegetable oils is catalyzed by ashes from the combustion of plant wastes such as coconut shells or fibers of a palm tree that contain K_2CO_3 or Na_2CO_3 as catalyst (*92*). Thus the methanolysis of palm oil by refluxing 2 h with MeOH in the presence of coconut shell ash gave 96-98% methyl esters containing only 0.8-1.0% soap. The ethanolysis of vegetable oils over the readily accessible ash catalysts gave lower yields and less pure esters than the methanolysis.

Several catalysts (CaO, K_2CO_3, Na_2CO_3, Fe_2O_3, MeONa, $NaAlO_2$, Zn, Cu, Sn, Pb, ZnO, and *Dowex 2X8* (anion exchange resin)) were tested (mainly at 60-63°C) for catalytic activity in the transesterification of low-erucic rapeseed oil with MeOH (*93*). The best catalyst was CaO on MgO. At 200°C and 68 atm, the anion exchange resin produced substantial amounts of fatty methyl esters and straight-chain hydrocarbons.

An enzymatic transesterification method utilizing lipases and methanol, ethanol, 2-propanol, and 2-methyl-1-propanol as alcohols gave alkyl esters of fatty acids (*94, 95*). This method eliminates product isolation and waste disposal problems.

Analysis of Transesterification Products. Hardly any chemical reaction, including transesterification, ever proceeds to completion. Therefore, the transesterified product, biodiesel, contains other materials. There are unreacted TGs and residual alcohol present as well as partially reacted mono- and diglycerides and glycerol co-product.

Glyceride mixtures were analyzed by TLC / FID (thin-layer chromatography / flame ionization detection) (*96*), which was also used in the studies on the variables affecting the yields of fatty esters from transesterified vegetable oils (*74*). Analysis of reaction mixtures by capillary GC determined esters, triglycerides, diglycerides and monoglycerides in one run (*97*). Free glycerol was determined in transesterified vegetable oils (*98*) Besides analyzing esters for sterols (*99-101*), which are often minor components in vegetable oils, and different glycerides (*102-103*), recently the previous GC method (*97*) was extended to include analysis of glycerol in one GC run (*104*). In both papers (*97, 104*), the hydroxy groups of the glycerides and glycerol were derivatized by silylation with *N*-methyl-*N*-trimethylsilyltrifluoroacetamide. A simultaneous analysis of methanol and glycerol was recently reported (*105*).

Other authors, using GC to determine the conversion of TGs to methyl esters, gave a correlation between the bound glycerol content determined by TLC/FID and the acyl conversion determined by GC (*106*). Glycerol has also been detected by high-performance liquid chromatography (HPLC) using pulsed amperometric detection, which offers the advantage of being more sensitive than refractometry and also suitable for detection of small amounts of glycerol for which GC may not be suitable (*107*).

Recently, an alternative method for determining the methyl ester content based on viscosity measurements, which agreed well with GC determinations, was reported (*108*). The method is reportedly more rapid than GC and therefore especially suitable for process control.

Properties of Vegetable Oil Esters.

Early 100 hr tests on transesterified sunflower oil initially showed the improved properties of esters for use in a diesel engine by reducing the viscosity of vegetable oils and solving engine problems (*29-31*).

Table IV compares the essential fuel properties of some esters. In all cases the viscosity decreases dramatically and is only about twice that of DF2. The CNs are also improved, now being higher than that of DF2.

The methyl and ethyl esters of soybean oil generally compared well with DF2 with the exception of gum formation which leads to problems with fuel filter plugging (*20*). Another study reports that methyl esters of rapeseed and high-linoleic safflower oils formed equal and lesser amounts of deposits than a DF standard while the methyl ester of high-oleic safflower oil formed more deposits (*55*). Methyl and ethyl esters of soybean oil were evaluated by 200 hr EMA (Engine Manufacturers Association) engine tests and compared to DF2. Engine performance with soybean esters differed little from that with DF. In that work, also a slight power loss was observed, together with an increase in fuel consumption due to the lower heating values of the esters. The emissions for the two fuels were similar, with the exception of NO_x which are higher for the esters as discussed above. Engine wear and fuel-injection system tests showed no abnormal characteristics for any of the fuels. Deposit amounts in the engine were comparable, however, the methyl ester showed greater varnish and carbon deposits on the pistons. Operating DI engines with neat soybean oil esters under certain conditions produced lubricating oil dilution which was not observed with an IDI engine (*109*). Lubricating oil dilution was estimated by Fourier-Transform infrared spectroscopy combined with a fiber optic probe when using rapeseed methyl ester as a fuel (*110*). The carbonyl absorption was used for quantitation.

Low-temperature Properties. As discussed above, one of the major problems associated with the use of biodiesel, including methyl esters, is its poor low-temperature properties, documented by relatively high cloud point (CP) and pour point (PP) (Tables III and IV). CPs and PPs of vegetable oils and their esters are a result of these materials being mixtures of various compounds. For example, as seen in Table I, saturated fatty compounds have significantly higher melting points than unsaturated fatty compounds and in a mixture they therefore crystallize at higher temperature than the unsaturates.

The CP, which occurs at a higher temperature than the PP, is the temperature at which a fatty material becomes cloudy due to formation of crystals and saturates solidifying. These solids can clog fuel lines. With decreasing temperature, more material solidifies and the compound approaches the pour point, at which it will no longer flow.

Besides CP (ASTM D2500) and PP (ASTM D97), two test methods exist for examining the low-temperature properties of diesel fuel (as discussed briefly in the section on "Biodiesel Standards"), the Low-Temperature Flow Test (LTFT; used in North America; ASTM D4539) and the Cold-Filter Plugging Point (CFPP; used in Europe). These methods have also been used to evaluate biodiesel. Low-temperature filterability tests are necessary because they correlate better with operability tests than CP or PP (*111*). Recent results showed that for fuel formulations containing at least 10% methyl esters, both LTFT and CFPP are linear functions of the CP (*112*). Additional statistical analysis showed a strong 1:1 correlation between LTFT and CP (*112*).

Five possible solutions to the low-temperature problems of esters have been investigated: blending with conventional DF, additives, branched-chain esters, bulky substituents in the chain, and winterization. Blending of esters is currently the preferred method for improving low-temperature properties and is discussed in the next section.

Numerous additives have been synthesized and reported mainly in the patent literature, which allegedly have the effect of lowering CP and PP. These additives are usually a variety of viscosity-modifying polymers such as carboxy-containing interpolymers (*113*), styrene-maleic anhydride copolymer (*114*), polymethacrylates (*114-115*), polyacrylates (*113-115*), nitrogen-containing polyacrylates (*113*), poly[alkyl (meth)acrylates] (*116*), ethylene-vinyl ester (acetate) copolymers (*117-120*), fumarate or itaconate polymers and copolymers (comb polymers) (*117-118*), polyoxyalkylene compounds (*113*). Polar nitrogen compounds (*117*) have also been reported as additives. Similar additives have also been tested for conventional diesel fuel (*7*). The beneficial effect of some additives appears to be limited, however, because they more strongly affect the PP than the CP, and the CP is more important than the PP for improving low-temperature flow properties (*121*).

Another route is the synthesis of fatty compound-derived materials with bulky substituents in the chain (*122*). The idea associated with these materials is that the bulky substituents would destroy the harmony of the solids which are usually oriented in one direction. However, these materials had only slight influence on the CP and PP.

The use of secondary alcohols in the transesterification reaction provides branched-chain esters such as isopropyl and 2-butyl instead of the methyl esters (*95, 123)* . These esters showed a lower crystallization onset temperature (T_{CO}) as determined by differential scanning calorimetry (DSC) for the isopropyl esters of SBO by 7-11°C and for the 2-butyl esters of SBO by 12-14°C (*123*). The CPs and PPs were also lowered by the branched-chain esters. However, economics would only permit the isopropyl soyate appear attractive as branched-chain ester, raising the price of a biodiesel blend containing 30% isopropyl soyate by $0.02/L while lowering the T_{CO} by 15°C.

In the winterization method (*121, 124*), solids formed during cooling of the vegetable oil are removed by filtration, thus leaving a mixture of more unsaturated fatty compounds with lower CP and PP. This procedure can be repeated to achieve the desired CPs and PPs. Saturated fatty compounds, which have higher CNs, are among the major compounds removed by winterization. Thus the CN of the biodiesel decreases. The T_{CO} of typical methyl soyate was lowered from 3.7°C to -7.1°C by winterization (*124*), but the yield was low (26%). Winterization of low-palmitate methyl soyate, however, gave a T_{CO} of -6.5°C with a yield of 86%. Winterization of typical methyl soyate diluted in hexane gave a T_{CO} of -5.8°C with 77% yield. In the latter method, crystal formation was greatly affected by the nature of the solvent, with acetone and chloroform being unsuitable for winterization.

In a paper on fatty acid derivatives for improving ignition and low-temperature properties (*125*), it was reported that tertiary fatty amines and amides were effective in enhancing the ignition quality of biodiesel without negatively affecting the low-temperature properties. In that paper, saturated fatty alcohols of chain lengths C_{12} and greater increased the PP substantially. Ethyl laurate was weakly decreased the PP.

Blending of Esters.

Blending conventional DF with esters (usually methyl esters) of vegetable oils is presently the most common form of biodiesel. The most common ratio is 80% conventional diesel fuel and 20% vegetable oil ester (also termed "B20," indicating the 20% level of biodiesel; see also list of biodiesel demonstration programs in Ref. 6). There have been numerous reports that significant emission reductions are achieved with these blends.

No engine problems were reported in larger-scale tests with, for example, urban bus fleets running on B20. Fuel economy was comparable to DF2, with the consumption of biodiesel blend being only 2-5% higher than that of conventional DF. Another advantage of biodiesel blends is the simplicity of fuel preparation which only requires mixing of the components.

Ester blends have been reported to be stable, for example, a blend of 20% peanut oil with 80% DF did not separate at room temperature over a period of 3 months (*126*). Stability was also found for 50:50 blends of peanut oil with DF (*43*).

A few examples from the literature may illustrate the suitability of blends of esters with conventional DF in terms of fuel properties. In transient emission tests on an IDI engine for mining applications (*62*), the soybean methyl ester used had a CN of 54.7, viscosity 3.05 mm^2/s at 40°, and a CP of -2°C. The DF2 used had CN 43.2, viscosity 2.37 mm^2/s at 40° and a CP of -21°C. A 70:30 DF2 : soybean methyl ester blend had CN 49.1, viscosity 2.84 mm^2/s at 40°C, and a CP of -17°C. The blend had 4% less power and 4% higher fuel consumption than the DF2, while the neat esters had 9% less power and 13% higher fuel consumption than DF2. Emissions of CO and hydrocarbons as well as other materials were reduced. NO_x emissions were not increased here, although higher NO_x emissions have been reported for blends (DI engines) (*43, 59*).

Irregularities compared to other ester blends were observed when using blends of the isopropyl ester of soybean oil with conventional DF (*127*). Deposits were formed on the injector tips. This was attributed to the isopropyl ester containing 5.2 mole-% monoglyceride which was difficult to separate fòrm the isopropyl ester.

Microemulsification.

The formation of microemulsions (co-solvency) is one of the four potential solutions for solving the problem of vegetable oil viscosity. Microemulsions are defined as transparent, thermodynamically stable colloidal dispersions in which the diameter of the dispersed-phase particles is less than one-fourth the wavelength of visible light. Microemulsion-based fuels are sometimes also termed "hybrid fuels," although blends of conventional diesel fuel with vegetable oils have also been called hybrid fuels (*128*). Some of these fuels were tested in engines including the 200 hr EMA test. A microemulsion fuel containing soybean oil, methanol, 2-octanol, and a cetane enhancer was the cheapest vegetable oil-based alternative diesel fuel ever to pass the EMA test.

The components of microemulsions can be conventional DF, vegetable oil, an alcohol, a surfactant, and a cetane improver. Water (from aqueous ethanol) may also be present in order to use lower-proof ethanol (*129*), thus increasing water tolerance of the microemulsions is important.

Microemulsions are classified as non-ionic or ionic, depending on the surfactant present. Microemulsions containing, for example, a basic nitrogen compound are termed ionic while those consisting, for example, only of a vegetable oil, aqueous ethanol, and another alcohol, such as 1-butanol, are termed non-ionic. Non-ionic microemulsions are often referred to as detergentless microemulsions, indicating the absence of a surfactant.

Viscosity-lowering additives were usually with C_{1-3}alcohols length while longer-chain alcohols and alkylamines served as surfactants. *n*-Butanol (CN 42) was claimed to be the alcohol most suitable for microemulsions, giving microemulsions more stable and lower in viscosity than those made with methanol or ethanol (*130*). Microemulsions with hexanol and an ionic surfactant had no major effect on gaseous emissions or efficiency. Emulsions were reported to be suitable as diesel fuels with viscosities close to that of neat DF. No additional engine tests were reported here (*130*).

Physical property studies of mixtures of TGs with aqueous ethanol and 1-butanol (*131*) showed that they form detergentless microemulsions. Mixtures of hexadecane, 1-butanol, and 95% ethanol were shown to be detergentless microemulsions. Evidence was presented in that paper that 1-butanol in combination with ethanol associates and interacts with water to form systems exhibiting microemulsion features.

Solubilization and microemulsification studies on TGs, especially triolein, with methanol in the presence of several even-numbered *n*-alcohols as surfactants showed that 1-octanol produced the microemulsions with the best water tolerance. Among the octanols, 1- and 4-octanol were superior to the 2- and 3- isomers. 1-Butanol and 1-tetradecanol gave microemulsions with the least water tolerance. The formation of molecular dispersions seemed more likely than the formation of nonaqueous microemulsions, but the addition of water produced systems that exhibited microemulsion properties (*132*). Studies on micellar solubilization of methanol with TGs and 2-octanol as co-surfactant gave the following sequence for water tolerance of three surfactant systems: tetradecyldimethylammonium linoleate > bis(2-ethylhexyl) sodium sulfosuccinate > triethylammonium linoleate. A nonaqueous microemulsion system formed from triolein / oleyl alcohol (9(*Z*)octadecen-1-ol) / methanol (*133*).

When studying different unsaturated fatty alcohols, it was reported that the viscosity is nearly independent of the configuration of the double bonds in the tailgroup structure. However, with increasing unsaturation in the tailgroup, viscosity decreased at constant methanol concentration. Generally, adding long-chain fatty alcohols substantially increased methanol solubility in non-aqueous triolein / unsaturated long-chain fatty alcohol / methanol solutions under most conditions. Physical property data were consistent with those for systems exhibiting co-solvent phenomena. However, for solutions with methanol concentration exceeding 0.444 vol frac, the results showed that solubilization of methanol within large aggregates was feasible (*134*). Mixed amphiphile systems investigating four unsaturated C_{18} fatty alcohols and five C_4 -C_{12} alkanols showed that large methanol-in-amphiphile aggregates resembling a microemulsion were feasible under limited conditions (*135*). These binary systems strongly affect miscibility between methanol and TG. Critical micelle concentration (CMC) studies showed that degree of unsaturation and double bond configuration significantly affected aggregation when using six unsaturated C_{18} fatty alcohols as amphiphiles (*136*). These compounds form large and polydisperse aggregates in methanol. The effect of solubilized soybean oil was studied. Viscosity results were consistent with those for microemulsions. Presumably soybean oil is solubilized by incorporation into large soybean oil-in-fatty alcohol aggregates in methanol solvent, resembling a nonaqueous detergentless microemulsion.

Microemulsions containing conventional diesel fuel. Fuel formulations containing conventional DF in emulsion with soybean oil have been subjected to engine testing. In an emulsion with ethanol (*137*), such a fuel burned faster with higher levels of premixed burning due to longer ignition delays and lower levels of diffusion flame burning than DF, resulting in higher brake thermal efficiencies, cylinder pressures, and rates of pressure rise. NO_x and CO emissions increased with these fuels, while smoke and unburned hydrocarbons decreased. A microemulsion consisting of 50 vol-% DF, 25 vol-% degummed, alkali-refined soybean oil, 5 vol-% 95% aqueous ethanol and 20 vol-% 1-butanol was studied by the 200 hr EMA (Engine Manufacturers Association) test (*138*). The engine running on this fuel completed the EMA test without difficulty. The microemulsion fuel caused less engine wear than conventional DF but produced greater amounts of carbon and lacquer on the injector tips, intake valves and tops of the cylinder liners besides the observation that engine performance degraded 5% at the end of the test. Another report on blends of alcohols with vegetable oils and conventional DF (the 40:40:20 and 30:40:30 DF/ degummed, dewaxed soybean oil / ethanol blends used in this study were not fully miscible and no surfactant system was used) confirmed that the performance of such fuels was comparable to conventional DF but the tests were too short-term to determine potential problems of carbon buildup, etc. (*139*).

Microemulsions for blending alcohols with diesel fuel employed unsaturated fatty acids. Saturated fatty acids were unsatisfactory because crystalline phases separated upon refrigeration (*129*). Addition of *N,N*-dimethylamino ethanol (DMAE) gave microemulsions with satisfactory viscosity. Two fuels were tested: (1) 66.7% DF2, 16.7% 95% ethanol, 12.5% soybean acids, and 4.1% DMAE (ionic); (2) 66.7% DF2, 11.1% 95% EtOH, and 22.2% 1-butanol (non-ionic). Both hybrid fuels gave acceptable performance, for example improved brake thermal efficiency and lower exhaust temperatures. Smoke and CO levels were reduced but the unburned hydrocarbons level increased. The detergentless microemulsion was superior to the ionic one in those SAE properties relevant to good engine performance. On the other hand, fundamental studies on properties of microemulsions such as rheology, density, water tolerance, and critical solution temperatures showed that the water tolerance of ionic systems was greater than that of the 1-butanol system (*138*). The relative viscosities of the detergentless microemulsion varied directly with the volume percent of the dispersed water phase while for the ionic system the relative viscosities varied with increasing volume percent of dispersed water by values greater than those predicted by theory (*140*).

Variations of the microemulsion technology have been reported in the patent literature not using vegetable oils but conventional DFs and the fatty ingredient being present only as part of a surfactant system in such emulsions. These microemulsions usually consisted of DF, water, an alcohol (or, combining the latter two components, an aqueous solution of an alcohol), and a system of surfactants. Several such microemulsions with a surfactant system comprising DMAE and a long-chain fatty substance (C_9-C_{22}) were patented (*141*). This microemulsion, which contains a fatty compound only in small amounts, showed a high tolerance for water, which enabled hybridizing diesel fuel with relatively high levels of aqueous alcohol and also showed low-temperature stability. Other systems were a cosurfactant combination of methanol and a fatty acid partially neutralized by a nitrogeneous base such as ammonia, ethanolamine, or *iso*-propanolamine (*142*) and, in a similar system, the use of ammonium salts of fatty acids as cosurfactants was patented (*143*).

Microemulsions with vegetable oils and without conventional DF are the most widely studied. A microemulsion comprising a vegetable oil, a lower (C_1-C_3) alcohol, water, and a surfactant system consisting of a trialkylamine or the reaction product of a trialkylamine with a long-chain fatty compound was reported (*144*). Addition of 1-butanol to the surfactant system was optional. In another patent (*145*), a microemulsion consisted of a vegetable oil, a C_1-C_3 alcohol, water, and 1-butanol as nonionic surfactant. These fuels had acceptable viscosity and compared favorably to DF2 in terms of engine performance. Another fuel composition consisted of a vegetable oil, methanol or ethanol, a straight-chain isomer of octanol, and optionally water (*146*), which again had properties such as high water tolerance, acceptable viscosity and performance properties comparable to DF2. Another patent (*147*) reported the formation of microemulsions from vegetable oil (preferably degummed; mainly rapeseed oil), water, and a surfactant such as an alkaline soap or a potassium salt of fatty acids. Another microemulsion composition was fatty esters, aqueous alcohol, and small amount of alkali metal soap with subsequent separation of the aqueous layer from the microemulsion (*148*).

Engine tests were performed on several microemulsions. A non-ionic microemulsion comprising of alkali-refined, winterized sunflower oil (53.3 vol-%), 95% aqueous ethanol (13.3 vol-%) and 1-butanol (33.4 vol-%) encountered incomplete combustion at low-load engine operation as major problem (*149*). Lubricating oil dilution was observed, followed by an abnormal increase in viscosity. Heavier carbon residues on the piston lands, in the piston ring grooves and in the intake ports were noted. Furthermore, premature injection-nozzle deterioration (needle sticking) was experienced. The tested microemulsion was not recommended for long-term use in a

DI engine, but further modifications in formulation might produce acceptable microemulsions.

Two other hybrid fuels were tested. One was non-ionic consisting of 53.3 vol-% soybean oil, 13.3 vol-% 95% aqueous ethanol and 33.4 vol-% 1-butanol (*150*), and the other was ionic composed of 52.3 vol-% soybean oil, 17.4 vol-% 95% aqueous ethanol, 20.5 vol-% 1-butanol, 6.54 vol-% linoleic acid, and 3.27 vol-% triethylamine. Generally, these fuels performed nearly as well as DF2 despite their lower CNs and less energy content, producing nearly as much engine power (non-ionic emulsion). The increased viscosity of the hybrid fuels produced a 16% increase in the mass of each fuel injection at maximum power, but the injections contained 6% less energy than those of DF2. There was a 6% gain in thermal efficiency.

Another paper reports using methyl *tert.*-butyl ether (MTBE), which is normally used as octane enhancer in gasoline, to homogenize mixtures of soybean or rape oil with ethanol (*151*). No engine tests were performed.

In two papers (*152-153*), emulsions of palm oil with diesel fuel and 5-10% water were tested to determine engine performance and wear characteristics on an IDI diesel engine under steady-state conditions and 20 h endurance tests. Engine performance and fuel consumption were comparable to conventional DF. Wear metal debris accumulation in the crankcase oil was lower than with conventional DF.

Pyrolysis.

Soybean oil pyrolyzed distillate, which consisted mainly of alkanes, alkenes, and carboxylic acids had a CN of 43, exceeding that of soybean oil (37.9) and the ASTM minimum value of 40 (*154*). The viscosity of the distillate was 10.2 cSt at 38°C, which is higher than the ASTM specification for DF2 (1.9-4.1 cSt) but considerably below that of soybean oil (32.6 cSt). Short-term engine tests were carried out on this fuel (*155*).

Used cottonseed oil from the frying process was decomposed with Na_2CO_3 as catalyst at 450° to give a pyrolyzate containing mainly C_{8-20} alkanes (70%) besides alkenes and aromatics (*156*). The pyrolyzate had lower viscosity, flash point, and PP than DF and equivalent calorific values. The CN of the pyrolyzate was lower.

Rapeseed oil methyl esters were pyrolyzed at 550 to 850°C and in nitrogen dilution (*157*). The major products were linear 1-alkenes, straight-chain alkenes, and unsaturated methyl esters. CO, CO_2, and H_2 were contained in the gas fraction. The C_{10-14} alkenes and short-chain unsaturated esters were optimally produced at 700°.

Catalytic conversion of vegetable oils using a medium severity refinery hydroprocess yielded a product in the diesel boiling range with a CN of 75-100 (*158*). The main liquid product was a straight-chain alkane. Other products of the process included propane, water, and CO_2.

Soybean, babassu and some less common vegetable oils were hydrocracked with a NiMo/γ-Al_2O_3 catalyst sulfided *in situ* with elemental sulfur under hydrogen pressure (*159*). Various alkanes, alkylcycloalkanes, and alkylbenzenes were observed. Oxygen in the oil feed was liberated as CO_2, H_2O, and CO. Decarboxylation was indicated by water and CO_2. C_{1-4} formation indicated acrolein decomposition. Differences between more saturated and unsaturated oils were observed. Besides NiMO/γ-Al_2O_3, an $NiSiO_2$

catalyst was studied (*160*) in the hydrocracking of vegetable oils at 10-200 bars hydrogen pressure and 623-673 K. The resulting product was a mixture of hydrocarbons, mainly alkanes, in the diesel fraction. Hydrogenolysis of palm oil over Ni/SiO_2 or over Co at 300° and 50 bar gave a nearly colorless oil, mainly C_{15-17} alkane (*161*). The same process gave soft solid with 80.4% C_{17} alkanes when applied to rapeseed oil. An octadecane model compound gave 50% conversion over Co/oil catalyst to C_{17} alkane as the main product.

Catalytic hydrocracking ($Rh-Al_2O_3$ catalyst) of soybean oil at 693 K and 40 bar hydrogen pressure gave liquid products which were distilled to gasoline and gas oil boiling-range hydrocarbons (*162*). Decarboxylation / decarbonylation was again noted.

Crude and partially hydrogenated soybean oil were decomposed by passage over solid acidic Al_2O_3 or basic MgO (*163*). The degree of unsaturation of the oil influenced product formation. Partially hydrogenated soybean oil yielded more hydrocarbons while crude soybean oil yielded a mixture of oxygenated products and hydrocarbons of lower mean molecular weight. The products derived from MgO cracking showed more unsaturates and aromatics than those from Al_2O_3 decomposition.

Kolbe electrolysis of the potassium salts of coconut fatty acids and acetic acid reportedly gave a liquid with good DF properties (*164*) and the products resembled those from pyrolytic procedures. This product contained 83% alkanes, mainly even-numbered compounds from C_{10-24}, with C_{12-18} being the most abundant.

Storage stability.

While most aspects of biodiesel discussed above have received considerable attention, relatively few papers (*165-167*) deal with the aspect of (storage) stability of biodiesel or fatty alkyl esters. The use of biodiesel is advantageous compared to conventional diesel fuel from the aspect of handling and storage safety because of the higher flash point of both vegetable oils and their methyl esters.

Generally, the stability of fatty compounds is influenced by factors such as presence of air, heat, traces of metal, peroxides, light, or structural features of the compounds themselves, mainly the presence of double bonds. The more conjugated or methylene-interrupted double bonds in a fatty molecule, the more susceptible the material is to oxidation and degradation.

Early storage tests gave the following decreasing order of stability for different refinement grades of various vegetable oils (*165*): soybean oil >> degummed soybean oil > refined soybean oil = refined sunflower oil > degummed sunflower oil = crude sunflower oil. The stability of the crude and degummed oils was significantly improved by the addition of diesel fuel (in 1:1 mixtures) but this did not improve the stability of refined oils. The storage stability of 1:1 mixtures were in the decreasing order of crude soybean oil ≥ crude sunflower oil > degummed soybean oil > degummed sunflower oil >> refined soybean oil > refined sunflower oil. A degummed oil / diesel blend with better stability characteristics than that of a refined oil / diesel blend could be prepared. Additionally, the purity of the degummed oils was sufficiently improved by the addition of diesel fuel to meet the required fuel specification.

A study on the stability of the methyl and ethyl esters of sunflower oil reports that ester fuels (biodiesel) should be stored in airtight containers, the storage temperature should be < 30 °C, that mild steel (rust-free) containers could be used, and that *tert.*-butylhydroquinone (TBHQ), an oxidation inhibitor, has a beneficial effect on oxidation stability (*166*). Methyl esters were slightly more stable than ethyl esters. Light caused only a small increase in the oxidation parameters of esters stored at the high temperature level. The changes in the samples were reflected by increasing acid and peroxide values in storage at 50 °C and increases in ultraviolet (UV) absorption.

Two parameters, namely temperature and the nature of the storage container, were claimed to have the greatest influence on the storage stability (*167*). Samples stored in the presence of iron behaved differently than those stored in glass. Higher temperature favored degradation of the hydroperoxide at a faster rate than when it was stored at room temperature. Secondary oxidation products were formed in greater amounts in the presence of iron (from the primary peroxides) while in glass the concentration of primary oxidation products is higher. Acidity values were also monitored in this work. Even for samples stored at 40 °C, the increase in free acids was within the limits of technical specifications. The free acids need to be controlled because they are mainly responsible for corrosion.

Other Sources of Biodiesel.

Animal fats. The most prominent animal fat to be studied for potential biodiesel use is tallow. Tallow contains a high amount of saturated fatty acids (Table II), and it has therefore a melting point above ambient temperature. Blends of tallow esters (methyl, ethyl, and butyl) with conventional DF were studied for this reason (*168*). Smoke emissions were reduced with the esters, particularly the butyl ester. Other features such as torque, power, and thermal efficiency did not deviate from conventional DF by more than 3% in any case. Specific fuel consumption was higher for the neat esters but only 1.8% higher for a 50:50 blend of butyl tallowate with conventional DF. A study on beef tallow and an inedible yellow grease both neat and a 1:1 (weight ratio) blend of tallow with DF in short-term engine tests with DI and IDI engines was carried out (*169*). The deposits were softer than those formed with reference cottonseed oil but still excessive. In a 200 h EMA test the deposits caused ring sticking and cylinder wear. Thus animal fats, like vegetable oils, were not suitable for long-term use unless modified.

Other researchers blended methyl tallowate with 35 vol-% ethanol to achieve the viscosity of petrodiesel and the fuel properties were closely related to that of No. 2 diesel fuel (*170*). In an investigation of blends of DF2 with methyl tallowate and ethanol (*171*), an 80:13:7 blend of DF2:methyl tallowate:ethanol reduced emissions the most without a significant drop in engine power output. The same authors determined numerous physical properties of blends of DF with methyl tallowate, methyl soyate and ethanol and found them to be similar to the pertinent properties of DF2.

Waste vegetable oils. Vegetable oils have many other applications, notably as food ingredients and cooking oils. Especially the latter use produces significant amounts of waste vegetable oils. These vegetable oils contain some degradation products of vegetable oils and foreign material. However, analyses of used vegetable oils claimed

(*172*) that the differences between used and unused fats are not very great and in most cases simple heating and removal by filtration of solid particles suffices for subsequent transesterification. The cetane number of a used frying oil methyl ester was given as 49 (*173*), thus comparing well with other materials, but little demand could be covered by this source. Biodiesel in form of esters from waste cooking oils was tested and it was reported that emissions were favorable (*174*). Used canola oil (only purified by filtration) was blended with DF2 (*175*). Fuel property tests, engine performance tests and exhaust emission values gave promising results. Filtered frying oil was transesterified under both acidic and basic conditions with different alcohols (methanol, ethanol, 1-propanol, 2-propanol, 1-butanol, and 2-ethoxyethanol) (*175*). The formation of methyl esters with base catalysis (KOH) gave the best yields. The methyl, ethyl, and 1-butyl esters obtained here performed well in short-term engine tests on a laboratory high-speed diesel engine.

Literature Cited

1. Nitske, W.R.; Wilson, C.M. *Rudolf Diesel: Pioneer of the Age of Power*; University of Oklahoma Press, 1965, p.139.
2. Shay, E.G. *Biomass Bioenergy* **1993**, *4*, 227-242.
3. Anonymous, *Industrial Uses of Agricultural Materials*, published by the Economic Research Service, USDA, September 1995, pp. 15-20.
4. Shearer, B. *Chemical Marketing Reporter*, March 25,1996, p. 10.
5. Ahouissoussi, N.B.C.; Wetzstein, M.E. *Industrial Uses of Agricultural Materials*, published by the Economic Research Service, USDA, September 1995, pp. 35-41.
6. Alternative Fuels Committee of the Engine Manufacturers Association *Biodiesel Fuels and Their Use in Diesel Engine Applications*, August 1995.
7. Lane, J.C. In *Kirk-Othmer, Encyclopedia of Chemical Technology*; Third Ed.; John Wiley & Sons: New York, NY, 1980, Vol. 11; pp. 682-689.
8. *Diesel Engine Reference Book*; Lilly L.R.C., Ed.; Butterworths: London, 1984, pp. 22/9-22/12.
9. *Biodiesel Report*; National Biodiesel Board: Jefferson City, MO, March 1996.
10. Ervin, C.A. *Federal Register*, **1995** (July 31), *60*, 38976-38977.
11. Goering, C.E.; Schwab, A.W.; Daugherty, M.J.; Pryde, E.H.; Heakin, A.J. *Trans. ASAE* **1982**, *25*, 1472-1477 & 1483.
12. Bagby, M.O.; Freedman, B.; Schwab, A.W. *ASAE Paper No.* 87-1583, 1987 Int. Winter Meeting ASAE.
13. Batel, W.; Graef, M.; Mejer, G.-J.; Möller, R.; Schoedder, F. *Grundl. Landtechnik* **1980**, *30*, 40-51.
14. Freedman, B; Bagby, M.O. *J. Am. Oil Chem. Soc.* **1989**, *66*, 1601-1605.
15. Freedman, B; Bagby, M.O., Khoury, H. *J. Am. Oil Chem. Soc.* **1989**, *66*, 595-596
16. Harrington, K.J. *Biomass* **1986**, *9*, 1-17.
17. Bagby, M.O. In *Proc. 9th Int. Conf. Jojoba Uses, 3rd Int. Conf. New Industr. Crops Prod.*; Princen, L.H., Rossi, C., Eds.; Assoc. Advancem. Industr. Crops. publ. 1996; pp. 220-224.
18. Knothe, G.; Bagby, M.O.; Ryan, T.W.. III; Callahan, T.J.; Wheeler, H.G. *Soc. Automot. Eng., [Spec. Publ.] SP* **1992**, SP-900, 37-63.

19. Ryan, T.W., III; Dodge, L.G.; Callahan, T.J. *J. Am. Oil Chem. Soc.* **1984**, *61*, 1610-1619.
20. Clark, S.J.; Wagner, L.; Schrock, M.D.; Piennaar, P.G. *J. Am. Oil Chem. Soc.* **1984**, *61*, 1632-1638.
21. Klopfenstein, W.E. *J. Am. Oil Chem. Soc.* **1985**, *62*, 1029-1031.
22. Freedman, B.; Bagby, M.O. *J. Am. Oil Chem. Soc.* **1990**, *67*, 565-571.
23. Douros, T.; Schindlbauer, H. *Erdoel, Erdgas, Kohle* **1987**, *103*, 228-230.
24. Freedman, B.; Bagby, M.O.; Callahan, T.J.; Ryan, T.W., III *Soc. Automot. Eng. Techn. Pap. Ser.* **1990**, Paper No. 900343.
25. Pande, S.G.; Hardy, D.R. *Fuel* **1990**, *69*, 437-442.
26. Ladommatos, N.; Goacher, J. *Fuel* **1995**, *74*, 1083-1093.
27. Krisnangkura, K. *J. Am. Oil Chem. Soc.* **1986**, *63*, 552-553.
28. *Handbook of Chemistry and Physics*; Weast, R.C.; Astle, M.J.; Beyer, W.H., Eds.; 66th Ed.; CRC Press: Boca Raton, FL, 1985-1986; pp. D-272 - D-278.
29. Bruwer, J.J.; van D. Boshoff, B.; Hugo, F.J.C.; du Plessis, L.M.; Fuls, J.; Hawkins, C.; van der Walt, A.N.; Engelbrecht, A., paper presented at the 1980 Symposium, South African Institute of Agricultural Engineers.
30. Bruwer, J.J.; van D. Boshoff, B.; Hugo, F.J.C.; Fuls, J.; Hawkins, C.; van der Walt, A.N.; Engelbrecht, A.; du Plessis, L.M., paper presented at the 1980 National Energy Symposium, American Society of Agricultural Engineers.
31. Bruwer, J.J.; v. d. Boshoff, B.; Hugo, F.J.C.; Fuls, J.; Hawkins, C.; v. d. Walt, A.N.; Engelbrecht, A.; du Plessis, L.M. *Comm. Eur. Communities, [Rep.] EUR* **1981**, EUR 7091, Energy Biomass Conf., 1st, 934-940.
32. Schwab, A.W.; Bagby, M.O.; Freedman, B. *Fuel* **1987**, *66*, 1372-1378.
33. Knothe, G.; Dunn, R.O.; Bagby, M.O. *INFORM* **1996**, *7*, 827-829.
34. Mittelbach, M. *Liq. Fuels, Lubr. Addit. Biomass, Proc. Altern. Energy Conf.*; Dale, B.E., Ed.; American Society of Agricultural Engineers, St. Joseph, MI, 1994.
35. Andrews, A.S.; Quick, G.R. *Energy Agric.* **1984**, *3*, 323-332.
36. Eskin, N.A.M.; McDonald, B.E.; Przybylski, R.; Malcolmson, L.J.; Scarth, R.; Mag, T.; Ward, K.; Adolph, D. In *Bailey's Industrial Oil and Fat Products*; Fifth Ed.; Hui, Y.H., Ed.; John Wiley & Sons, 1996; pp. 1-95 **(p. 6)**.
37. *Diesel Technology*; National Academy Press: Washington, D.C., 1982; pp. 61-157.
38. *Diesel Technology*; National Academy Press; Washington, D.C., 1982; pp. 314-344 (Appendix B).
39. Krahl, J.; Vellguth, G.; Graef, M.; Munack, A.; Stalder, K.; Bahadir, M. *Biomass Energy Environ., Agric. Ind., Proc. Eur. Biomass Conf., 8th*; Chartier, P., Ed.; Elsevier: Oxford, 1995 (Meeting Date 1994); Vol. 2; pp. 1186-1191.
40. *Diesel Engine Reference Book*; Lilly, L.R.C., Ed.; Butterworths: London, 1984; pp. 18/9-18/10.
41. Mills, G.A; Howard, A.G. *J. Inst. Energy* **1983**, *56*, 131-137.
42. Geyer, S.M.; Jacobus, M.J.; Lestz, S.S. *Trans. ASAE* **1984**, *27*, 375-381.
43. Mittelbach, M.; Tritthart, P.; Junek, H. *Energy Agric.* **1985**, *4*, 207-215.
44. Wenzel, G.; Schulze-Lammers, P. *Fat. Sci. Technol.* **1995**, *97*, 475-481.
45. Cruz, J.M.; Ogunlowo, A.S.; Chancellor, W.J.; Goss, J.R. *Resour. Conserv.* **1981**, *6*, 69-74.

46. Bacon, D.M.; Brear, F.; Moncrieff, I.D.; Walker, K.L. In *Beyond the Energy Crisis*; Fazzolare, R.A.; Smith, C.B., Eds.; Pergamon Press: Oxford, 1981, Vol. III; p. 1525-1533.
47. Bettis, B.L.; Peterson, C.L.; Auld, D.L.; Driscoll, D.J.; Peterson, E.D. *Agron. J.* **1982**, *74*, 335-339.
48. Peterson, C.L.; Wagner, G.L.; Auld, D.L. *Trans. ASAE* **1983**, *26*, 322-327, 332.
49. Pryor, R.W.; Hanna, M.A.; Schinstock, J.L.; Bashford, L.L. *Trans. ASAE* **1983**, *26*, 333-337.
50. Korus, R.A.; Mousetis, T.L.; Lloyd, L. *ASAE Publ.* **1982**, (4-82, Veg. Oil Fuels), 218-223.
51. Darcey, C.L.; LePori, W.A.; Yarbrough, C.M.; Engler, C.R. *Trans. ASAE* **1983**, *26*, 1626-1632.
52. Vellguth, G. *Grundl. Landtechnik* **1982**, *32*, 177-186.
53. Clevenger, M.D.; Bagby, M.O.; Goering, C.E.; Schwab, A.W.; Savage, L.D. *Trans. ASAE* **1988**, *31*, 381-388.
54. Rewolinski, C.; Shaffer, D.L. *J. Am. Oil Cem. Soc.* **1985**, *62*, 1120-1124.
55. Peterson, C.L.; Korus, R.A; Mora, P.G.; Madsen, J.P. *Trans. ASAE*, **1987**, *30*, 28-35.
56. Krahl, J.; Vellguth, G.; Munack, A.; Stalder, K.; Bahadir, M. *Motortechn. Z.* **1996**, *57*, 24-31.
57. Stalder, K.; Gerhard, V.; Krahl, J.; Munack, A. *Landtechnik* **1995**, *50*, 266-267.
58. Ali, Y.; Hanna, M.A.; Leviticus, L.I. *Bioresour. Technol.* **1995**, *52*, 185-195.
59. Last, R.J.; Krüger, M.; Dürnholz, M. *Soc. Automot. Eng. [Spec. Publ.] SP* **1995**, SP-1093, 9-21.
60. J.F. McDonald, Purcell, D.L.; McClure, B.T.; Kittelson, D.B. *Soc. Automot. Eng. [Spec. Publ.] SP* **1995**, SP-1093, 191-207.
61. Zubik, J.; Sorenson, S.C.; Goering, C.E. *Trans. ASAE*, **1984**, *27*, 1251-1256.
62. Purcell, D.L.; McClure, B.T.; McDonald, J.; Basu, H.N. *J. Am. Oil Chem. Soc.* **1996**, *73*, 381-388.
63. Scholl, K.W.; Sorenson, S.C. In *Alternate Fuels: A Decade of Success and Promise*; Bata, R.M., Ed.; Society of Automotive Engineers, Warrendale, PA, 1994; pp. 555-567.
64. Knothe, G.; Bagby, M.O.; Ryan, T.W., III; Callahan, T.J. J. Am. Oil Chem. Soc. **1991**, *68*, 259-267 N 6564
65. Knothe, G.; Bagby, M.O.; Ryan, T.W., III; Wheeler, H.G.; Callahan, T.J. *J. Am. Oil Chem. Soc.* **1992**, *69*, 341-345.
66. Clothier, P.Q.E.; Aguda, B.D.; Moise, A.; Pritchard, H.O. *Chem. Soc. Rev.* **1993**, *22*, 101-108.
67. Vaitilingom, G.; Higelin, P.; Andrzejewski, J.; Sapinski, A. *Entropie* **1991**, *27*, 39-43.
68. Poirier, M.A.; Steere, D.E.; Krogh, J.A.; U.S. Patent 5,454,842; October 3, 1995.
69. Msipa, C.K.M.; Goering, C.E.; Karcher, T.D. *Trans. ASAE* **1983**, *26*, 1669-1672.
70. Ziejewski, M.; Goettler, H.; Pratt, G.L. *SAE Techn. Pap. Ser.* **1986**, Paper No. 860301.

71. Schmidt, A.; Staetter, W.; Marhold, A.; Zeiner, W.; Joos, G. *Erdoel, Erdgas, Kohle* **1992**, *108*, 415-418.
72. Isiğigür, A.; Karaomanoğlu; Aksoy, H.A.; Hamdallahpur, F.; Gülder, Ö. L. *Appl. Biochem. Biotechnol.* **1993**, *39-40*, 89-105.
73. Goering, C.E. *Combustion of Agriculturally Derived, Nonpetroleum Fuel in Diesel Engines*, Final Report, CR Grant No. 59-2171-0-2-090-0, U.S. Dept. Agric.
74. Freedman, B.; Pryde, E.H.; Mounts, T.L. *J. Am. Oil Chem. Soc.* **1984**, *61*, 1638-1643.
75. Freedman, B.; Butterfield, R.O., Pryde, E.H. *J. Am. Oil Chem. Soc.* **1986**, *63*, 1375-1380.
76. Mulangala, M.; Tshiamala, K. *Entropie* **1981**, *98*, 97-98.
77. Freedman, B.; Pryde, E.H. *ASAE Publ.* **1982**, (4-82, Veg. Oil Fuels), 117-122.
78. Mittelbach, M.; Wörgetter, M.; Pernkopf, J.; Junek, H. *Energy Agric.* **1983**, *2*, 369-384.
79. Ahn, E.; Koncar, M.; Mittelbach, M.; Marr, R. *Sep. Sci. Technol.* **1995**, *30*, 2021-2033.
80. Işigigür, A.; Karaosmanoğlu, F.; Aksoy, H.A. *Appl. Biochem. Biotechnol.* **1994**, *45-46*, 103-122.
81. Stage, H. *Fett Wiss. Technol.* **1988**, *90*, 28-32.
82. Enchelmaier, H.; Rasehorn, H.J.; German Patent 4,238,195; May 19, 1994.
83. Germani, M.M.G.; German Patent 4324875; February 3, 1994.
84 Gaskoks-Vertriebs GmbH; Austrian Patent 387,399; January 10, 1989.
85. Junek, H.; Mittelbach, M.; Andreae, F.; German Patent 3,727,981, March 3, 1988.
86. Hofmann, P.; German Patent 3,512,497; October 9, 1986.
87. Hock, O.S.; May, C.Y.; Yoo, C.K.; Australian Patent 626,014, July 23, 1992.
88. Kaufmann, H.P.; Grandel, F.; Grothues, B. *Fett Seifen Anstrichm.* **1958**, *60*, 919-930
89. Graille, J.; Lozano, P.; Pioch, D.; Geneste, P.; *Oleagineux* **1986**, *41*, 457-464.
90. Harrington, K.J.; D'Arcy-Evans, C. *J. Am. Oil Chem. Soc.* **1985**, *62*, 1009-1013.
91. Graille, J.; Lozano, P.; Pioch, D.; Geneste, P.; Guida, A. *Oleagineux* **1982**, *37*, 421-424.
92. Graille, J.; Lozano, P.; Pioch, D.; Geneste, P. *Oleagineux* **1985**, *40*, 271-276.
93. Peterson, G.R.; Scarrah, W.P. *J. Am. Oil Chem. Soc.* **1984**, *61*, 1593-1597.
94. Nelson, L.; Foglia, T.A.; Marmer, W.N. *INFORM* **1995**, *6*, 509-510.
95. Nelson, L.A.; Foglia, T.A.; Dunn, R.O.; Marmer, W.N. submitted for publication.
96. Freedman, B.; Pryde, E.H.; Kwolek, W.F. *J. Am. Oil Chem. Soc.* **1984**, *61*, 1215-1220.
97. Freedman, B.; Kwolek, W.F.; Pryde, E.H. *J. Am. Oil Chem. Soc.* **1986**, *63*, 1370-1375.
98. Mittelbach, M. *Chromatographia* **1993**, *37*, 623-626.
99. Plank, C.; Lorbeer, E. *Fett Wiss. Technol.* **1994**, *96*, 379-386.
100. Plank, C.; Lorbeer, E. *J. Chromatogr., A* **1994**, *683*, 95-104.
101. Plank, C.; Lorbeer E. *J. High Resolut. Chromatogr.* **1993**, *16*, 483-487.
102. Plank, C.; Lorbeer E. *J. High Resolut. Chromatogr.* **1992**, *15*, 609-612.
103. Bondioli, P.; Mariani, C.; Lanzani, A.; Fedeli, E. *Riv. Ital. Sostanze Grasse* **1992**, *69*, 7-9.

104. Plank, C.; Lorbeer, E. *J. Chromatogr., A* **1995**, *697*, 461-468.
105. Mittelbach, M.; Roth, G.; Bergmann, A. *Chromatographia* **1996**, *42*, 431-434.
106. Cvengroš, J.; Cvengrošová, Z. *J. Am. Oil Chem. Soc.* **1994**, *71*, 1349-1352.
107. Lozano, P.; Chirat, N.; Graille, J.; Pioch, D. *Fresenius' J. Anal. Chem.* **1996**, *354*, 319-322.
108. De Filippis, P.; Giavarini, C.; Scarsella, M.; Sorrentino, M. *J. Am. Oil Chem. Soc.* **1995**, *72*, 1399-1404.
109. Siekmann, R.W.; Pischinger, G.H. *Proc. Vegetable Oil Diesel Fuel, Seminar III*, Agric. Rev. Manuals, publ. by U.S. Dept. Agric., Agric. Res. Serv., 1983.
110. Sadeghi-Jorabchi, H.; Wood, V.M.E.; Jeffery, F.; Bruster-Davies, A.; Loh, N.; Coombs, D. *Spectrosc. Eur.* **1994**, *6*, 16, 18, 20-1.
111. Owen, K.; Coley, T. *Automotive Fuels Handbook*; Soc. Automot. Eng.: Warrendale, PA, 1990; pp. 353-403.
112. Dunn, R.O.; Bagby, M.O. *J. Am. Oil Chem. Soc.* **1995**, *72*, 895-904.
113. Lal, K.; Dishong, D.M.; Dietz, J.G.; Eur. Pat. Appl. EP 604,125; June 29, 1994.
114. Lal, K.; U.S. Patent 5,338,471; August 16, 1994.
115. Demmering, G.; Schmid, K.; Bongardt, F.; Wittich, L.; German Patent 4,040,317; June 25, 1992.
116. Böhmke, U.; Pennewiss, H.; Eur. Pat. Appl. EP 543,356; May 26,1993.
117. Davies, B.W.; Lewtas, K.; Lombardi, A.; PCT Int. Appl. WO 94 10,267; May 11, 1994.
118. Lewtas, K.; Block, D.; PCT Int. Appl. WO 93 18,115; September 16, 1993.
119. Bormann, K.; Gerstmeyer, A.; Franke, H.; Stirnal, G.; Wagner, K.D.; Flemmig, B.; Kosubeck, K.; Fuchs, W.; Voigt, R.; Welker, J.; Viehweger, U.; Wehner, K.; German (East) Patent DD 287,048; February 14, 1991.
120. Müller, M.; Pennewiss, H.; Janssen, D.; Eur. Pat. Appl. EP 406,684, January 9, 1991.
121. Dunn, R.O.; Shockley, M.W.; Bagby, M.O. *J. Am. Oil Chem. Soc.*, in press.
122. Shockley, M.W.; Dunn, R.O.; Bagby, M.O. submitted for publication.
123. Lee, I.; Johnson, L.A.; Hammond, E.G. *J. Am.Oil Chem. Soc.* **1995**, *72*, 1155-1160
124. Lee, I.; Johnson, L.A.; Hammond, E.G. *J. Am. Oil Chem. Soc.* **1996**, *73*, 631-636.
125. Stournas, S.; Lois, E.; Serdari, A. *J. Am. Oil Chem. Soc.* **1995**, *72*, 433-437.
126. Goodrum, J.W.; Law, E.S. *Trans. ASAE* **1982**, *25*, 897-900.
127. Zhang, Y.; Van Gerpen, J.H. *Soc. Automot. Eng., [Spec. Publ.] SP* **1996**, SP-1160, 1-15.
128. Goering, C,E.; Schwab, A.W.; Campion, R.M.; Pryde, E.H. *ASAE Publ.* **1982**, (4-82, Veg.Oil Fuels), 279-286.
129. Boruff, P.A.; Schwab, A.W.; Goering, C.E.; Pryde, E.H. *Trans. ASAE* **1982**, *25*, 47-53.
130. Googin, J.M.; Compere, A.L.; Griffith, W.L. *Energy Res.* **1983**, *3*, 173-186.
131. Schwab, A.W.; Nielsen, H.C.; Brooks, D.D.; Pryde, E.H. *J. Dispers. Sci. Technol.* **1983**, *4*, 1-17.
132. Schwab, A.W.; Pryde, E.H. *J. Dispers. Sci. Technol.* **1985**, *6*, 563-574.
133. Dunn, R.O.; Schwab, A.W.; Bagby, M.O. *J. Dispers. Sci. Technol.* **1992**, *13*, 77-93.

134. Dunn, R.O.; Schwab, A.W.; Bagby, M.O. *J. Dispers. Sci. Technol.* **1993**, *14*, 1-16.
135. Dunn, R.O.; Bagby, M.O. *J. Am. Oil Chem. Soc.* **1994**, *71*, 101-108.
136. Dunn, R.O.; Bagby, M.O. *J. Am. Oil Chem. Soc.* **1995**, *72*, 123-130.
137. Faletti, J.J.; Sorenson, S.C.; Goering, C.E. *Trans. ASAE* **1984**, *27*, 322-325.
138. Goering, C.E.; Fry, B. *J. Am. Oil Chem. Soc.* **1984**, *61*, 1638-1643.
139. Braun, D.E.; Stephenson, K.Q. *ASAE Publ.* **1982**, (4-82, Veg. Oil Fuels), 294-302.
140. Schwab, A.W.; Fattore, R.S.; Pryde, E.H. *J. Dispers. Sci. Technol.* **1982**, *3*, 45-60.
141. Schwab, A.W.; U.S. Patent 4,451,265; May 29, 1984.
142. Schon, S.G.; Hazbun, E.A.; U.S. Patent 5,004,479; April 2, 1991.
143. Sexton, M.D.; Smith, A.K.; Bock, J.; Robbins, M.L.; Pace, S.J.; Grimes, P.G.; Eur. Pat. Appl. EP 475,620, March 18, 1992.
144. Schwab, A.W.; Pryde, E.H.; U.S. Patent 4,451,267; May 29, 1984.
145. Schwab, A.W.; Pryde, E.H.; U.S. Patent 4,526,586; July 2, 1985.
146. Schwab, A.W.; Pryde, E.H.; U.S. Patent 4,557,734; December 10, 1985.
147. Martin, J.; Vanhemelryck, J.-L.; Eur. Pat. Appl. EP 587,551; March 16, 1994.
148. Hunter, H.F.; U..S. Patent 5,380,343; January 10, 1995.
149. Ziejewski, M.; Kaufman, K.R.; Schwab, A.W.; Pryde, E.H. *J. Am. Oil Chem. Soc.* **1984**, *61*, 1620-1626
150. Goering, C.E.; Schwab, A.W.; Campion, R.M.; Pryde, E.H. *Trans. ASAE* **1983**, *26*, 1602-1604, &1607.
151. Kato, M.; Tanaka, H.; Ueda, H.; Moriya, S.; Yaginuma, F.; Isshiki, N. *Sekiyu Gakkaishi* **1992**, *35*, 115-118.
152. Masjuki, H.; Abdulmain, M.Z.; Sii, H.S.; Chua, L.H.; Seow, K.S. *J. Energy, Heat Mass Transfer* **1994**, *16*, 295-304.
153. Sii, H.S.; Masjuki, H.; Zaki, A.M. *J. Am. Oil Chem. Soc.* **1995**, *72*, 905-909.
154. Schwab, A.W.; Dykstra, G.J.; Selke, E.; Sorenson, S.C.; Pryde, E.H. *J. Am. Oil Chem. Soc.* **1988**, *65*, 1781-1786.
155. Niehaus, R.A.; Goering, C.E.; Savage, L.D., Jr.; Sorenson, S.C., paper no. 85-1560 presented at the 1985 ASAE Winter Meeting, Chicago, IL.
156. Zaher, F.A.; Taman, A.R. *Energy Sources* **1993**, *15*, 499-504.
157. Billaud, F.; Dominguez, V.; Broutin, P.; Busson, C. *J. Am. Oil Chem. Soc.* **1995**, *72*, 1149-1154.
158. Stumborg, M.; Soveran, D.; Craig, W.; Robinson, W.; Ha, K. *Energy Biomass Wastes* **1993**, *16*, 721-738.
159. Da Rocha Filho, G.N.; Brodzki, D.; Djéga-Mariadassou, G. *Fuel* **1993**, *72*, 543-549.
160. Gusmão, J.; Brodzki, D.; Djéga-Mariadassou, G.; Frety, R. *Catal. Today*, **1989**, *5*, 533-544.
161. Cecchi, G.; Bonfand, A. *Rev. Fr. Corps Gras* **1987**, *34*, 397-401.
162. Nunes, P.P.; Brodzki, D.; Bugli, G.; Djéga-Mariadassou, G. *Rev. Inst. Fr. Pet.* **1986**, *41*, 421-431.
163. Dos Anjos, J.R.S.; Gonzalez, W.D.A.; Lam, Y.L.; Frety, R. *Appl. Catal.* **1983**, *5*, 299-308.

164. Sumera, F.C.; Sadain, S. *Philipp. J. Sci.* **1990**, *119*, 333-345.
165. Du Plessis, L.M. *CHEMSA* **1982**, *8*, 150-154.
166. Du Plessis, L.M.; de Villiers, J.B.M.; van der Walt, W.H. *J. Am. Oil Chem. Soc.* **1985**, *62*, 748-752.
167. Bondioli, P.; Gasparoli, A.; Lanzani, A.; Fedeli, E.; Veronese, S.; Sala, M. *J. Am. Oil Chem. Soc.* **1995**, *72*, 699-702.
168. Sims, R.E.H. *Trans. ASAE* **1985**, *28*, 716-721.
169. LePori, W.A.; Engler, C.R.; Johnson, L.A; Yarbrough, C.M. *In Liq. Fuels Renewable Resour., Proc.Altern. Energy Conf.*; Cundiff, J.S.. Ed.; ASAE: St. Joseph, MI, 1992; pp. 89-98.
170. Ali, Y.; Hanna, M.A. *Bioresour. Technol.* **1994**, *47*, 131-134.
171. Ali, Y.; Hanna, M.A.; Borg, J.E. *Bioresour. Technol.* **1995**, *52*, 237-243.
172. Mittelbach, M.; Pokits, B.; Silberholz, A. In *Liq. Fuels Renewable Resour., Proc. Altern. Energy Conf.*; Cundiff, J.S., Ed.; ASAE: St. Joseph, MI, 1992; pp. 74-78.
173. Reed, T.B.; Graboski, M.S.; Gaur, S. *Energy Biomass Wastes* **1991**, *15*, 907-914.
174. Karaosmanoğlu, F.; Isiğigür, A.; Hamdallahpur, F.; Gülder, Ö.L.; Aksoy, H.A. In *Renewable Energy: Technol. Environ., Proc. World Renewable Energy Congr., 2nd*; Sayigh, A.A.M., Ed.; Pergamon, Oxford, 1992, Vol.3; pp. 1455-1459.
175. Nye, M.J.; Williamson, T.W.; Deshpande, S.; Schrader, J.H.; Snively, W.H.; Yurkevich, T.P.; French, C.L. *J. Am. Oil Chem. Soc.* **1983**, *60*, 1598-1601.
176. Applewhite, T.H. In *Kirk-Othmer, Encyclopedia of Chemical Technology*; Third Ed.; John-Wiley & Sons: New York, NY, 1980, Vol. 9; pp. 795-811.
177. Ali, Y.; Hanna, M.A.; Cuppett, S.L. *J Am. Oil Chem. Soc.* **1995**, *72*, 1557-1564.
178. Isiğigür, A.; Karaosmanoğlu, F.; Aksoy, H.A.; Hamdallahpur, F.; Gülder, Ö.L. *Appl. Biochem. Biotechnol.* **1994**, *45-46*, 93-102.
179. Kaufman, K.R.; Ziejewski, M. *Trans. ASAE* **1984**, *27*, 1626-1633.
180. Avella, F.; Galtieri, A.; Fiumara, A. *Riv. Combust.* **1992**, *46*, 181-188.
181. Gunstone, F.D.; Harwood, J.L.; Padley, F.B. *Lipid Handbook*; Second Ed.; Chapman & Hall: London, 1994.

Chapter 11

A New Perspective on Hydrogen Production by Photosynthetic Water Splitting

James Weifu Lee and Elias Greenbaum[1]

Chemical Technology Division, Oak Ridge National Laboratory, Oak Ridge, TN 37831–6194

Present energy systems are heavily dependent on fossil fuels. This will eventually lead to the foreseeable depletion of fossil energy resources and, according to some reports, global climate changes due to the emission of carbon dioxide. In principle, hydrogen production by biophotolysis of water can be an ideal solar energy conversion system for sustainable development of human activities in harmony with the global environment. In photosynthetic hydrogen production research, there are currently three main efforts: (1) direct photoevolution of hydrogen and oxygen by photosynthetic water splitting using the ferredoxin/hydrogenase pathway, (2) dark hydrogen production by fermentation of organic reserves such as starch that are generated by photosynthesis during the light period, and (3) Two-stage hydrogen production in a combined fermentative and light-driven algae/bacteria system. In this chapter, the advantages and challenges of these approaches for hydrogen production are discussed in relation to a new opportunity brought by our recent discovery of a new photosynthetic water-splitting reaction [*Nature*, **373**, 438-441 (1995); *Science*, **273**, 364-367 (1996)], which, theoretically, has twice the energy efficiency of conventional water splitting via the two-light-reaction Z-scheme of photosynthesis.

[1]Corresponding author

Hydrogen is a versatile, clean, and environmentally acceptable energy carrier. It can be produced by photolysis of water, an inexpensive and inexhaustible raw material. Photolysis can be performed using either inorganic systems such as semiconductors or living organisms such as cyanobacteria or green microalgae. It is now clear that green algae are probably better for H_2 production than cyanobacteria, since the latter use the more energy-intensive enzyme, adenosine triphosphate (ATP)–requiring nitrogenase, for production of H_2. Based on a recent feasibility analysis (*1*), H_2 production by green algae can be more cost-effective than semiconductor photovoltaic electronics. The discussion in this article is focused on H_2 production by photosynthetic water splitting.

Fundamentals of Photosynthesis and H_2 Production

Photosynthesis is the fundamental biological process that converts the electromagnetic energy of sunlight into stored chemical energy that supports essentially all life on Earth. In green algae as in higher plants, photosynthesis occurs in a specialized organelle, the chloroplast. Light energy captured by the photosynthetic reaction centers is stored predominately by reduction of CO_2, using water as the source of electrons. As illustrated in Figure 1, the key components of the photosynthetic apparatus involved in light absorption and energy conversion are embedded in thylakoid membranes inside the chloroplast. They are two chlorophyll (chl)–protein complexes, Photosystem I (PSI) with a reaction center, P700, and Photosystem II (PSII) with another distinct reaction center, P680. According to the current and prevailing concept of oxygenic photosynthesis, the Z-scheme, first proposed by Hill and Bendall (*2*) and now described in many textbooks (*3–7*), PSII can split water and reduce the plastoquinone (PQ) pool, the cytochrome (Cyt) b/f complex, and plastocyanin (PC), while PSI can reduce ferredoxin (Fd)/nicotinamide adenine dinucleotide phosphate ($NADP^+$) and oxidize PC, the Cyt b/f complex, and the PQ pool. As a result, the electrons derived from water splitting are transferred to Fd/$NADP^+$, which provides the reducing power for reduction of CO_2 to carbohydrate in the stromal region of the chloroplast by a series of enzymatic reactions collectively called the Calvin cycle. Electron transport in the membrane is coupled with proton transport from the stroma into the lumen, generating a proton gradient across the thylakoid membrane. The proton gradient drives phosphorylation through the coupling factor CF_0-CF_1 to make essential ATP for the reduction of CO_2. This is the common description of oxygenic photosynthesis.

In many green algae, such as *Chlamydomonas*, there is a hydrogenase that can be induced under anaerobic conditions (*8, 9*). The hydrogenase can catalyze the reduction of protons to produce H_2 using electrons from the reduced Fd as shown in Figure 1. Since protons are also produced by water splitting at PSII, the net result of this Fd/hydrogenase pathway is simultaneous photoevolution of H_2 and O_2, using water as the substrate and light energy as the driving force.

Under anaerobic conditions and darkness, H_2 may be produced by fermentative metabolic degradation of organic reserves such as starch (Figure 1). This fermentative metabolic process has been well studied (*10, 11*). Since the substrate of the fermentative pathway is generated photosynthetically by reduction of CO_2, the net result of the

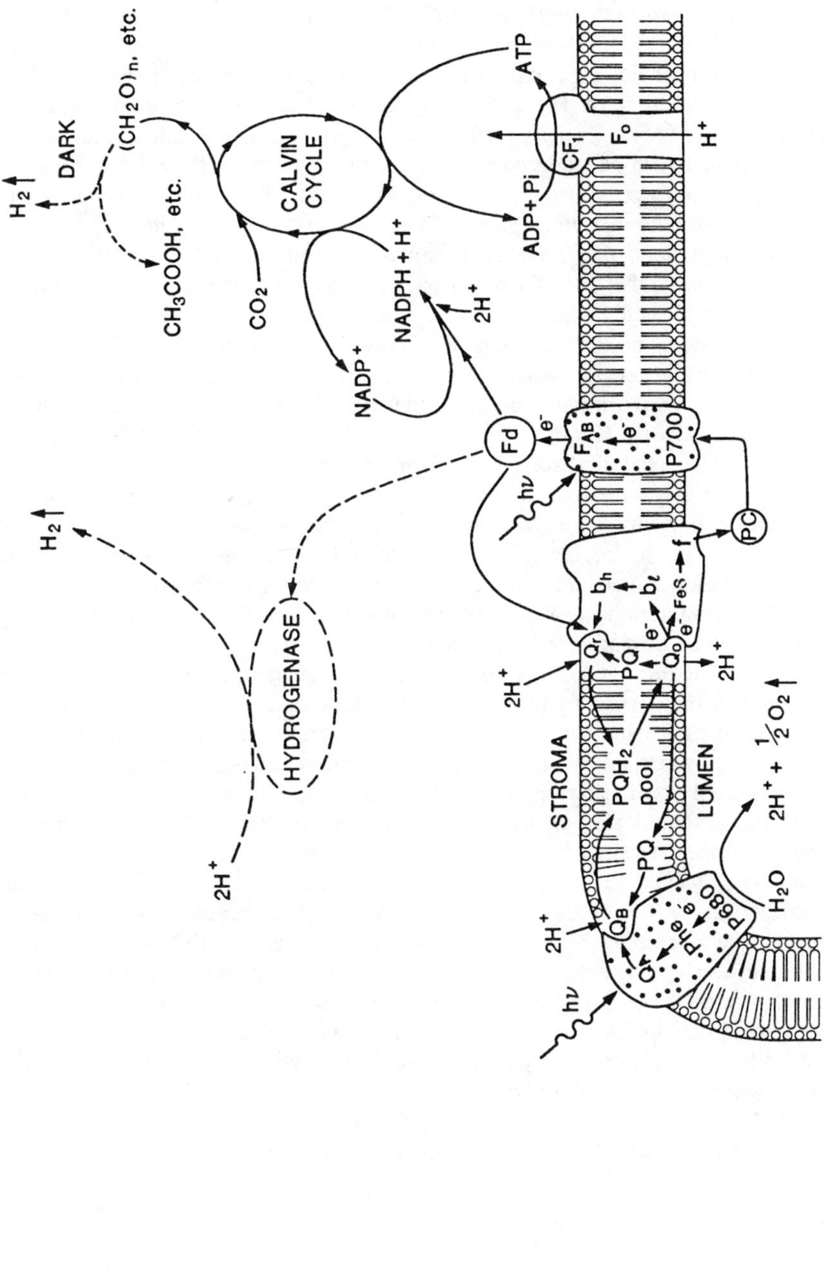

Figure 1. Conventional photosynthetic pathway based on the Z-scheme.

sequential process, photosynthesis and fermentation, is still splitting water to H_2 and O_2, with CO_2 as an intermediate.

Hydrogen Production by Direct Photosynthetic Water Splitting

The discovery of photosynthetic H_2 production is based on the classic work of Gaffron and Rubin in 1942 (*12*). However, only since 1973, at the time of the energy crisis, has photosynthetic H_2 production been investigated as a potential source of energy (*13–41*). Sustained photoevolution of H_2 and O_2 by microalgae was first demonstrated by Greenbaum in 1980 (*15*). Under anaerobic conditions, sustained photoevolution of H_2 and O_2 in microalgae can be readily demonstrated using a reactor-flow-detection system (Figure 2). Figure 3 presents a typical measurement of H_2 and O_2 production in *Chlamydomonas* in a helium atmosphere using the reactor-flow-detection system. The data clearly demonstrate that photoevolution of H_2 and O_2 can occur stably with a stoichiometric ratio of H_2 to O_2 of nearly 2:1 as expected for water splitting. Photoevolution of H_2 and O_2 can be sustained for weeks.

The advantage of the simultaneous H_2 and O_2 photoevolution is that it can potentially have a high energy conversion efficiency since electrons energized by the light reactions are used directly in the reduction of protons to produce H_2 by the Fd/hydrogenase pathway (Figure 1). However, H_2 production by this mechanism requires gas product separation since both H_2 and O_2 are produced simultaneously in the same volume. Furthermore, until an O_2-insensitive hydrogenase is developed (*42*, *43*), the O_2 concentration in the algal suspension has to be kept low to maintain H_2 production since the hydrogenase is sensitive to O_2. Therefore, an efficient and inexpensive technique to separate and remove gas products is needed.

Another important aspect is that the photoevolution of H_2 and O_2 in microalgae is often saturated at a relatively low actinic intensity. This is probably due to three factors: (1) accumulation of a back-proton gradient because of the limited permeability of the thylakoid membrane to protons and the loss of ATP utilization due to the inactivation of the Calvin cycle, (2) partial inactivation of PSII activity owing to loss of CO_2 binding at a regulatory site on the PSII reaction center in the absence of CO_2, and (3) the normal nonlinear response of the light saturation curve of photosynthesis. Therefore, the potential still exists to improve the efficiency of photosynthetic H_2 production. Further research is needed to eliminate these limiting factors. The first limitation may be solved on a short-term basis by using an appropriate proton uncoupler that dissipates the proton gradient across the thylakoid membrane (*44*), whereas the second could be overcome by eliminating the requirement of CO_2 binding through molecular engineering. The third limitation can, in principle, be overcome by reducing the antenna size of the photosynthetic reaction centers.

To avoid gas product separation and to increase efficiency, we have previously proposed a PSI and PSII reactor system that can potentially produce H_2 and O_2 in two separate compartments (*32*). This reactor system is based on the structure and function of isolated PSI and PSII reaction centers and on the catalytic activity of metallic platinum and osmium for H_2 production (*45–49*). As illustrated in Figure 4, O_2 and

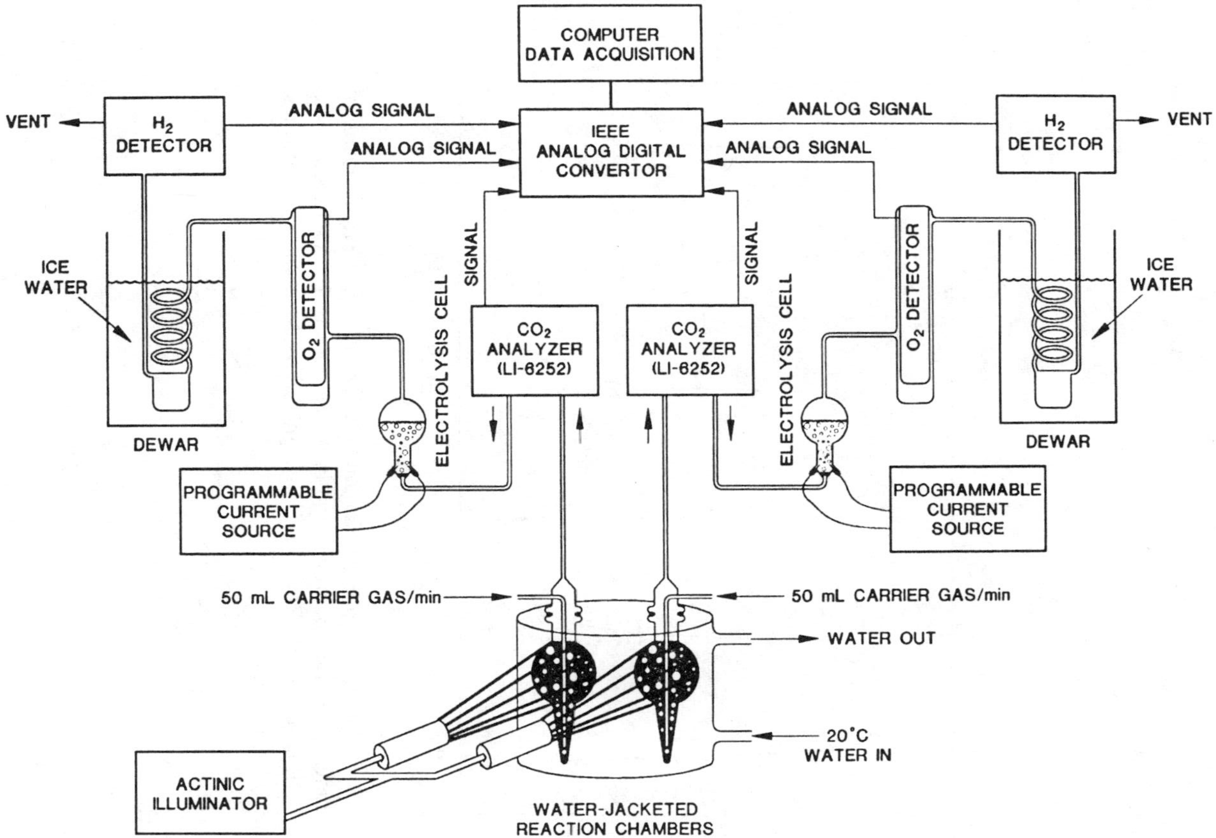

Figure 2. Schematic illustration of a dual-reactor-flow system for simultaneous detection of H_2, O_2, and CO_2. The dual-reactor system permits two independent experiments to be performed simultaneously.

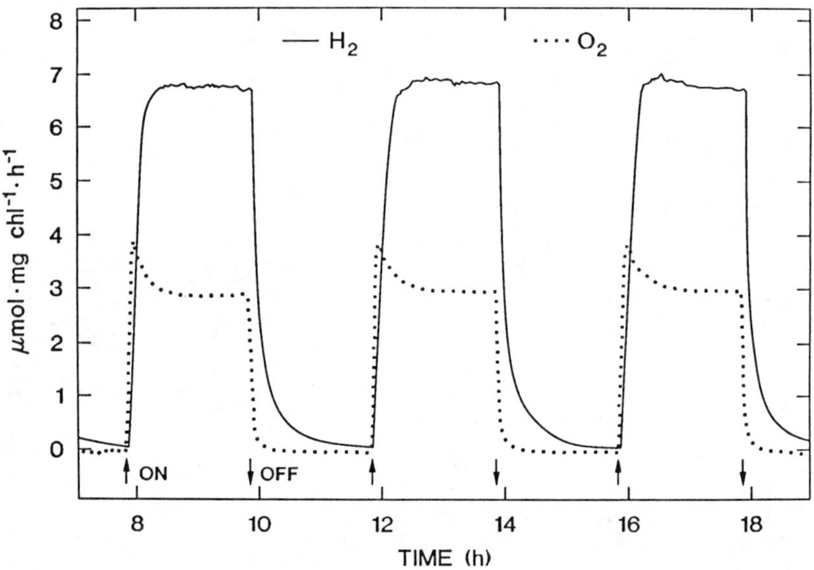

Figure 3. Sustained photoassimilation of H_2 and O_2 in *Chlamydomonas* 137c under anaerobic conditions and in the absence of CO_2. (Reproduced with permission from Ref. *35* Copyright 1996 Macmillan Magazines Limited)

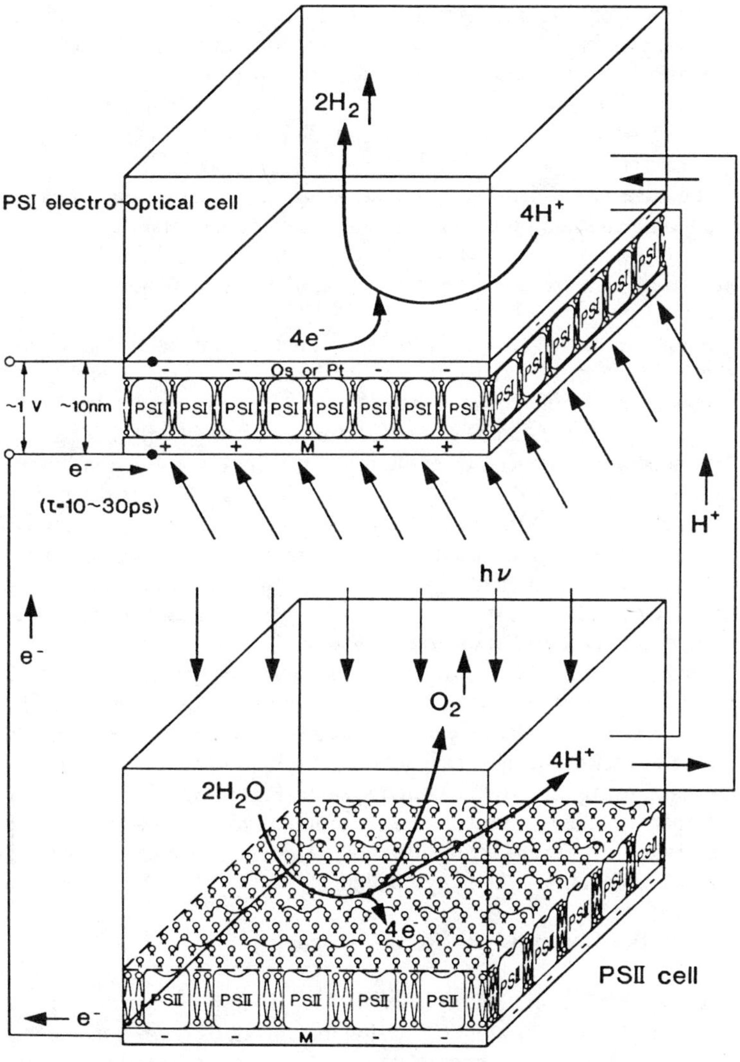

Figure 4. A photosynthetic reactor system made of PSI and PSII electro-optical cells for production of H_2 and O_2 in separate compartments. (Reproduced with permission from Ref. *32* Copyright 1996 Human Press Inc.)

protons are produced by water splitting by an array of PSII reaction centers in the PSII compartment. The electrons acquired from water splitting are wired to the reducing side of a PSI array where the electrons are energized again by PSI photochemistry. The PSI-energized electrons are then used to evolve H_2 by platinum- (or osmium-) catalyzed reduction of protons that come from the PSII compartment through a proton-conducting channel. As described previously, this system should be able to operate continuously since the number of protons and electrons generated can be balanced with the number consumed. We believe this is an important direction for future research. In this laboratory, research progress has been made in this direction (*50–52*). Recently, we have constructed a two-dimensional spatial array of PSI reaction centers on a gold surface at nanometer scale by a platinization anchoring technique (*52*).

Hydrogen Production by Dark Fermentation of Photosynthetic Product

An advantageous feature of fermentative H_2 production is its temporal separation from photosynthesis (Figure 1). That is, O_2 evolution and CO_2 photoassimilation by photosynthesis occur during the day, whereas fermentative H_2 production by degradation of photosynthetic product (starch) can occur during the night. By taking advantage of this temporal separation between O_2 evolution (day) and H_2 production (night), one can potentially develop an algal H_2 production technology that avoids the problem of gas product separation. This approach is a major project by scientists in Japan (*22*, *29*, *33*, *34*), and important progress has been made. Miura et al. (*34*) have demonstrated algal fermentative H_2 production that is temporally separated from photosynthesis (CO_2 fixation and O_2 evolution), using a combination of green algae and photosynthetic bacteria comprising over 100 L of green algae.

The challenge, as always, in this fermentative H_2 production approach is the efficiency. Figure 5 presents a typical measurement of photosynthesis and fermentative H_2 production in wild-type *Chlamydomonas* in a helium atmosphere in the presence of CO_2 under cycles of 12 h of moderate actinic illumination (PAR, 200 $\mu E \cdot m^{-2} \cdot s^{-1}$) and 12 h of darkness, using the flow-detection system (Figure 2). From the data, it can be clearly seen that the rate of fermentative H_2 production is very slow—less that 5% of the rate of oxygen evolution during the day. Therefore, enhancing the rate of dark fermentative H_2 production is the key challenge.

New Opportunities Brought by a Recent Discovery

In this laboratory, our current research has focused on improving the energy efficiency of photosynthetic hydrogen production. Using mutants of *Chlamydomonas* that lack PSI but contain PSII, we have demonstrated a new type of photosynthesis: that is, photoevolution of O_2 and H_2 and photoassimilation of CO_2 by PSII light reaction alone (*35*, *36*). This work builds on the original demonstration by Biochenko et al. (*53*) of hydrogen and oxygen transients in PSI-deficient mutants of *Chlamydomonas*. Based on studies of the electron transport pathway (Lee and Greenbaum, 1996, unpublished), the newly discovered water-splitting reaction for H_2 and O_2 production (reaction 1) or for

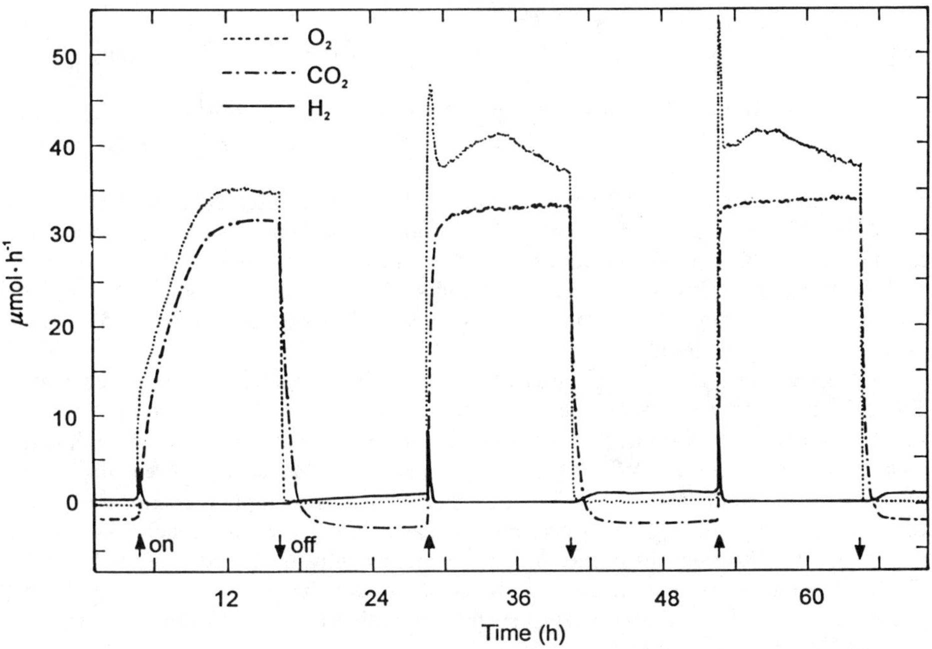

Figure 5. Photoassimilation of CO_2 and evolution of H_2 and O_2 by *Chlamydomonas* 137c in the presence of 700 ppm CO_2 in helium.

CO_2 fixation and O_2 evolution (reaction 2) may require only half the number of photons of conventional Z-scheme photosynthetic reactions 3 and 4.

Newly Discovered PSII Photosynthesis:

$$H_2O + 2\ h\nu \dashrightarrow H_2 + 1/2\ O_2 \qquad \Delta G^\circ = -\ 115\ \text{kJ/mol} \qquad (1)$$
$$\text{Energy efficiency} = 67.4\%\ (\lambda = 680\ \text{nm})$$

$$CO_2 + H_2O + 4\ h\nu \dashrightarrow 1/6(C_6H_{12}O_6) + O_2 \qquad \Delta G^\circ = -\ 224\ \text{kJ/mol} \qquad (2)$$
$$\text{Energy efficiency} = 68.2\%\ (\lambda = 680\ \text{nm})$$

Conventional Z-scheme Photosynthesis:

$$H_2O + 4\ h\nu \dashrightarrow H_2 + 1/2\ O_2 \qquad \Delta G^\circ = -\ 446\ \text{kJ/mol} \qquad (3)$$
$$\text{Energy efficiency} = 33.7\%\ (\lambda = 680\ \text{nm})$$

$$CO_2 + H_2O + 8\ h\nu \dashrightarrow 1/6(C_6H_{12}O_6) + O_2 \qquad \Delta G^\circ = -\ 927\ \text{kJ/mol} \qquad (4)$$
$$\text{Energy efficiency} = 34.1\%\ (\lambda = 680\ \text{nm})$$

Both reactions 1 and 2 have a significantly large negative value of ΔG°. They should be able to occur spontaneously. Therefore, although the discovery is surprising and novel, it still obeys the laws of thermodynamics. Since reactions 1 and 2 require half the number of photons of reactions 3 and 4, the discovery can potentially lead to H_2 production and/or CO_2 fixation technology with twice the energy conversion efficiency of conventional Z-scheme photosynthesis.

The demonstration of photosynthesis by a single light reaction proved that a single light reaction can span the potential difference between water oxidation and proton reduction for sustained evolution of H_2 and O_2, which was previously thought to be difficult to achieve (*54*). Therefore, we can now propose a new reactor system containing biometallocatalysts that requires only a single type of photochemical reaction center (Figure 6) but is able to perform the same function as the reactor system in Figure 3. As illustrated in Figure 6, when water is split to O_2 and protons by PSII, electrons from the reducing side of PSII should, neglecting resistive loss, be able to reduce protons on a platinum catalyst surface to evolve H_2 in a separate compartment without PSI.

At low light intensity and under ideal laboratory conditions, the maximum sunlight to H_2 energy conversion efficiency for Z-scheme photosynthesis has been measured to be about 10% (*55*). From a practical point of view, application of PSII photosynthesis can potentially double the sunlight conversion efficiency from 10 to 20% (*35*). This potentially higher efficiency can put photosynthetic H_2 production in a much more competitive position vis-à-vis other solar technologies. Moreover, since PSII photosynthesis can also photoassimilate CO_2, it should also be able to improve the energy efficiency of photosynthesis (CO_2 fixation) in general. Therefore, the discovery also provides a new opportunity to improve energy efficiency for production of H_2 by the photosynthesis/fermentation combined system of Miura et al. (*34*).

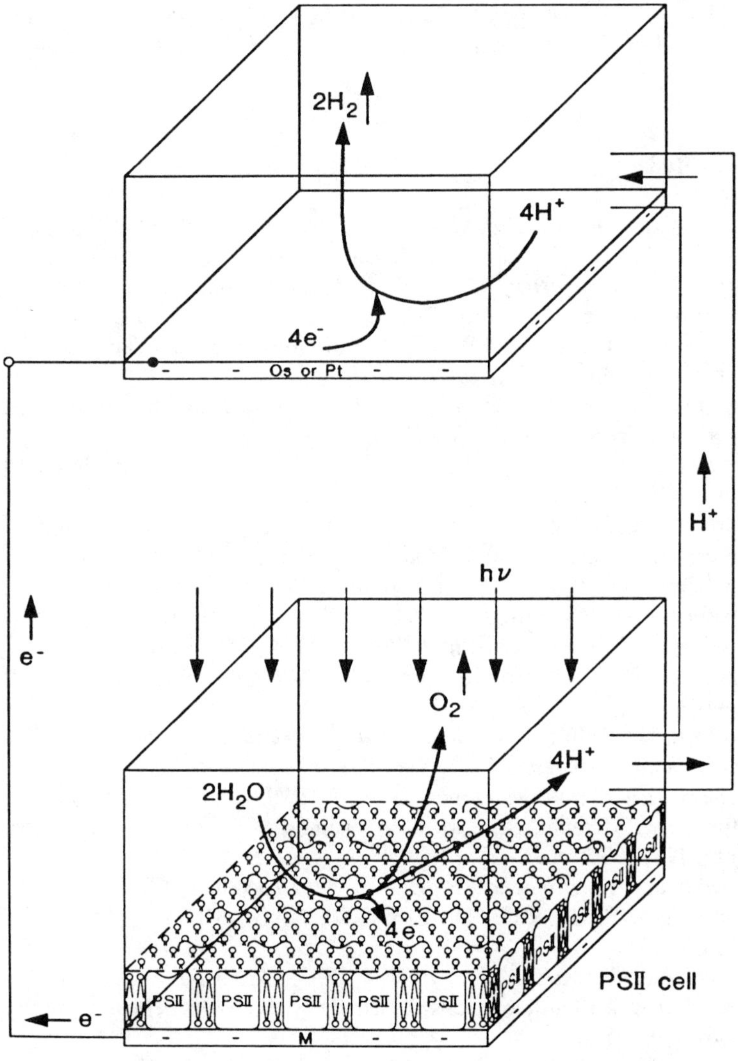

Figure 6. A proposed biometallocatalytic reactor system for production of H_2 and O_2 in separate compartments by a single light reaction (PSII).

Acknowledgments

This research was supported by the U.S. Department of Energy, the Pittsburgh Energy Technology Center, and the ORNL Laboratory Director's R&D Fund. Oak Ridge National Laboratory is managed by Lockheed Martin Energy Research Corp., for the U.S. Department of Energy under contract DE-AC05-96OR22464.

Literature Cited

1. Benemann, J. R. *Xth Proc. World Hydrogen Energy Conf.* **1995**, pp. 931–40.
2. Hill, R.; Bendall, F. *Nature* **1960**, *186*, 136–137.
3. Stryer, L. *Biochemistry*, 4th ed.; W. H. Freeman and Company: New York, NY, 1995; pp. 617–618, 658–664, 676–677.
4. Salisbury, F. B.; Ross, C. W. *Plant Physiology*, 3rd ed.; Wadsworth Publishing Company: Belmont, CA, 1985; pp. 191–194.
5. Goodwin, T. W.; Mercer, E. I. *Introduction to Plant Biochemistry*, 2nd ed.; Pergamon Press: Oxford, UK, 1986; pp. 117–124.
6. Alberts et al. *Molecular Biology of the Cell*; Garland Publishing, Inc.: New York, NY 1983; pp. 517–519.
7. Darnell, J.; Lodish, H.; Baltimore, D. *Molecular Cell Biology*; Scientific American Books, Inc.: New York, NY 1986; pp. 901–903.
8. Roessler, P.G.; Lien, S. *Plant Physiol.* **1984** *76*, 1086–1089.
9. Boichenko, V. A.; Hoffmann, P. *Photosynthetica* **1994**, *30*, 527–552.
10. Gfeller, R. P.; Gibbs, M. *Plant Physiol.* **1984**, *75*, 212–218.
11. Chen, C. Ph.D. Dissertation, Brandeis University, Waltham, MA, 1992, pp. 99–109.
12. Gaffron, H.; Rubin, J. *J. Gen. Physiol.* **1942**, *26*, 219–240.
13. Greenbaum, E. *Science* **1977**, *196*, 879–878.
14. Bishop, N. I.; Frick, M.; Jones, L. W. In *Biological Solar Energy Conversion*; Mitsui, A.; Miyachi, S.; San, P. A; Tamura, S., Eds.; Academic Press: New York, NY 1977; pp. 2–22.
15. Greenbaum, E., *Biotechol. Bioeng. Symp. 10* **1980**, 1–13.
16. Bamberger, E. S.; King, D.; Erbes, D.; Gibbs, M. *Plant Physiol.* **1982**, *69*, 1268–1273.
17. Greenbaum, E.; Ramus, J. *J. Phycol.* **1983**, *19*, 53–57.
18. Greenbaum, E.; Guillard, R. R. L; Sunda, W. G. *Photochem. Photobiol.* **1983**, *37*(6), 649–655.
19. Hoopen, H. T.; Snellink, L. J.; Gemert, J. V; Fuchs, A.; Roels, J. A. *Prog. Ind. Microbiol.* **1984**, *20*, 191–196.
20. Greenbaum, E. *Photobiochem. Photobiophys.* **1984**, *8*, 323–332.

21. Aparicio, P. J.; Azuara, M. P.; Ballesteros, A.; Fernadez, V. M. *Plant Physiol.* **1985**, *78*, 803–806.
22. Miura, Y.; Ohta, S.; Mano, M.; Miyamoto, K., *Agric. Biol. Chem.* **1986**, *50*(11), 2837–2844.
23. Mahro, B.; Grimme, L. H. *Arch. Microbiol.* **1986** *144*, 25–28.
24. Greenbaum, E. In *Algae And Human Affairs*; Lembi, C. A.; Waaland, R. J., Eds.; Cambridge University Press: Cambridge, MA, 1988, pp. 283–304.
25. Graves, D. A.; Reeves, M. E.; Greenbaum, E. *Plant Physiol.* **1988**, *87*, 603–608.
26. Graves, D. A.; Tevault, C. V.; Greenbaum E. *Photobiochem. Photobiophys.* **1989**, *50*, 571–576.
27. Greenbaum, E. *Adv. Hydrogen Technol.* **1990**, *2*, 743–754.
28. Graves, D. A.; Spradlin, G. M.; Greenbaum, E. *Photobiochem. Photobiophys.* **1990**, *52*, 585–590.
29. Miura, Y.; Saitoh, C.; Matsuoka, K. *Biosci. Biotech. Biochem.* **1992**, *56*, 751–754.
30. Cinco, R. M.; MacInnis, J. M.; Greenbaum, E. *Photosyn. Res.* **1993**, *38*, 27–33.
31. Lee, J. W.; Blankinship, S. L.; Greenbaum, E. *Appl. Biochem. Biotechnol.* **1995** *51/52*, 379–385.
32. Lee, J. W.; Greenbaum, E. *Appl. Biochem. Biotechnol.* **1995**, *51/52*, 295–305.
33. Miura, Y. *Process Biochem.* **1995**, *30*, 1–7.
34. Miura, Y.; Akano, T.; Fukatsu, K.; Miyasaka, H.; Mizoguchi, T.; Yagi, K.; Maeda, I.; Ikuta, Y.; Matsumoto, H. *Energy Convers. Mgmt* **1995**, *36*, 903–906.
35. Greenbaum, E.; Lee, J. W.; Tevault, C. V.; Blankinship, S. L.; Mets, L. J. *Nature* **1995**, *376*, 438–441.
36. Lee, J. W.; Tevault, C. V.; Owens, T. G.; Greenbaum, E. *Science* **1996**, *273*, 364–367.
37. Lien, S.; San, P. A. "An Inquiry into Biophotolysis of Water to Produce Hydrogen" NSF RANN Grant GI 40253, Indiana University, Boomington, IN.
38. Weaver, P. F.; Lien, S.; Seibert, M., *Sol. Energy* **1980**, *24*, 3–45.
39. Rao, K. K; Hall, D. O.; Vlachopoulos, N.; Grätzel, M.; Evans, M. C. W.; Seibert, M. *J. Photochem. Photobiol. B: Biol.* **1990**, *5*, 379–389.
40. Hall, D. O.; Rao, K. K. *Chimicaoggi* **1989**, March, 41–47.
41. Hall, D. O.; Rao, K. K.; Park, I. H. *Proc. 8th Intl. Conf. Photochem. Convers. Storage Sol. Energy* **1990**, Palermo, Italy.
42. McBride, A. C.; Lien, S.; Togasaki, R. K.; San P. A. In *Biological Solar Energy Conversion*; Mitsui, A.; Miyachi, S.; San, P. A.; Tamura, S., Eds.; Academic Press: New York, NY, 1977.
43. Ghirardi, M. L.; Togasaki, R. K.; Seibert, M. *Appl. Biochem. Biotechnol.* **1996**, in press.
44. Nicholls, D. G.; Ferguson, S. J. *Bioenergetics*, 2nd ed.; Academic Press Inc.: San Diego, CA, 1992, pp. 31–33.
45. Greenbaum, E. *Science* **1985**, *230*, 1373–1375.
46. Greenbaum, E. *J. Phys. Chem.* **1988**, *92*, 4571–4574.
47. Greenbaum, E. *J. Phys. Chem.* **1990**, *94*, 6151–6153.

48. Greenbaum, E. *J. Phys. Chem.* **1992**, *96*, 514–516.
49. Lee, J. W.; Tevault, C. V.; Blankinship, S. L.; Collins, R. T.; Greenbaum, E. *Energy Fuels* **1994**, *8*(3), 770–773.
50. Lee, J. W.; Lee, I.; Laible, P. D.; Owens, T. G.; Greenbaum, E. *Biophys. J.* **1995**, *69*, 1–8.
51. Lee, I.; Lee, J. W.; Warmack, R. J.; Allison, D. P.; Greenbaum, E. *Proc. Natl. Acad. Sci. USA* **1995**, *92*, 1965–1969.
52. Lee, J. W.; Lee, I.; Greenbaum, E. *Biosensors Bioelectronics* **1996**, *11*, 375–387.
53. Biochenko, V. A.; Allakhverdiev, S. I.; Ladygin, V. G.; Klimov, V. V., *Dokl Akad Nauk SSSR* **1986**, *290*, 995–998.
54. Bolton, J. R. *Science* **1978**, *202*, 705–711.
55. Greenbaum, E. *Biophys. J.* **1988**, *54*, 365–368.

CHEMICALS

Chapter 12

Lactic Acid Production and Potential Uses: A Technology and Economics Assessment

Rathin Datta and Shih-Perng Tsai

Waste Management and Bioengineering Section, Energy Systems Division, Argonne National Laboratory, Argonne, IL 60439

Lactic acid has been an intermediate-volume specialty chemical (world production ~50,000 tons/yr) used in a wide range of food processing and industrial applications. Lactic acid has the potential of becoming a very large volume, commodity-chemical intermediate produced from renewable carbohydrates for use as feedstocks for biodegradable polymers, oxygenated chemicals, environmentally friendly "green" solvents, plant growth regulators, and specialty chemical intermediates. The recent announcements of plant expansions and building of new development-scale plants for producing lactic acid and/or polymer intermediates by major U.S. companies, such as Cargill, Chronopol, A.E. Staley, and Archer Daniels Midland (ADM), attest to this potential. In the past, efficient and economical technologies for the recovery and purification of lactic acid from crude fermentation broths and the conversion of lactic acid to the chemical or polymer intermediates had been the key technology impediments and main process cost centers. The development and deployment of novel separations technologies, such as electrodialysis (ED) with bipolar membranes, extractive distillations integrated with fermentation, and chemical conversion, can enable low-cost production with continuous processes in large-scale operations. The use of bipolar ED can virtually eliminate the salt or gypsum waste produced in the current lactic acid processes. Thus, the emerging technologies can use environmentally sound processes to produce environmentally useful products from lactic acid. The process economics of some of these processes and products can also be quite attractive. In this paper, potential products and recent technical advances in lactic and polylactic acid processes are discussed. The technical accomplishments at Argonne National Laboratory (ANL) and the future directions of this program at ANL are discussed.

Lactic acid (2-hydroxypropionic acid), $CH_3CHOHCOOH$, is the most widely occurring hydroxycarboxylic acid. It was first discovered in 1780 by the Swedish chemist Scheele. Lactic acid is a naturally occurring organic acid that can be produced by fermentation or chemical synthesis. It is present in many foods, both naturally or as a product of in-situ microbial fermentation (as in sauerkraut, yogurt, buttermilk, sourdough breads, and many other fermented foods). Lactic acid is also a major metabolic intermediate in most living organisms — from anaerobic prokaryotes to human beings.

Although lactic acid has been ubiquitous in nature and has been produced by fermentation or chemical synthesis for over 50 years, it has not been a large-volume chemical. Its worldwide production volume by 1995 had grown to approximately 50×10^3 tons/yr with only a few major producers — CCA biochem b.v. of the Netherlands and its subsidiaries in Brazil and Spain, ADM in Decatur, Illinois, as the primary manufacturers. Sterling Chemicals, Inc., in Texas City, used to be a major producer but has recently announced the closure of its plant and exit from the business. Musashino in Japan has been a smaller manufacturer. CCA and ADM uses carbohydrate feedstocks and fermentation technology, while Sterling and Musashino use a chemical technology. Thus, lactic acid was considered a relatively mature fine chemical in that only its use in new applications (such as a monomer in plastics or as an intermediate in synthesis of high-volume oxygenated chemicals) would cause a significant increase in its anticipated demand (*1*).

For lactic acid to enter these applications, economical, efficient, and environmentally sound manufacturing processes are needed for its production. In the past, efficient and economical technologies for the recovery and purification of lactic acid from crude fermentation broths and conversion of lactic acid to the chemical or polymer intermediates had been the key technology impediments and main process cost centers. The development and deployment of novel separations technologies, such as electrodialysis (ED) with bipolar membranes, extractive distillations integrated with fermentation, and chemical conversion, can enable low-cost production with continuous processes in large-scale operations. The use of bipolar ED can virtually eliminate the salt or gypsum waste produced in the current lactic acid processes. Thus, the emerging technologies can use environmentally sound processes to produce environmentally useful products from lactic acid. Recent announcements of new lactic acid production plants by major chemical and agriprocessing companies may usher new technologies for the efficient, low-cost manufacture of lactic acid and its derivatives for new applications (*2-5*).

Current Uses and Manufacturing Technologies

The major use of lactic acid is in food and food-related applications, which, in the U.S., accounts for approximately 85% of the demand. The rest (~15%) of the uses are for nonfood industrial applications. As a food acidulant, lactic acid has a mild acidic taste in contrast to other food acids. Lactic acid is nonvolatile, odorless, and is classified as GRAS (generally recognized as safe) for use as a general purpose food additive by the FDA in the U.S. and other regulatory agencies elsewhere. It is a very good preservative and pickling agent for sauerkraut, olives, and pickled vegetables. It is used as acidulant/flavoring/pH buffering agent or inhibitor of bacterial spoilage in a wide variety of processed foods, such as candy, breads and bakery products, soft drinks, soups, sherbets, dairy products, beer, jams and jellies, mayonnaise, and processed eggs — often in conjunction with other acidulants (*6*). An emerging new use for lactic acid or its salts is in the disinfection and packaging of carcasses, particularly those of poultry and fish, where the addition of aqueous solutions of lactic acid and its salts during the processing increased shelf life and reduced the growth of anaerobic spoilage organisms such as *Clostridium botulinum* (*7-8*).

A large fraction (> 50%) of the lactic acid for food-related uses goes to produce emulsifying agents used in foods — particularly for bakery goods. These emulsifying agents are esters of lactate salts with longer chain fatty acids, and the four important products are calcium and sodium, stearoyl-2-lactylate, glyceryl lactostearate, and glyceryl lactopalmitate. Of the stearoyl lactylates, the calcium salt is a very good dough conditioner, and the sodium salt is both a conditioner and an emulsifier for yeast-leavened bakery products. The glycerates and palmitates are used in prepared cake mixes, other bakery products, and in liquid shortenings. In prepared cake mixes, the palmitate improves cake texture, whereas the stearate increases cake volume and permits mixing tolerances (*6*). The manufacture of these emulsifiers requires heat-stable lactic acid — hence, only the synthetic or the heat-stable fermentation grades are used for this application.

Technical-grade lactic acid has long been in use in the leather tanning industry as an acidulant for deliming hides and in vegetable tanning. In various textile finishing operations and acid dying of wool, technical-grade lactic acid was used extensively. Cheaper inorganic acids are now more commonly used in these applications. The future availability of lower cost lactic acid and the increasing environmental restrictions on waste salt disposal may reopen these markets for lactic acid.

Lactic acid is currently used in a wide variety of small-scale, specialized industrial applications where the functional specialty of the molecule is desirable. Some examples are pH adjustment of hardening baths for cellophane that is used in food packaging, terminating agent for phenol-formaldehyde resins, alkyd resin modifier, solder flux, lithographic and textile printing developers, adhesive formulations, electroplating and electropolishing baths, detergent builders (with maleic anhydride to form carboxymethoxysuccinic acid-type compounds). Because of the current high cost and low volume of production, these applications account for only 5-10% of the consumption of lactic acid (*6, 9*).

Lactic acid and ethyl lactate have long been used in pharmaceutical and cosmetic applications and formulations, particularly in topical ointments, lotions, parenteral solutions, and biodegradable polymers for medical applications (such as surgical sutures, controlled-release drugs, and prostheses). A substantial part of pharmaceutical lactic acid is used as the sodium salt for parenteral and dialysis applications. The calcium salt is widely used for calcium-deficiency therapy and as an effective anti-caries agent. As humectants in cosmetic applications, the lactates are often superior to natural products and more effective than polyols (*6, 9*). Ethyl lactate is the active ingredient in many anti-acne preparations. The use of the chirality of lactic acid for synthesis of drugs and agrichemicals is an opportunity for new applications for optically active lactic acid or its esters. The chiral synthesis routes to R (+) phenoxypropionic acid and its derivatives using S (-) lactate ester as a chiral synthon has been described (*6*). These compounds are used in herbicide production. Another use as an optically active liquid crystal whereby lactic acid is used as a chiral synthon has been recently described (*10*). These advances could open new small-volume specialty chemical opportunities for optically active lactic acid and its derivatives.

Lactic acid can be manufactured by either (1) chemical synthesis or (2) carbohydrate fermentation — both are used for commercial production. In the U.S., lactic acid is manufactured synthetically by means of the lactonitrile route by Sterling Chemicals, Inc. In Japan, Musashino Chemical Co. used this technology for all of Japan's production. CCA Biochemical b.v. of the Netherlands uses carbohydrate fermentation technology in plants in Europe and Brazil and markets worldwide. Prior to 1991, the annual U.S. consumption of lactic acid was estimated at 18,500 metric tonnes, with domestic production of approximately 8,600 tonnes, by Sterling Chemical and the rest imported from Europe and Brazil. The worldwide consumption was estimated at approximately 40,000 tonnes/yr.

(1) Chemical Synthesis. The chemical-synthesis routes produce only the racemic lactic acid. The commercial process is based on lactonitrile, which used to be a by-product from acrylonitrile synthesis. It involves base catalyzed addition of hydrogen cyanide to acetaldehyde to produce lactonitrile. This is a liquid-phase reaction and occurs at atmospheric pressures. The crude lactonitrile is then recovered and purified by distillation and is hydrolyzed to lactic acid by using either concentrated hydrochloric or sulfuric acid, producing the corresponding ammonium salt as a by-product. This crude lactic acid is esterified with methanol, producing (1) methyl lactate, which is recovered and purified by distillation and hydrolyzed by water under acid catalysts to produce lactic acid, which is further concentrated, purified, and shipped under different product classifications, and (2) methanol, which is recycled (equations 1-3).

$$CH_3\,CHO + HCN \xrightarrow{\text{catalyst}} CH_3\,CHO\,HCN \qquad (1)$$

$$CH_3\,CHO\,HCN + H_2O + \frac{1}{2}H_2\,SO_4 \rightarrow CH_3\,CHOH\,COOH + \frac{1}{2}(NH_4)_2\,SO_4 \qquad (2)$$

$$CH_3\,CHOH\,COOH + CH_3\,OH \rightarrow CH_3\,CHOH\,COOCH_3 \uparrow + H_2O$$

$$CH_3\,CHOH\,COO\,CH_3 + H_2O \rightarrow CH_3\,CHOH\,COOH + CH_3OH \uparrow \qquad (3)$$

Other possible chemical synthesis routes for lactic acid include base catalyzed degradation of sugars; oxidation of propylene glycol; reaction of acetaldehyde, carbon monoxide, and water at elevated temperatures and pressures; hydrolysis of chloropropionic acid (prepared by chlorination of propionic acid), and nitric acid oxidation of propylene, among others. None of these routes have led to technically and economically viable processes (*9, 11*).

(2) Carbohydrate Fermentation. The fermentation technology can make a desired stereoisomer of lactic acid. The existing commercial production processes use homolactic organisms, such as *Lactobacillus delbrueckii, L. bulgaricus, L. leichmanii.* A wide variety of carbohydrate sources can be used (molasses, corn syrup, whey, dextrose, cane, or beet sugar). The use of a specific carbohydrate feedstock depends on its price, availability, and purity. Proteinaceous and other complex nutrients required by the organisms are provided by corn steep liquor, yeast extract, and soy hydrolysate, for example. Excess calcium carbonate is added to the fermenters to neutralize the acid produced and produce a calcium salt of the acid in the broth. The fermentation is conducted as a batch process, requiring 4 to 6 days to complete. Lactate yields of approximately 90% (w/w) from a dextrose equivalent of carbohydrate are obtained. Keeping the calcium lactate in solution is desirable so that it can be easily separated from the cell biomass and other insolubles, and this limits the concentration of carbohydrates that can be fed in the fermentation and the concentration of lactate in the fermentation broth, which is usually around 10% (w/v). The broth containing calcium lactate is filtered to remove cells, carbon treated, evaporated, and acidified with sulfuric acid to convert the salt into lactic acid and insoluble calcium sulfate, which is removed by filtration. The filtrate is further purified by carbon columns and ion exchange and evaporated to produce technical and food-grade lactic acid, but not a heat-stable product, which is required for the stearoyl lactylates, polymers, and other value-added applications. The technical-grade lactic acid can be esterified with methanol or ethanol, and the ester is recovered by distillation, hydrolyzed by water, evaporated, and the alcohol is recycled. This separation process produces a highly pure product, which, like the synthetic product, is water white and heat stable (equations 4-7).

$$C_6\,H_{12}\,O_6 + Ca(OH)_2 \xrightarrow{\text{Fermentation}} (2 \bullet CH_3\,CHOH\,CO\bar{O})\,CA^{++} + 2H_2O \quad (4)$$

$$(2 \bullet CH_3\,CHOH\,CO\bar{O})\,Ca^{++} + H_2SO_4 \longrightarrow 2\,CH_3\,CHOH\,COOH + CaSO_4 \downarrow \quad (5)$$

$$CH_3\,CHOH\,COOH + CH_3\,OH \longrightarrow CH_3\,CHOH\,COO\,CH_3 \uparrow + H_2O \quad (6)$$

$$CH_3\,CHOH\,COOCH_3 + H_2O \longrightarrow CH_3\,CHOH\,COOH + CH_3\,OH \uparrow \quad (7)$$

Some of the major economic hurdles and process cost centers of this conventional carbohydrate fermentation process are in the complex separation steps that are needed to recover and purify the product from the crude fermentation broths. Furthermore, approximately one ton of gypsum by-product is produced and needs to be disposed of for every ton of lactic acid produced by the conventional fermentation and recovery process. These factors had made large-scale production by this conventional route economically and ecologically unattractive.

Potential Products and Markets

The future growth opportunities for lactic acid are in its use as a feedstock for potentially large-volume applications. In Table I, these applications are classified into four categories — biodegradable polymers, oxygenated chemicals, "Green" chemicals/solvents, and plant-growth regulators. The overall size of this opportunity, both in terms of mass/volume and product sales value, is substantial. For the U.S. markets, this could be approximately 6.4-8.4 x 10^9 lb/yr (2.9-4.0 x 10^6 tons/yr), with sales volume between approximately \$4.3-6.2 x 10^9/yr. The volume and selling price projections for the new products (i.e., the degradable plastics, "green" chemicals, and derivatives) are made on the basis of several published studies by Battelle and others and some internal Argonne estimates. It should be noted that the high volumes can be reached only when the prices are within the acceptable ranges (Table I) and vice versa. The list in Table I is by no means comprehensive nor would all these products (particularly the oxychemicals) be derived from lactic acid in the near future. It should be noted, however, that recently a large U.S. agriprocessing company, Cargill, has announced a potential plant of 250 x 10^6 lb/yr by 1997/1998 (*5*), substantiating that large-volume, economical manufacturing of lactic acid may be feasible with new technologies and for new or existing products.

Polymers of lactic acids are biodegradable thermoplastics. A fairly wide range of properties are obtainable by copolymerization with other functional monomers, such as glycolide, caprolactone, and polyether polyols. The polymers are transparent, which is important for packaging applications. They offer a good shelf life because they degrade slowly by hydrolysis, which can be controlled by adjusting the composition and molecular weight. These have potential uses in a wide variety of consumer products, such as paper coatings, films, moulded articles, foamed articles, and fibers. Some of the published information on some of the properties of lactic copolymers that approach those of large-volume, petroleum-derived polymers (such as polystyrene, flexible polyvinyl chloride [PVC], and vinylidene chloride) are summarized in the article of Lipinsky and Sinclair (*1*). There are numerous patents and articles on lactic acid polymers and copolymers, their properties, potential uses, and processes that date back to the early work by Carothers at DuPont. A discussion of this work is beyond the scope of this article. Several reference articles and patents (*12-15*) can provide the reader with a basis for further information.

Table I. Lactic Acid — Potential Products, Volumes, and Value

Product	*Uses*	*U.S. Market*[a] *Volume* *(10^9 lb/yr)*	*Selling*[a] *Price* *($/lb)*	*Value* *($\$10^6$/yr)*
Degradable Plastics	Packaging, films	0.3-2.0[b]	0.5-0.8[b]	150-1,600
Oxychemicals:				
Propylene Glycol	Polymers, food deicers, humectants	1.0	0.60	600
Acrylates	Polymers, plastics films, coatings	1.4	0.85	1,200
Propylene Oxide	Polymers, plastics	3.5	0.64	2,200
"Green" Chemicals/Solvents:				
Esters	Non-toxic non-volatile solvents/cleaners	0.1-0.4[c]	1.0	100-400
Ester/ Derivatives	Specialty chemicals	0.05-0.1[c]	1.5	75-150
		6.4-8.4		4,325-6,150

[a] Market volumes and prices are for 1995 (C&EN and CMR) unless otherwise stated.

[b] Estimates from Battelle, SRI, Cargill (1993) announcement.

[c] Argonne's estimates.

Among the other new product opportunities, the use of lactate esters as "green" solvents is substantial because they are high-boiling (nonvolatile), nontoxic, and degradable compounds. With increasing consumer and political consciousness with environmentally sound products, the use of these solvents as replacements for other solvents or cleaners could be a very important expansion opportunity for lactic acid. The estimates of the volume of these (Table I) are based on typical volumes and lower prices than several intermediate volume non-volatile solvents, such as n-methyl pyrrolidone, di-basic esters and such. Low-molecular-weight polymers of L-lactic acid (degree of polymerization [dp] 2-10) have been recently discovered to stimulate plant growth in a variety of crops and fruits when applied at a low level (*16-17*). These findings may lead to specialized products and formulations that would incorporate L-polylactic acid as or into controlled release or degradable mulch films for large-scale agricultural applications.

Recent Technological Advances

Technological advances in the major process components — fermentation, primary purification, and secondary purification and polymerization/chemical conversion of lactic acid and its derivatives — have recently occurred. These and other advances would enable low-cost, large-volume, and environmentally sound production of lactic acid and its derivative products.

In fermentation, high (> 90%) yield from carbohydrate, such as starch, is feasible, together with high product concentration (90 g/L, 1 M). Stable strains with good productivity (> 2 g/L h) that utilize low-cost nutrients (such as corn steep liquor) are available. Furthermore, the fermentation is anaerobic and thus has low power and cooling needs. All of these make the fermentation step very facile and inexpensive.

Recent advances in membrane-based separation and purification technologies, particularly in micro- and ultrafiltration and electrodialysis, have led to the inception of new processes for lactic acid production. These processes would, when developed and commercialized, lead to low-cost production of lactic acid, with a reduction of nutrient needs and without creating the problem of by-product gypsum (*18-21*). Desalting electrodialysis has been shown to need low amounts of energy to recover, purify, and concentrate lactate salts from crude fermentation broths (*19*). The recent advances in water-splitting electrodialysis membranes enable the efficient production of protons and hydroxyl ions from water and can thus produce acid and base from a salt solution (*20-21*). These advances have led to the development of proprietary technologies for lactic acid production from carbohydrates without producing salt or gypsum by-products (*19-20*). In recently issued patents to Datta and Glassner (*19-20*), an efficient and potentially economical process for lactic acid production and purification is described. By using an osmotolerant strain of lactic acid bacteria and a configuration of desalting electrodialysis, water-splitting electrodialysis, and ion-exchange purification steps, a concentrated lactic acid product containing less than 0.1% of proteinaceous impurities could be produced from a carbohydrate fermentation. The electric power requirement for the electrodialysis steps was approximately 0.5 kWh/lb (~1 kWh/kg) lactic acid. The process produces no by-product gypsum, only a small amount of by-product salt from the ion-exchange regeneration. Such a process can be operated in a continuous manner, can be scaled up for large-volume production, and forms the basis for commercial developments for several companies that have announced intention to be commercial producers of lactic acid and its derivative products (*3-4*).

Another recent entrant, Ecochem, a DuPont-ConAgra partnership, had developed a recovery and purification process that produces a by-product ammonium salt instead of insoluble gypsum cake and attempted to commercialize the process and sell this by product as a low-cost fertilizer. A 10^3-ton/yr demonstration-scale plant was recently completed to prove the process and develop products and markets for polymers and derivatives. However, due to poor choice of feedstock (whey) and purification process technology, this attempt failed. The polymer patent portfolio of DuPont was acquired by Chronopol, Inc. which is building a demonstration-scale plant for lactic acid polymer production.

The utilization of the purified lactic acid to produce polymers and other chemical intermediates requires the development of secondary purification and integration of catalytic chemical conversion process steps with the lactic acid production processes. Examples of such process steps would be dilactide production for polymerization to make high-molecular-weight polymers or copolymers and hydrogenolysis to make propylene glycol — a large-volume intermediate chemical. In the past, very little effort was devoted to develop efficient and potentially economical processes for such integrations, because only small-volume, high-margin specialty polymers for biomedical applications or specialty chemicals were the target products.

Recently, several advances in catalysts and process improvements have occurred and proprietary technologies have been developed that may enable the commercialization of integrated processes for large-scale production in the future. In a recent patent issued to Gruber et al. of Cargill, Inc. (*22*), the development of a continuous process for manufacture of lactide polymers with controlled optical purity from purified lactic acid is described. The process uses a configuration of multistage evaporation followed by polymerization to a low-molecular-weight prepolymer, which is then catalytically converted to dilactide, and the purified dilactide is recovered in a distillation system with partial condensation and recycle. The dilactide can be used to make high-molecular-weight polymers and copolymers. The process has been able to use fermentation-derived lactic acid, and the claimed ability to recycle and reuse the acid and prepolymers could make such a process very efficient and economical (*22*). In recent patents issued to Bhatia et al. of DuPont, Inc. (*23-30*), processes to make cyclic esters, dilactide, and glycolide from their corresponding acid or prepolymer are described. This process uses an inert gas, such as nitrogen, to sweep away the cyclic esters from the reaction mass and then recovers and purifies the volatilized cyclic ester by scrubbing with an appropriate organic liquid and finally separates the cyclic ester from the liquid by precipitation or crystallization and filtration of the solids. Very high purity lactide with minimal losses due to racemization have been claimed to be produced by this process. Recycle and reuse of the lactic moiety in the various process streams have been claimed to be feasible (*30*). DuPont's patents have been acquired by Chronopol and both Cargill and Chronopol are developing their processes to commercial scale; their goal is large-scale production of biodegradable polymers in the future.

Hydrogenolysis reaction technology to produce alcohol from organic acids or esters has also advanced recently — new catalysts and processes yield high selectivity and rates and operate at moderate pressures (*31-33*). This technology has been commercialized to produce 1,4 butanediol, tetrahydrofuran, and other four-carbon chemical intermediates from maleic anhydride. In the future, such technologies could be integrated with low-cost processes for the production of lactic acid to make propylene glycol and other intermediate chemicals (*34*).

Technical Accomplishments and Future Directions at Argonne

In the past two years, under a U.S. Department of Energy-sponsored project at ANL, several important technical advances have been achieved and demonstrated at the laboratory scale. Notably, these advances have occurred in fermentation, primary purification, and polymer synthesis. In fermentation, high product yield (95%) from starch by means of an enzymatic saccharification/fermentation process with high lactate concentration (100 g/L) and good productivity (3 g/L•h) have been achieved. The ED-based primary purification process has been operated in the laboratory in short-term feasibility experiments to obtain flux and power data for design and economics. A proprietary method to produce a high-molecular-weight copolymer of polylactic acid with other copolymers has been developed at the laboratory scale. Methods to modify and test the degradability of polylactic acid have been developed. Furthermore, the development of secondary purification processes and specialty products derived from lactic acid with targeted properties have been initiated.

The ANL program of oxychemicals and polymer feedstock production from carbohydrate-derived lactic acid is schematically shown in Figure 1. The fermentation and primary purification process to make purified lactic acid has been developed and demonstrated at ANL and elsewhere. The program is now focusing on developing efficient and economical secondary purification processes to make esters that can serve as the key intermediate for the production of a host of other chemicals, polymers, and specialty derivatives. The products and the processes to be developed or integrated are shown in Figure 1. This matches several of the target products listed in Table I that can be derived from lactic acid.

Thus, a rational program targeted at development of economical processes for key intermediates of lactic acid and its derivative products has been the primary focus at Argonne.

Conclusions

A wide range of products with U.S. market size exceeding 6×10^9 lb/yr and product values exceeding $\$4 \times 10^9$/yr could be potentially manufactured from lactic acid.

Degradable and environmentally sound products will provide the initial impetus for development and deployment of new lactic acid technologies and products.

Several major U.S. agriprocessing/chemical companies have built demonstration-scale plants and have identified the trends in the environmentally sound products and processes; consequently, they have plans for major large-scale plants in the future.

Novel separations processes that have recently emerged can enable large-scale and economical production of purified lactic acid without waste gypsum or salt by-product.

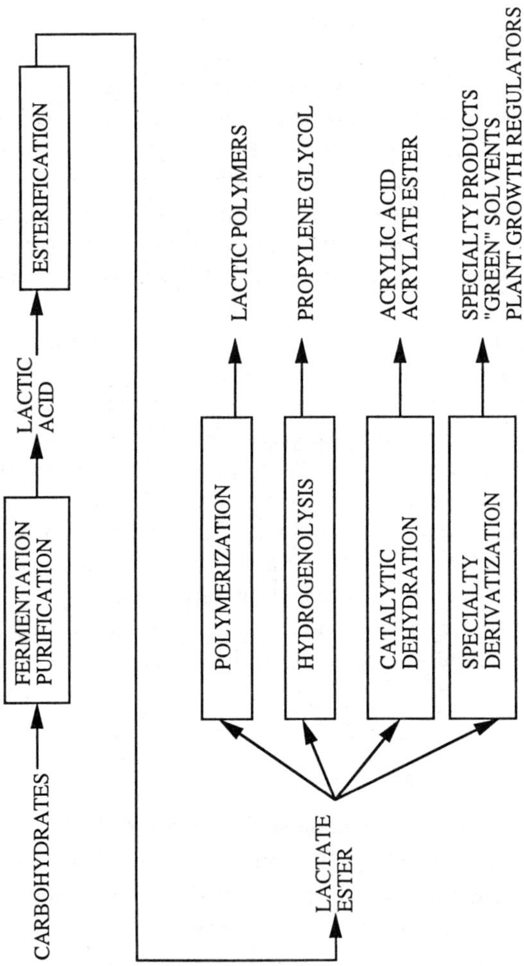

Figure 1. ANL Program of Oxychemicals and Polymer Feedstock Production from Carbohydrate Derived Lactic Acid

Several novel processes are being deployed for facile production of lactic polymer feedstock from lactic acid.

A wide variety of polymers/copolymers with many potential consumer uses could be derived as these products and processes are brought on-stream.

With the new technologies, the manufacturing costs and economics of lactic acid and its derivatives have an attractive potential in large-scale systems.

The lactic acid program at ANL has achieved several important milestones mainly in fermentation and methods of copolymer development.

The technical strategy of the program is to develop novel and economical technologies for key intermediates and products (beyond the degradable polymers) that have a wide range of potential applications.

Acknowledgments

Work supported by the U.S. Department of Energy, Assistant Secretary for Energy Efficiency and Renewable Energy, under Contract W-31-109-Eng-38..

Literature Cited

1. Lipinsky, E.S.; Sinclair, R.G., *Chemical Engineering Progress*, **1986**, *82*, 8:26-32.
2. Anon., *Lactic Acid Market Readies for ADM Entry*, Chem. Marketing Reporter 22, Oct. 15, 1990.
3. Anon., *Lactic Prospects May Ride on the Degradable Plastics Use*, Chem. Marketing Reporter 5, Oct. 16, 1989.
4. Anon., *A.E. Staley Announces Lactic Acid Production Plans*, company announcement/press release May 1996.
5. Anon., *Cargill Latest to Step Up in Lactates Field*, Chem. Marketing Reporter 9, Oct. 21, 1991, Ag Industrial Materials and Products, pp. 14, Sept. 1993, "Cargill & CCA Announce Joint Production Plant for Lactic Acid," May 1996.
6. *Lactic Acid and Lactates*, Product Bulletin, Purac, Inc., Arlington Heights, Ill. 1989.
7. U.S. Patent 4,888,191, Dec. 19, 1989, Anders, R.J.; Cerveny, J.G.; Milkowski, A.L. (to Oscar Mayer Foods Corp., U.S.A.).
8. U.S. Patent 5,017,391, May 21, 1991, Anders, R.J.; Cerveny, J.G.; Milkowski, A.L. (to Oscar Mayer Foods Corp., U.S.A.).
9. Van Ness, J.H., *Hydroxy Carboxylic Acid*s, in Kirk-Othmer Encyclopedia of Chemical Technology, **1981**, 3rd Ed., *13*:80-103, J. Wiley & Sons, N.Y.
10. U.S. Patent 4,812,259, Mar. 14, 1989, Yoshinaga, K.; Katagiri, K.; Shinjo, K. (to Canon Kabushiki Kaisha, Japan).
11. Holten, C.H.; Muller, A.; Rehbinder, D., *Lactic Acid*, *Verlag Chemie*, International Research Association, Copenhagen, Denmark, 1971.
12. Dave, P., et al., in Polymer Preparation, **1990**, *31*(1):442-443, American Chemical Soc.

13. U.S. Patents 4,045,418, & 4,057,537, 1977, Sinclair, R.G. (to Gulf Oil, USA).
14. Zhu, J., et al., *Homopolymers and Copolymers of Glycolide and Lactide*, in Proceedings of C-MRS Intl. Symp. **1990**, *3*:387-390.
15. Nakamura, T., et al., *Surgical Applications of Biodegradable Films Prepared from Lactide-epsilon-Caprolactone Copolymers*, Advanced Biomaterials 7, **1987**, Biomater. Clin. Appl. 759-764.
16. U.S. Patent 4,813,997, Mar. 21, 1989, Kinnersley, A. et al. (to CPC International, U.S.A.).
17. U.S. Patent 5,059,241, Oct. 22, 1991, Young, D.C. (to Union Oil Co., U.S.A.).
18. Yao, P.X.; Toda, K., J. Gen. Appl. Microbiol., **1990**, 36:111-125.
19. U.S. Patent 4,885,247, Dec. 5, 1989, Datta, R. (to Michigan Biotechnology Institute, U.S.A.).
20. Eur. Pat. Appl. Publication No. 0 393 818 A1, Feb. 20, 1990, Glassner, D.; Datta, R. (to Michigan Biotechnology Institute, U.S.A.).
21. Mani, K.N., J. Membrane, Sci., **1991**, 58:117-138, and AQUATECH Systems Product Brochure, **1985**, Allied Signal Corp.
22. U.S. Patent 5,142,023, Aug. 25, 1992, Gruber, P.R. et al. (to Cargill, Inc., US).
23. U.S. Patent 4,835,293, May 30, 1989, Bhatia, K.K. (to E.I. DuPont de Nemours Co., U.S.A.).
24. U.S. Patent 5,023,349, June 11, 1991, Bhatia, K.K. (to E.I. DuPont de Nemours Co., U.S.A.).
25. U.S. Patent 5,023,350, June 11, 1991, Bhatia, K.K. (to E.I. DuPont de Nemours Co., U.S.A.).
26. U.S. Patent 5,043,458, Aug. 27, 1991, Bhatia, K.K. (to E.I. DuPont de Nemours Co., U.S.A.).
27. U.S. Patent 5,091,544, Feb. 25, 1992, Bhatia, K.K. (to E.I. DuPont de Nemours Co., U.S.A.).
28. U.S. Patent 5,117,008, May 26, 1992, Bhatia, K.K.; Drysdale, N.E.; and Kosak, J.R. (to E.I. DuPont de Nemours Co., U.S.A.).
29. U.S. Patent 5,136,057, Aug, 4, 1992, Bhatia, K.K. (to E.I. DuPont de Nemours Co., U.S.A.).
30. U.S. Patent 5,138,074, Aug, 11, 1992, Bhatia, K.K.; Bellis, H.E. (to E.I. DuPont de Nemours Co., U.S.A.).
31. German Patent, DE 3443277, A1, June 5, 1985, Kipax, J.; Harris, N.; Turner, K.; Anthony, B.; Rathmell, C. (to Davy Mckee [London] Ltd.).
32. U.S. Patent, 4,777,303, Oct. 11, 1988, Kitson, M.; Williams, P.S. (to B.P. Chemicals [London] Ltd.).
33. German Patent 1,257,766, Jan. 4, 1968, Mueller, W. et al. (to Lentia GMBH, Chemie U. Pharm. Erzeugnisse-Ind.).
34. Datta, R., *Technological Advances & Needs for Large Scale Bioprocessing & Separations*, Proc. of Corn Utilization Conf. 4, National Corn Growers Assn., St. Louis, Mo., June 1992; & Datta, R.; Tsai, S.P., *Technology & Economic Assessment of Lactic Acid Production & Uses*, Corn Utilization Conference 6, National Corn Growers Assn., St. Louis, Mo., June 1996.

Chapter 13

Production of Succinate by Anaerobic Microorganisms

R. R. Gokarn, M. A. Eiteman, and J. Sridhar

Department of Biological and Agricultural Engineering, Driftmier Engineering Center, University of Georgia, Athens, GA 30602

Numerous anaerobic microorganisms synthesize succinic acid as a fermentation product. This chapter reviews the literature for succinate producing organisms and compares the growth and succinate production of two widely differing anaerobic bacteria, *Fibrobacter succinogenes* and *Clostridium coccoides*. *F. succinogenes* degrades simple sugars such as glucose as well as cellulosic materials such as pulped shredded office paper. The principal products after 90 hours from 10 g/L pulped paper are succinate (3.2 g/L) and acetate (0.58 g/L), with lower concentrations of formate (0.070 g/L). *C. coccoides* degrades simple sugars only, and after 24 hours the principal products from 5 g/L glucose are acetate (3.0 g/L), succinate (0.57 g/L) and lactate (0.58 g/L).

Succinic acid is a four-carbon aliphatic dicarboxylic acid having $pKa_1 = 4.2$ and $pKa_2 = 5.6$. The depronated form succinate can be produced by many anaerobic microorganisms at their operating conditions, usually near neutral pH. Succinic acid can be used to manufacture specialty chemicals including tetrahydrofuran, 1,4-butanediol, maleic anhydride, adipic acid, and dimethyl succinate. Its derivatives are used in the food, pharmaceutical, cosmetics and polymer industries. Anaerobic processes from renewable resources are particularly appealing for the synthesis of succinate becuase of their high yields and straght-forward scale-up requirements. This chapter reviews the literature on anaerobic processes for succinate production as well as comparing the succinate production by two widely different microorganisms.

Rumen Bacteria

The rumen is a highly competitive microbial ecosystem of primarily anaerobic bacteria, fungi and protozoa. These microorganisms ferment cellulose, starch and various other carbohydrates into numerous low molecular weight products. The principal acid products from rumen fermentations are acetate, propionate and butyrate. The production of propionate in the rumen involves cross feeding between succinate-producing microorganisms and species that decarboxylate succinate to propionate and carbon dioxide (1,2). Therefore, even though succinate itself is not a product of the entire rumen ecosystem, numerous anaerobes have been isolated which synthesize succinate as a primary end product.

Fibrobacter succinogenes. *Fibrobacter succinogenes* (previously named *Bacteroides succinogenes*) is the predominant cellulolytic bacterial species found in the rumen. Hungate (3) isolated these organisms from bovine rumen and characterized them as obligately anaerobic, gram negative, cellulolytic, non-motile, non-sporeforming, rod-shaped bacteria whose morphology may change to lemon-shaped when cultivated in the laboratory. Strains of this organism have also been isolated from mice caeca (4), pig caeca (5), gut of horses (6), mice (7), langur monkeys (8) and several African ruminants (9). A more recent study (10) comparing the 16S ribosomal ribonucleic acid sequence demonstrated that this organism differs from other *Bacteroides* species, and hence the organism was renamed *Fibrobacter*.

Most *F. succinogenes* strains utilize glucose, cellobiose, maltose, dextrins, lactose, pectin or cellulose, while some strains also use starch as a carbon source (3). *F. succinogenes* requires one branched (i.e., isobutyrate or α-methyl butyrate) and one linear (i.e., valeric, caproic, heptaonic or caprylic) volatile fatty acid for the synthesis of long chain fatty acids and aldehydes incorporated into phospholipids (11,12). *F. succinogenes* has an absolute requirement for biotin (13) and for ions such as Na^+, K^+, Ca^{2+}, Mg^{2+} and PO_4^{3-} (2). *F. succinogenes* synthesizes almost all of its cellular nitrogenous compounds from exogenous ammonia even when large amounts of amino acids and nucleotide precursors are present in the media. Although glutamine or asparagine may substitute for ammonia, ammonia is preferred when multiple nitrogen sources are present (14). Like all rumen bacteria which produce succinate as a major end product (15,16), *F. succinogenes* has an absolute requirement for carbon dioxide. Initiation of the growth of *F. succinogenes* is achieved at 0.02 to 0.05% of carbon dioxide, while optimal growth is observed when the carbon dioxide concentration is above 0.1% (15). *F. succinogenes* also fixes carbon dioxide during succinate production, with carbon dioxide incorporated in the carboxyl group of succinate (16).

F. succinogenes can degrade highly structured, crystalline cellulose such as cotton fibers (17). The cellulose degrading enzymes and mode of cellulose degradation by this organism have been extensively studied (18–20). In order to carry out cellulose degradation, cells must have intimate contact with cellulose fibers, since the cellulase enzyme endo-β-1,4-glucanase is membrane bound (18). This requirement for contact makes the available gross surface area of the substrate a major determinant factor of hydrolytic rate (21). Endoglucanase activity is about seven times greater when the organisms are grown on cellulose than when grown on cellobiose or glucose, suggesting that the enzyme system may be regulated by a catabolite repression mechanism (18). However, Hiltner and Dehority (22) found that the presence of glucose or cellobiose does affect cellulose digestion when pH is controlled, an observation which seems to contradict the catabolic repression hypothesis. In addition to cellulose degradation, *F. succinogenes* also degrades hemicellulose (18,23). However, the organism cannot utilize as a substrate for growth the pentoses which are released during hemicellulose degradation (24,25). The inability of *F. succinogenes* to utilize pentoses is attributed to the lack of key enzymes such as xylose permease, xylose isomerase and xylulokinase (26).

F. succinogenes has restricted ranges of redox potential and pH. The redox potential range for cell viability is -290 to +175 mV, and the most prevalent morphology at highest redox potential is greatly elongated cells (27). *F. succinogenes* has a pH range for growth of 6.1 to 6.9 (28), and cell wash out occurs at a pH of 6.0 when the organisms are grown on cellobiose in a chemostat (29). The highest cell yield on cellobiose and cellulose occurs at the lower pH limit (28,29). The inability of the organisms to grow at lower pH may be due to inhibition of the glucose transport system or due to low substrate affinity (28,30). When grown on microcrystalline cellulose, *F. succinogenes* has a maximum specific growth rate of 0.076 h^{-1} and a maintenance requirement of 0.04–0.06 g cellulose/g cells (28), while on cellobiose or cellodextrins *F. succinogenes* has a maximum specific growth rate of about 0.44 to 0.48 h^{-1} (31).

Figure 1 summaries the biochemical pathway for succinate production by *F. succinogenes* (34). The degradation products of cellulose, glucose and cellobiose, are transported into the cell by a highly specific active transport system. The glucose transport system is energized by a proton gradient, while the cellobiose transport system is energized by a sodium ion gradient (32). *F. succinogenes* possesses fructose 1,6-biphosphate aldolase (33) and glyceraldehyde-3-phosphate dehydrogenase (34). Oxaloacetate formation from phosphoenolpyruvate is accompanied by carbon dioxide fixation and is catalyzed by GDP-dependent phosphoenolpyruvate-carboxykinase. Reduction of oxaloacetate results in malate formation with NADPH or NADH acting as the electron donor. Even though fumarase activity has not been demonstrated, Miller (34) proposed that conversion of malate to fumarate is catalyzed by fumarase. Fumarate is then reduced to succinate by a flavin-dependent membrane-bound fumarate reductase (34). During the fumarate reduction, cytochrome *b* acts as an electron carrier, and the step may result in ATP generation via electron transport linked phosphorylation. This hypothesis is supported by observations of higher growth yields than can be explained solely by substrate level phosphorylation (27), and of decreased growth rates in the presence of electron uncouplers (35). Conversion of pyruvate to acetyl-CoA is accompanied by the reduction of FMN with carbon dioxide evolution. The formation of acetyl phosphate from acetyl-CoA is catalyzed by phosphotransacetylase, and acetate production from acetyl phosphate yields ATP. Studies with partially isolated phosphoenolpyruvate-carboxykinase indicate that this enzyme is active only in presence of bicarbonate, GDP and the Mn^{2+} ion (34).

***Ruminococcus flavefaciens*.** Sijpesteijn described *Ruminococcus flavefaciens* as a gram positive, non-motile, anaerobic, cellulolytic, streptococci of 0.8–0.9 µm diameter (36). The cells can exist singly, in pairs or may form a chain. *R. flavefaciens* ferments xylans, cellobiose or cellulose, while the fermentations of glucose, xylose or other simple carbohydrates is restricted to only a few strains (36,37). *R. flavefaciens* is active on amorphous cellulose (17) and also breaks down hemicellulose, but most strains cannot utilize pentoses as an energy source (38). A distinct feature of *R. flavefaciens* is the production of a yellow pigment when grown on cellulose. The temperature range for growth is 30 to 45°C, with 39°C being the optimum. *R.*

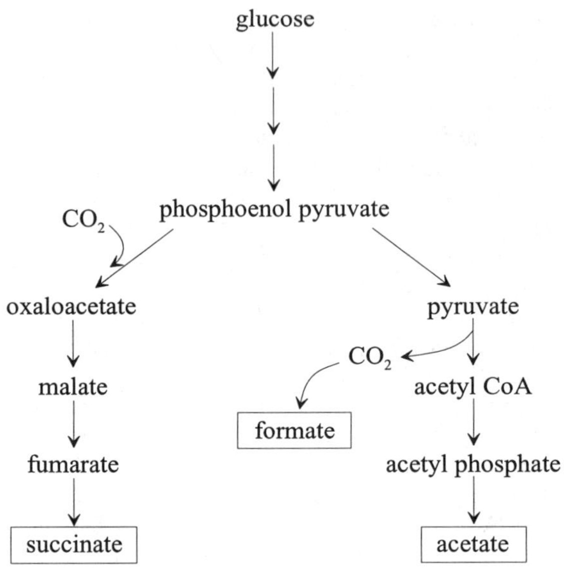

Figure 1 Biochemical pathway for succinate production by *F. succinogenes.*

flavefaciens produces reducing sugars when grown in excess cellulose, although usual fermentation products include succinate, acetate and at least trace formate and lactate (36).

R. flavefaciens has a requirement for branched fatty acids, which are incorporated into lipids and amino acids (39). *R. flavefaciens* can utilize ammonia as a sole nitrogen source (14). Biotin is required by all strains, while vitamin B_{12} is required by some strains. *p*-Aminobenzoic acid has been shown to stimulate growth (13). *R. flavefaciens* has an absolute requirement for carbon dioxide: 0.05 to 0.1% carbon dioxide for growth initiation and above 0.1% for optimal growth (15).

R. flavefaciens possesses enzymes which are active on cellulose, hemicellulose and pectin. Most of these enzymes are believed to be cell wall associated, since cell attachment to cellulose fibers is necessary for plant wall degradation. *R. flavefaciens* has active exo-1,4-β-glycosidase enzymes which generate cellobiose and cellotriose from cellulose and xylobiose and xylotriose from xylan. *R. flavefaciens* also exhibits low levels of aryl β-glucosidase and aryl β-xylosidase activity (40). Presence of soluble sugars such as cellobiose seems not to affect cellulose digestion, although cellulose digestion may be influenced by a pH decrease when soluble sugars are rapidly fermented (22). *R. flavefaciens* has a poor affinity for cellobiose which may result in its poor utilization. With cellobiose in a continuous culture, cell wash out occurs at a pH of about 6.1, and cell yield decreases abruptly (29). An increase in growth rate results in a shift toward more acetate and formate with less succinate. On cellulose the organism has a low maintenance requirement of 0.07 g cellulose/g cell per hour (41).

For one *R. flavefaciens* strain isolated from sheep, succinate is the major product of glucose fermentation in the presence of carbon dioxide, but in the absence of carbon dioxide the fermentation shifts to a homolactic pattern (42). The pathway for succinate and acetate production by *R. flavefaciens* is believed to be identical to that of *F. succinogenes* shown in Figure 1 (42–44). Succinate formation is accompanied by fixation of carbon dioxide, which is incorporated in the carboxyl group. Formate can form from free carbon dioxide or from pyruvate (43). PEP-carboxykinase requires GDP and the bicarbonate ion as cosubstrates, and this enzyme is most effective in converting phosphoenolpyruvate to oxaloacetate in the presence of Mn^{2+} (44).

Ruminobacter amylophilus. Hamlin and Hungate (45) first isolated *Ruminobacter amylophilus* (*Bacteroides amylophilus*) from bovine rumen and characterized the species as an obligately anaerobic, gram negative, non-sporeforming, non-motile bacterium (45). A more recent study of the 16S ribosomal ribonucleic acid sequence showed that this organism differs from other *Bacteroides* species and hence was renamed *Ruminobacter* (46). *R. amylophilus* cells are rod-shaped, 0.9–1.6 µm, but may also exhibit larger irregular shapes. The organism has subsequently been isolated by Blackman and Hobson (47), Bryant and Hobson (48), Caldwell et al. (16), and Blackman (49). *R. amylophilus* uses only starch and maltose as substrates, and the organism's population in the rumen is increased when the animal is fed a high starch diet (36). Fermentation products include acetate, succinate, formate and trace ethanol and lactate (45). The organisms grow at a pH of 6.5–7.8 and temperature range of 35 to 45°C (45). *R. amylophilus* growth occurs within a wide redox potential range of -320 mV to +250 mV, although the specific growth rate decreases above 0 mV (27). Above +200 mV more lactate is produced at the expense of succinate (27).

R. amylophilus has an absolute requirement for Na^+, and this requirement cannot be replaced by K^+, Li^+, Cs^+ or Rb^+ (50). In addition to the Na^+ ion, *R. amylophilus* requires K^+, PO_4^{3-} and trace Mg^{2+} (50). The Na^+ and K^+ ions affect the growth rate and growth yield, while PO_4^{3-} affects only the growth yield (50). *R. amylophilus* also has an absolute requirement of carbon dioxide for growth; however, bicarbonate is a suitable substitute (16). Growth is initiated at a carbon dioxide concentration between 4.5×10^{-3} M and 9×10^{-3} M, while optimal growth is achieved at 1.2×10^{-3} M (16).

R. amylophilus possesses starch-degrading enzymes such as amylase and amylopectinase. The attachment between the cell and starch molecule is mediated by a protein or protein complex (51). *R. amylophilus* possesses enzymes of the Embden-Meyerhof-Parnas pathway (52). The presence of fumarate reductase enzyme suggests that succinate is produced by the reduction of fumarate (16). The production of succinate involves fixation of carbon dioxide which is incorporated as the carboxyl group (16). The pathway for succinate production is thought to be identical as the pathway for *F. succinogenes* shown in Figure 1.

Succinimonas amylolytica. *Succinimonas amylolytica* was isolated from bovine rumen (53). The organism is a gram negative, motile, anaerobic, non-sporeforming, short, rounded to coccoid bacterium 1.0 to 1.5 µm by 1.2 to 3 µm. *S. amylolytica* grows in a temperature range of 30 to 37°C and utilize glucose, maltose, starch or dextrin as a substrate. Fermentation products include succinate, acetate and trace propionate. The organisms grow well in media containing trypticase and yeast extract, and do not grow in the absence of either bicarbonate or carbon dioxide (53). The concentration of carbon dioxide required for the initiation of growth and optimal growth is at least 0.1% (15). *S. amylolytica* also requires acetate and other volatile fatty acids or casein hydrolysate (48). Even though *S. amylolytica* is normally associated with starch digestion in the rumen, the species possesses a wide range of glycoside hydrolases which aid in the utilization of plant cell wall degradation products (54).

Succinivibrio dextrinosolvens. Bryant and Small (55) first isolated *Succinivibrio dextrinosolvens* from a bovine rumen and described the species as an anaerobic, non-sporeforming, gram negative, mobile, helicoidal rod-shaped bacterium, 0.3 to 0.5 μm by 1 to 5 μm. The organisms can metabolize glucose, fructose, L-arabinose, D-xylose, galactose, maltose, sucrose, dextrins or pectins (55). *S. dextrinosolvens* cannot degrade cellulose, hemicellulose or starch, although the organism possesses a wide range of monosaccharide-generating glycoside hydrolases (54,55). Glucose fermentation yields principally acetate and succinate and is accompanied by significant carbon dioxide uptake (55). Formate is a minor product, while some strains also produce trace lactate (55). Similar organisms isolated from an ovine rumen by Wilson (56) were later also termed *S. dextrinosolvens* strains by Bryant (36).

S. dextrinosolvens has an absolute requirement for naphthoquinone, menadione or vitamin K_5, with naphthoquinone resulting in best growth (57). *S. dextrinosolvens* possesses several nitrogen-assimilating enzymes such as urease, glutamate dehydrogenase and glutamine synthetase. Under ammonia limiting conditions the organism uses the ATP-driven glutamine synthetase system, while under excess ammonia the glutamate dehydrogenase enzyme is utilized (58). In addition to ammonia, *S. dextrinosolvens* requires an exogenous supply of amino acids to satisfy nitrogen requirements (48). In the absence of carbon dioxide, limited growth occurs after an extended lag phase (15). For optimal growth, a carbon dioxide concentration above 0.1% is required (55). *S. dextrinosolvens* requires volatile fatty acids and Na^+ ion (which cannot be replaced by K^+, Li^+, Cs^+ or Rb^+). The Na^+ concentration affects both the growth rate and the growth yield of *S. dextrinosolvens* (59).

The catabolic end products of glucose fermentation by *S. dextrinosolvens* are affected by the growth rate. Increased growth rate results in decreased succinate and acetate production and increased lactate formation (60). The pathway of succinate production appears to be identical to other rumen organisms already described (see Figure 1) with fixed carbon dioxide incorporated into the carboxyl group of succinate. *S. dextrinosolvens* has been shown to produce formate from free carbon dioxide present in the media (60).

Prevotella ruminocola. Bryant et al. (53) first isolated *Prevotella ruminicola* (previously named *Bacteroides ruminicola*) from bovine rumen and characterized the species as gram negative, non-motile, rod-shaped 0.8–1 μm by 0.8–3 μm, with slightly tapered, rounded ends (53). The organism recently was renamed *Prevotella* based on its genetic material (61). Several subspecies of *P. ruminicola* may be distinguished on the bases of morphology, substrates fermented and nutrient requirement. Most strains belonging to *P. ruminicola* subsp. *ruminicola* can utilize xylose, glucose and maltose. Some strains have the ability to hydrolyze starch and to utilize arabinose, sucrose and dextrins. Xylans and pectins are rapidly fermented by this particular subspecies (53). *P. ruminicola* subsp. *ruminicola* lacks the enzyme superoxide dismutase which is present in the subspecies *brevis* (62). Cells of *P. ruminicola* subsp. *brevis* are coccoid-to-oval shaped and do not require hemin (48,53). Most strains of this subspeices cannot utilize xylose and xylans, but can use pentoses as carbon sources. They also have the ability to hydrolyze starch and utilize maltose and sucrose (53).

P. ruminicola requires volatile fatty acids and acetate for growth. Use of casein hydrolysate can stimulate the growth of these organisms (48). As all other succinate producing rumen bacteria, *P. ruminicola* requires carbon dioxide for growth (63). The initiation of growth is achieved at 0.02–0.05% carbon dioxide, while above 0.1% optimal growth occurs (15). *P. ruminicola* exhibits proteolytic activity and possesses at least three different active proteinases (64).

P. ruminicola produces acetate, succinate and formate as products of sugar fermentation (53). In the presence of vitamin B_{12}, strain 23 of *P. ruminicola* can also produce propionate (65). In this case the formation of propionate occurs via the direct reductive pathway (acrylate pathway) (66). *P. ruminicola* has low affinity for maltose, sucrose and cellobiose in comparison to glucose (67). The maintenance coefficient for *P. ruminicola* when grown on glucose is 0.135 g glucose /g cells • h (68). Glucose toxicity is observed with *P. ruminicola* strain B_14 (69). A pH below 5.7 halts the growth of *P. ruminicola*; however, growth is not significantly affected by pH changes above 5.7 (70).

P. ruminicola B_14 grows rapidly in a batch culture with a doubling time of 1.65 h (63). The conversion of glucose to phosphoenolpyruvate occurs via the Embden-Meyerhof pathway. The routes of synthesis for succinate and acetate are identical to previously described organisms (see Figure 1), with carbon dioxide again incorporated into the carboxyl group of succinate (63, 71).

Wolinella succinogenes. *Wolinella succinogenes* (previous named as *Vibrio succinogenes*) was isolated by Wolin et al. (72) from bovine rumen and characterized as a curved rod-shaped, motile, anaerobic bacterium approximately 0.6 by 0.3 μm in size (72). A more recent study of the ribosomal ribonucleic acid content and evidence of the organism's inability to ferment sugars resulted in their renaming to *Wolinella succinogenes* (73). *W. succinogenes* contains cytochromes which impart pink color to the cells. These organisms conserve energy by oxidation-reduction reaction in which hydrogen or formate acts as an electron donor and fumarate, malate, asparagine, nitrate, elemental sulfur or nitrous oxide acts as an electron acceptor (72,74,75). These oxidation-reduction reactions can be represented as (72, 75):

$$\text{formate} + H^+ + \text{fumarate} \rightarrow CO_2 + \text{succinate}$$
$$H_2S + \text{fumarate} \rightarrow \text{succinate} + S$$
$$\text{formate} + H^+ + S \rightarrow CO_2 + H_2S$$

Major fermentation products are carbon dioxide and succinate when these organism grow on fumarate and formate (72). Nitrous oxide can be reduced to nitrogen, nitrate to nitrite or ammonia, and elemental sulfur to hydrogen sulfide (74, 75).

W. succinogenes can use oxygen as an electron acceptor, but only at a low partial pressure of oxygen (72). At such low oxygen concentrations, hydrogen peroxide generated is degraded by peroxidase, but at high oxygen concentrations, hydrogen peroxide is hypothesized to accumulate and inhibit growth (76). *W. succinogenes* has a requirement for some succinate when grown on formate and nitrate. However, the presence of succinate does not appear to be necessary when the organisms are grown on fumarate, which is itself reduced to succinate (77).

The reduction of fumarate to succinate is associated with generation of ATP via electron transport phosphorylation (78). *W. succinogenes* has a doubling time of 3.2 h when grown on formate and fumarate, with growth yield of 4.8 g dry cell/mol formate (79). When grown on hydrogen sulfide and fumarate, the doubling time is 3.8 h and 6.0 g dry cell/mol fumarate is achieved. Use of formate and elemental sulfur results in a 1.2 h doubling time and a growth yield of 3.5 g dry cell/mol formate (75).

W. succinogenes can synthesize its cellular components from fumarate. Pyruvate acts as an intermediate for carbohydrates, nucleotides, phospholipids and for most of the amino acids. Glutamate is derived from α-ketoglutarate and further acts as a intermediate for the synthesis of amino acids belonging to its family (79).

Non-Rumen Bacteria

Anaerobiospirillum succiniproducens. Succinate is also produced by a variety of non-ruminal anaerobic bacteria. *Anaerobiospirillum succiniciproducens* was isolated from beagles (dogs), and the bacteria are gram negative, anaerobic, motile, spiral-shaped of 0.6 to 0.8 μm by 3 to 8 μm size (80). These organisms grow in a temperature range of 25 to 40°C, although growth is slow at 25°C. Most *A. succiniciproducens* strains utilize glucose, lactose, or sucrose, while some hydrolyze starch and utilize fructose, dextrose or raffinose. The major fermentation products are succinate and acetate with lactate and formate as minor products (80). *A. succiniciproducens* has an absolute requirement for carbon dioxide and grows well under 0.1 atmosphere partial pressure of carbon dioxide (81,82). Tryptophan is believed to be necessary for utilization of dextrose and corn steep liquor (82).

A. succiniciproducens metabolizes glucose using the pathway shown in Figure 2 (81). The pathways for succinate and acetate production from glucose are identical to those for several rumen anaerobes previously described (see Figure 1). In this case, pyruvate is also the precursor for lactate and ethanol. The formation of lactate from pyruvate is catalyzed by a NADH-dependent lactate dehydrogenase enzyme (81). Pyruvate converted to acetyl-CoA may also be reduced to acetaldehyde then to ethanol. Environmental conditions such as pH and CO_2-HCO_3^- concentration affect the catabolic end products of glucose fermentation. At high pH (7.2) and low CO_2-HCO_3^- concentration, lactate production is favored, lactate dehydrogenase and alcohol dehydrogenase activities are detected and low phosphoenolpyruvate-carboxykinase activity is observed in the cell extract. At low pH (6.2) and high CO_2-HCO_3^- concentration, succinate is the major product, lactate dehydrogenase and alcohol dehydrogenase activities are absent while high phosphoenolpyruvate-carboxykinase activity is observed (81).

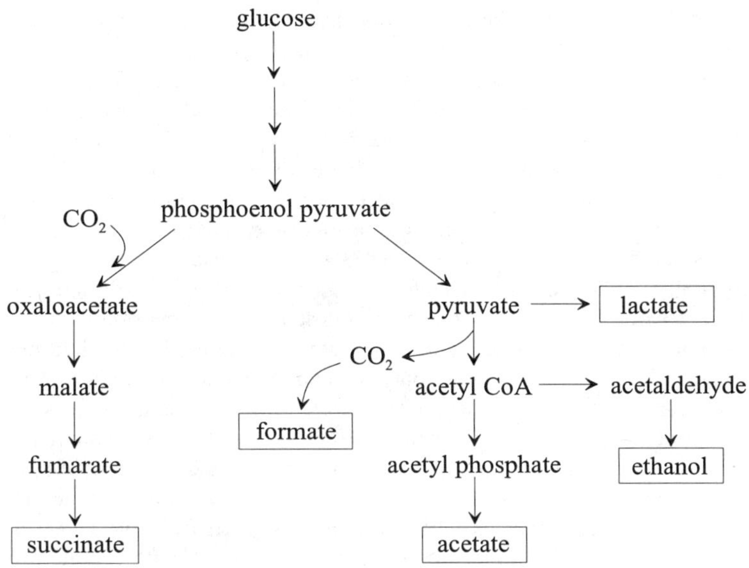

Figure 2 Biochemical pathway for succinate production by *A. succiniproducens.*

Phosphoenolpyruvate-carboxykinase, a key enzyme in succinate pathway, has been isolated and purified from *A. succiniciproducens* (83). The enzyme has a pH range for optimal activity of 6.5–7.1 with an isoelectric point of 4.9. This enzyme differs from that isolated from *F. succinogenes*, using ADP as a cosubstrate instead of GDP. The enzyme has an absolute requirement for a divalent metal ion. In presence of Mn^{2+} or Co^{2+} alone the enzyme activity is 7 to 8 times greater than in the presence of Mg^{2+} alone, while its highest activity is in the presence of either Mn^{2+} or Co^{2+} and Mg^{2+} (83).

Several patents (82,84–86) have been issued for the production of succinate using *A. succiniciproducens* with dextrose and corn steep liquor as substrates. Optimal conditions for this process are 39°C, a pH range of 5.8–6.6 with a carbon dioxide partial pressure of greater than 0.1 atmosphere. Using this process 20 g/L to 40 g/L of succinate production has been demonstrated.

***Clostridia* species.** *Clostridia* are the most widely dispersed of the anaerobic bacteria that produce succinate. Woods (87) noted that the importance of the *Clostridia* for biotechnology was recognized at the beginning of the century when *Clostridium acetobutylicum* was used industrially to produce acetone, butanol, and ethanol. *Clostridia* comprise approximately hundred anaerobic species which can utilize polysaccharides and proteins to produce industrially important products through fermentation, and some *Clostridia* can produce succinate from different substrates.

Clostridium thermosuccinogenes, a species of thermophilic anaerobic bacteria, has been isolated from several sources, including beet pulp at a sugar refinery, soil around a Jerusalem artichoke, fresh cow manure, and mud at a tropical pond in a botanical garden (88). *C. thermosuccinogenes* ferments fructose, glucose or inulin to produce succinate, formate, acetate, ethanol, lactate and hydrogen. The formation of succinate as a major fermentation product distinguishes *C. thermosuccinogenes* from other thermophilic clostridia (88), although *C. thermosaccharolyticum* and *C. thermocellum* produce trace amounts of succinate (89,90). *C. aminophilum* can act on amino acid carbon sources to produce traces of succinate and lactate, although the major fermentation products are ammonia, acetate and butyrate (91). *C. hobsonii* comb. nov. isolated from an anaerobic cattle waste digester ferments glucose to produce ethanol, acetate, formate, lactate and succinate (92). *C. aldrichii* sp. nov. isolated from a wood fermenting anaerobic digester acts on cellobiose to produce succinate, acetate, propionate, isobutyrate, butyrate, isovalerate, lactate, hydrogen and carbon dioxide (93). Acetate, formate, butyrate and lactate are the major products of *C. innocuum* strain Co15-23 with succinate being a minor fermentation product (94). *C. kluyveri* ferments pyruvate to succinate (95,96), and the proposed pathway is shown in Figure 3. The amount of succinate produced by the fermentation has been directly correlated to the quantity of carbon dioxide used in the fixation reactions (95).

For *C. coccoides*, a species which was first isolated from the feces of mice, succinate is the major fermentation product from PYFG broth with acetate produced in moderate amounts. Like the other *Clostridia*, this species is obligately anaerobic, non-motile, sporeforming, gram-positive, with coccobacilli to rod-shaped morphology (97). *C. coccoides* has also been isolated from the fecal microflora of elderly persons in rural and urban areas in Japan (98), and from the fecal microflora of laboratory mice, rats, hamsters and rabbits (99).

Propionibacteria. Several *Propionibacteria* produce succinate as one of their fermentation products. Succinate is produced in the range of 7.9–26.1 mmol/100 mmol glucose with several *Propionibacteria* species (100). Carbon dioxide is utilized in the formation of succinate during the fermentation of glucose and glycerol by *Propionibacterium* species (101,102). *P. freudenreichii* grows anaerobically on lactate (103) to produce succinate in trace amounts (104). A study of the metabolism of aspartate using *P. freudenreichii* showed that 41.5 mM succinate, 40.3 mM acetate, 40.6 mM propionate and 39.8 mM carbon dioxide were produced from 82 mM lactate and 41 mM aspartate. The metabolism of aspartate to succinate and ammonium ion by *P. freudenreichii* is influenced by the pathway by which lactate is fermented to propionate, acetate and carbon dioxide (Figure 4) (105).

P. freudenreichii grown on the cheese originally made with *Lactobacillus bulgaricus* produces between 18–39 mmol succinate/kg cheese (106). Succinate synthesized by *Propionibacteria* is formed by one of two pathways, one that generates carbon dioxide or one that consumes carbon dioxide. The first pathway involves the enzyme carboxytransphosphorylase catalyzing a carbon dioxide fixation step, which results in succinate formation from propionate (107). The second pathway involves the metabolism of aspartate to succinate during the fermentation of lactate by *P. freudenreichii* subsp. *shermanii* (105,108). In the presence of added propionate, *P. freudenreichii* consumes aspartate according to the following equation (109):

$$3 \text{ aspartate} + 1 \text{ propionate} \rightarrow 3 \text{ succinate} + 1 \text{ acetate} + 1\ CO_2 + 3NH_3$$

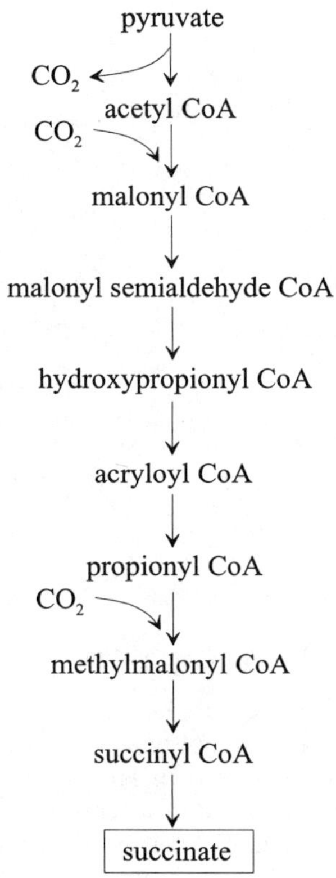

Figure 3 Biochemical pathway for succinate production by *C. kluyveri.*

Growth of *P. acidipropionici* on lactose or glucose results in production of larger quantities of succinate than when the organism is grown on lactate. At a pH of 6.6, the yields of succinate production on lactose, glucose and lactate are 0.114, 0.073 and 0.009 g/g, respectively, with the production of succinate reduced at lower pH (110). Hsu and Yang also reported that the yields of succinate from growth of *P. acidipropionici* on lactose are dependent on pH (111).

Lactobacilli. *Lactobacillus pentosus* grown on limited glucose in the presence of citrate results in the production of acetate, formate, lactate and succinate as the major fermentation products (112). The proposed pathway involves the splitting of citrate into oxaloacetate and acetate by citrate lyase. Oxaloacetate and lactate are metabolized, respectively, to generate succinate and acetate by the pathway shown in Figure 5. The overall reaction is represented by:

$$2 \text{ lactate} + 1 \text{ citrate} \rightarrow 3 \text{ acetate} + 2 \text{ formate} + 1 \text{ succinate}$$

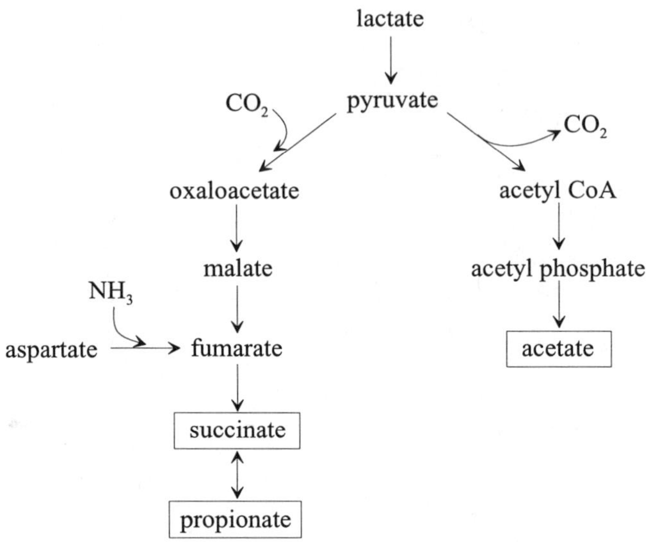

Figure 4 Biochemical pathway for succinate production by *P. freudenreichii*.

In one study, 9.5 mmol of citrate and 8 mmol of glucose were transformed into 27 mmol of acetate and 9 mmol of succinate (112). The presence of nitrate and nitrite prevented the formation of formate, acetate and succinate.

A number of *Lactobacilli* isolated from fermented cane molasses in Thailand produce succinate in de Man-Rogosa-Sharpe broth (113). Among the strains that produced succinate were 23 of 39 *Lactobacillus reuteri* strains, 6 of 18 *L. cellobiosus* strains and 1 of 6 unidentified strains. Diammonium citrate was found to be a precursor of the succinate. Other *Lactobacilli* which produce succinate include *L. brevis* (114) which converts tartarate to succinate through oxaloacetate, malate and fumarate, and *L. crispatus* (115) which produces succinate from glucose. A few *Lactobacilli* strains isolated from ciders and perries have been shown to produce succinate from malate (116).

L. plantarum degrades L-lactate in the presence of citrate to form formate, acetate and succinate along with carbon dioxide (117). In the absence of citrate, *L. plantarum* metabolizes lactate to acetate and formate. Evidence suggests that oxaloacetate formed from citrate acts as an electron acceptor for the production of succinate, a result consistent with those obtained for the mannitol fermentation by *L. plantarum* to produce succinate (118).

Bacteriodes. Species of *Bacteriodes* are strict anaerobes isolated from gastrointestinal tracts, including some rumen (119). A study of the role of carbon dioxide in the metabolism of glucose of *Bacteriodes fragilis* (120) showed that carbon dioxide has no effect on the growth rate or cell yield at concentrations above 30% (equivalent to an available CO_2-HCO_3^- of 25.5 mM). A slight decrease in the growth rate and cell yield occurs at 20% and 10% carbon dioxide. When CO_2-HCO_3^- concentrations are

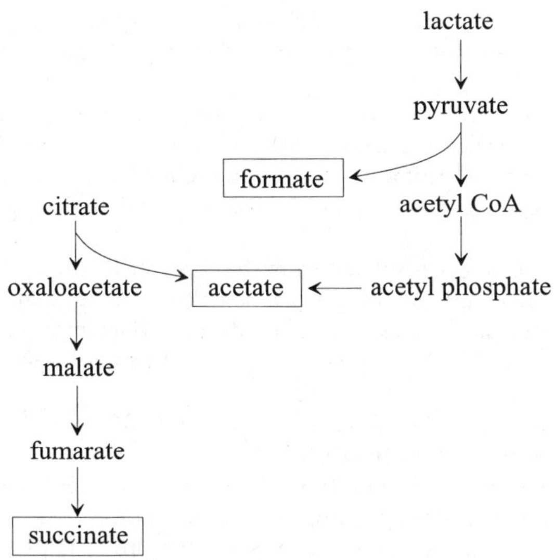

Figure 5 Biochemical pathway for succinate production by *L. pentosus*.

below 10 mM, the lag phase lengthens and a decrease in maximal growth rate and cell yield is observed. *B. fragilis* has been extensively studied in continuous culture to determine the effect of carbon dioxide on the growth rate at a constant physiological state (120). At 100% carbon dioxide or 100% nitrogen, decreasing the dilution rate favors propionate and acetate production over succinate, D-lactate, L-malate and formate. When grown in 100% nitrogen, propionate is formed at a greater concentration than succinate. Except at lowest dilution rates, the reverse is true when grown in 100% carbon dioxide. In 100% nitrogen, the maximum dry cell yield is 67.9 g cells/mol glucose while in 100% carbon dioxide this yield is 59.4 g/mol (120).

In one study, *Bacteriodes ovatus* grown on starch (2 g/L) or arabinogalactan (5 g/L) was used, respectively, to compare carbon-limited with nitrogen-limited conditions. Succinate produced under nitrogen limitation ranged from 9–14 mM compared with 1–7.1 mM under carbon limitation. The production of succinate and propionate is dependent on dilution rate, with higher dilution rates favoring propionate (121). A similar relationship between propionate/succinate and growth rate has been demonstrated for *B. thetaiotaomicron* (122) and *B. fragilis* (120).

Macy and Probst (119) reviewed the biology of gastrointestinal *Bacteriodes*. *B. ureolyticus* produces succinate (123,124), *B. melaninogenicus* grows on aspartate to produce succinate (125), and *B. oralis* forms succinate, acetate and formate (126,127). As noted earlier, organisms originally named *B. succinogenes*, *B. amylophilus* and *B. ruminicola* have been respectively renamed *Fibrobacter succinogenes*, *Ruminobacter amylophilus* and *Prevotella ruminicola*.

Other bacteria. *Desulfobacterium cetonicum* can degrade acetone to form succinate from isocitrate by isocitrate lyase (128). *D. dehalogenans* isolated from pond water can grow on pyruvate in the presence of 3-chloro-4-hydroxyphenylacetate or fumarate as an electron acceptor produce 4-hydroxyphenylacetate and succinate in equimolar amounts (129).

Acetonema longum is a hydrogen-oxidizing and carbon dioxide-reducing acetogenic bacterial species isolated from the gut of the termite *Pterotermes occidentis*. The organism consumes rhamnose to produce acetate and butyrate as the major fermentation products and propionate, succinate and 1,2-propanediol as minor products (130).

Anaerobic methanogens isolated from the feces of rats fed with high-fiber or fiber-free diets generate acetate, propionate and butyrate as major products and lactate, succinate and formate as minor products. Succinate produced by the fermentations is 1.3 mmol/g dry matter for the high-fiber diet and 5.5 mmol/g dry matter for the fiber-free diet (131).

Different strains of *Klebsiella oxytoca* grown on 10% xylose (666 mM) for 96 hours produce 41–59 mM succinate (132).

Peptostreptococcus productus strain Co8-4 produces acetate and succinate as major fermentation products (94). *Peptostreptococcus micros* strain Co33-6 produces acetate, lactate and succinate as its major fermentation products (94).

Fusobacterium russii strain Co21-3 produces acetate and butyrate as major products and succinate as a minor fermentation product (94).

Eubacterium sp B86, *Peptostreptococcus* sp 610, *Enterococcus faecalis* sp 84 and *Veillonella ratti* sp 36 isolated from conventional rat microflora ferment starch to produce succinate, acetate and propionate (133).

Anaerovibrio burkinabensis sp. nov., a strictly anaerobic bacterium isolated from rice fields soil by using lactate as the sole carbon and energy source, ferments fumarate, malate and aspartate to produce succinate (134). *A. lipolytica* isolated from both ovine and bovine rumen and grown on a linseed oil/rumen fluid agar media breaks down glycerol into propionate and succinate (135).

The myxobacteria *Cytophaga succinicans* have been shown to degrade glucose in the presence of carbon dioxide to produce succinate, acetate and formate in a 3:2:1 ratio (136).

Fungi and Yeast

A very few fungi and yeast produce succinate. The fermentation of filter paper cellulose by several anaerobic fungi has been studied in the absence and presence of methanogenic bacteria. In the absence of methanogens, *Neocallimastix* sp. strain L2, *N. frontalis* RE1, *N. patriciarum* CX, *Piromonas communis* P and *Sphaeromonas communis* FG10 have been found, respectively, to produce 0.48, 0.59, 0.39, 0.81 and 0.26 mol succinate/10 mol hexose. In the presence of methanogenic bacteria, the production of succinate is reduced significantly for all the fungi, a result attributed to the process of hydrogen transfer between the fungi and bacteria (137).

Muratsubaki elucidated the pathway by which succinate is produced by the anaerobic growth of *Saccharomyces cerevisiae* on glucose (138). The activity of fumarate reductase, which catalyzes the conversion of fumarate to succinate, is three times greater under anaerobic conditions than under aerobic conditions. However, succinate dehydrogenase activity is completely lost after ten hours of fermentation.

These observations indicate that for this organism the citric acid cycle has been modified to become a reductive pathway leading to succinate production during the anaerobic growth of *S. cerevisiae* on glucose (138).

Downstream Processing of Succinate

Although not strictly related to the production of succinate by anaerobic microorganisms, the downstream processing of this compound ultimately is of paramount importance for any industrial process, and could indeed influence the selection of microorganism. A few processes have been studied for the recovery of succinate from fermentation media, and each of these has its unique advantages and disadvantages. In general, processes developed heretofor have focused on the characteristic of charge that distinguishes succinate from many other components in fermentation media.

One method to recover succinate is by extraction, and recent articles describe extraction of various organic acids (139–146). Extraction processes for succinate and other negatively charged solutes usually involve the transfer of the solute into an organic phase by the use of a positively charged extractant, such as long-chain tertiary amines. The selectivity of separation is very good with amine extractants, because they have favorable equilibration chemistry with deprotonated acids (140,141). A potential problem is water coextraction (142), which reduces the extraction yield of the acid. Of course, the succinate-amine complex in an envisioned extraction process must still be reextracted (stripped) from the organic phase, a process which might, for example, involve a strong base. An advantage of the process is that extraction is a mature process, with numerous designs and devices available to implement the process. A potential problem is the toxicity of the amine extractant if the extraction is carried out *in situ*. If a water-soluble volatile tertiary amine such as trimethylamine is used to back-extract the succinate, the solution may be partially evaporated to produce the acid product in crystalline form (143).

Another recovery method is adsorption, whereby an ion exchange resin adsorbs the negatively charged ions from the fermentation media (145,146). To avoid clogging the adsorption bed with cells, a membrane separation is required. By the appropriate selection of ion exchange resin, the succinate along with other organic acids can very effectively be removed from the fermentation media. The next step in the process is removal of the succinate from the ion exchange column and regeneration of the ion exchange material. This process may require the use of both a strong base to strip the succinate and a strong acid both to convert the succinate into succinic acid and to regenerate the column. Alternatively, the acid can be back-extracted into a trimethylamine solution, as described in the previous paragraph.

A related, general extractive means to recover carboxylic acids from fermentation media is through the use of liquid membranes (147), and an example is the recovery of lactic acid (148). In supported liquid membranes, a porous polymeric membrane impregnated with the extractant is situated between the feed and the stripping solution. Often additional compounds such as surfactants are added to enhance the transport rate of the extracted species. Liquid membranes can suffer from instability, relatively high cost and comigration of other components from the fermentation media.

A unique method to recover succinic acid is by electrodialysis. This process has been the subject of patents (84,85,149) and a review article (150). Transportation of other mobile ions (and cells if the media is not ultrafiltered) in the fermentation media such as sodium, potassium, chloride, phosphate, sulfate, etc. can pose a problem for the efficiency of the electrodialysis. Essentially the method utilizes electric charge first by "desalting" electrodialysis to concentrate sodium succinate, and then by "water-splitting" electrodialysis to remove preferentially sodium and hydroxide ions from the salt stream yielding a precipitated succinic acid product. When electrodialysis was applied to a succinate fermentation by *A. succiniproducens*, a solid product of 99.9% purity was obtained (150).

Methods

Fibrobacter succinogenes. *Fibrobacter succinogenes* S85 (provided by P. J. Weimer, U. S. Dairy Forage Research Center, Madison, Wisconsin, USA) was cultivated in a defined basal medium of following composition (pH 6.5): $K_2HPO_4{\cdot}3H_2O$, 0.292 g/L; KH_2PO_4, 0.24 g/L; $(NH_4)_2SO_4$, 0.48 g/L; NaCl, 0.48 g/L; $MgSO_4{\cdot}7H_2O$, 0.123 g/L; $MnCl_2{\cdot}2H_2O$, 0.098 g/L; $CaCl_2{\cdot}2H_2O$, 64 mg/L; Na_2CO_3, 4 g/L; cysteine HCl, 0.6 g/L; isobutyric acid, 57.6 mg/L; isovaleric acid, 55.8 mg/L; n-valeric acid, 56.4 mg/L; 2-methyl butyric acid, 54 mg/L; resazurin, 1 mg/L; thiamine HCl, 0.44 mg/L; calcium D-pantothenate, 0.44 mg/L; nicotinamide, 0.44 mg/L; riboflavin, 0.44 mg/L; pyridoxine, 0.44 mg/L; biotin 0.1 mg/L; vitamin B_{12}, 0.004 mg/L; folic acid 0.002 mg/L; tetrahydrofolic acid, 0.002 mg/L. All fermentations of *F. succinogenes* were incubated at 39°C, and samples anaerobically withdrawn from each using a sterile syringe and needle and stored at -20 °C for later analyses.

Batch Fermentations. Batch fermentations of 500 mL liquid volume containing 10.0 g/L of a single substrate were conducted in 550 mL gas washing bottles. Substrates used were glucose (Sigma Chemical Co., St. Louis, Missouri, USA) and pulped office paper. Pulped office paper (500 g) was prepared by first shredding paper (Xerox 4200, Xerox Corp., Palo Alto, California, USA) in an office shredder and then pulping with 8.5 L of water for 75 minutes at room temperature in a laboratory pulper. Prior to its use as a substrate, pulped office paper was spun dried at room temperature in a cloth drier to about 60% water content. For both substrates, a culture (49 mL) grown on microcrystalline cellulose was used as inoculum. Duplicate gas washing bottles were used for each substrate, and each washing bottle had a continuous low flow of oxygen-free carbon dioxide.

Biotin Studies. Biotin studies (100 mL liquid volume) using microcrystalline cellulose (Avicel PH102, FMC Corp., Philadelphia, Pennsylvania, USA) were carried out in 160 mL serum bottles (Wheaton, Millville, New Jersey, USA) capped with butyl rubber stoppers to maintain anaerobic conditions (O_2-free CO_2). For these experiments the media biotin concentrations were 0.001 mg/L (low biotin) and 0.1 mg/L (high biotin).

The basal medium (88 mL) without vitamins was anaerobically pipetted into serum bottles containing 1.0 g of microcrystalline cellulose. After autoclaving, 2 mL of filter-sterilized biotin-free vitamin solution and the required volume of filter-sterilized biotin solution were anaerobically added, and the bottles were inoculated with 10 mL of a culture which had been grown on microcrystalline cellulose (in the presence of 0.01 mg/L biotin).

Carbon Dioxide Studies. Carbon dioxide studies (100 mL liquid volume) using 10.0 g/L glucose were also carried out in 160 mL serum bottles capped with butyl rubber stoppers to maintain anaerobic conditions (in this case, O_2-free N_2). For these experiments Na_2CO_3 was used as a source of carbon dioxide and all media and solutions were prepared under nitrogen.

The basal medium (83 mL) without vitamins and Na_2CO_3 was anaerobically pipetted into serum bottles containing solid Na_2CO_3 to achieve final Na_2CO_3 concentrations of 2 g/L, 5 g/L, 10 g/L or 20 g/L. After autoclaving, 2 mL of filter-sterilized vitamin solution and 5 mL of 20% glucose solution were anaerobically added to each serum bottle. The pH of the media was adjusted to 6.4–6.6 by anaerobically adding 20% H_2SO_4. Serum bottles were then inoculated with 10 mL of culture grown on microcrystalline cellulose.

Clostridium coccoides. *Clostridium coccoides* (ATCC 29236) was cultivated in a medium of the following composition (pH 7.0): glucose, 5.0 g/L; peptone, 2.0 g/L; yeast extract, 2.0 g/L; cysteine HCl, 0.5 g/L; $K_2HPO_4 \cdot 3H_2O$, 8.0 g/L; $NaH_2PO_4 \cdot H_2O$; $MnCl_2 \cdot 4H_2O$, 0.10 g/L; $MgSO_4 \cdot 7H_2O$, 0.05 g/L, $CaCl_2 \cdot 2H_2O$; 0.01 g/L; Na_2CO_3, 0.40 g/L, hemin, 5 mg/L. Batch fermentations of 250 mL liquid volume were conducted in 300 mL gas washing bottles, with a continuous low flowrate of carbon dioxide. All fermentations of *C. coccoides* were incubated at 37°C, and samples were anaerobically withdrawn and stored in a fashion identical to that described for *F. succinogenes*. Optical density was measured at 620 nm (Beckman DU-650).

Analyses. A Shimadzu HIC-6A ion chromatography system with a 5 μL sample loop was used in this work. The system consisted of a Model LC-6A pump, a Model CTO-6AS oven, a Model SCL-6B controller, a Model SPD-6AV UV/visible detector, and analytical data were computer stored on Chromapak C-R4A software. The UV detector was operated at 210 nm, and the flowrate of the eluent was 0.6 mL/min.

A Coregel 64H (Interaction Chromatography, San Jose, CA) ion-exclusion column (300 × 7.8 mm ID) of 10 μm particle size was used for acid analyses. A Coregel 64H guard column protected the column, which was stored in 1 mN H_2SO_4.

Sulfuric acid (Fisher Scientific, Fair Lawn, New Jersey, USA) and deionized water (Modulab Analytical UF Polishing System, Continental Water Systems, Atlanta, Georgia, USA) were used for the preparation of eluents. Optimal concentration of eluent (7.0–16.0 mN H_2SO_4) and optimal temperature (40–60°C) were selected to separate the specific fermentation products.

Glucose concentration in the medium was determined by DNS method (151).

Results and Discussion

Fibrobacter succinogenes. Batch fermentation studies conducted to compare the substrates of simple sugar (glucose) and complex cellulose (pulped office paper) resulted in the production of acetate and succinate as shown in Figure 6 (small

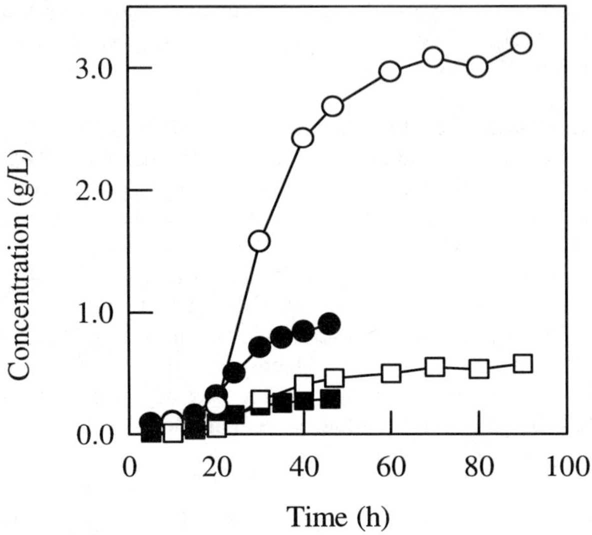

Figure 6 Succinate (○,●) and acetate (□,■) concentration during *F. succinogenes* fermentation of glucose (filled symbols) or pulped paper (hollow symbols).

amounts of formate and trace quantities of fumarate and pyruvate were also detected in the media). For all fermentations, these substrates were only incompletely utilized. For example, only 23% of glucose was consumed before growth ceased. From glucose, succinate was the primary fermentation product with a final concentration of about 0.90 g/L (i.e., at the cessation of product formation, 46 h). In contrast, the final concentration of acetate was 0.29 g/L, while the final formate concentration was 0.068 g/L. The maximum succinate productivity was 23.9 mg/L·h, while for acetate this maximum rate was 8.0 mg/L·h. These productivities were much lower than those achieved with *A. succiniproducens* grown on dextrose under optimal conditions (82,84–86). As the biochemical pathway (Figure 1) indicates, carbon dioxide must be consumed during the synthesis of formate and succinate and generated during acetate synthesis. From a mass balance, the carbon dioxide consumption during glucose fermentation averaged 0.18 g/L (gaseous carbon dioxide was supplied during the fermentation).

As noted in the introducton, *F. succinogenes* is also able to ferment cellulose to succinate. From this complex substrate, the final concentration (at 90 h) of succinate was 3.2 g/L, acetate was 0.58 g/L and formate was 0.070 g/L (formate data not shown in figure). During the fermentation of pulped paper, the maximum succinate productivity achieved was 60.6 mg/L·h, while the maximum rate of acetate production was 10.3 mg/L·h. Carbon dioxide consumption during the fermentation of paper pulp averaged 0.84 g/L.

Interestingly, the acetate and succinate productivities and their final concentrations were much greater when paper was the substrate than when glucose was the substrate. Another significant difference between these substrates was the difference in the relative production of succinate and acetate. In the case of glucose, the maximum succinate-to-acetate molar ratio was above 2.0 only at the onset of the fermentation, and this ratio leveled off at 1.6 after 24 h. From paper pulp the maximum succinate-to-acetate ratio of 3.0 occurred between 40 h and 60 h, and this ratio decreased to 2.8 by the end of the fermentation. *F. succinogenes* has a unique characteristic of growing and synthesizing succinate more quickly and on the complex cellulosic substrate than on the simple sugar, glucose.

The effect of biotin on organic acid production by *F. succinogenes* with microcrystalline cellulose as the substrate is shown in Figure 7 (data not shown for formate, which was present below 0.01 g/L). In the presence of a low biotin concentration (0.001 mg/L), the final succinate concentration (at 90 h) was 1.3 g/L, and the final acetate concentration was 0.34 g/L. In the presence of a high biotin concentration (0.1 mg/L), the final succinate concentration was 2.03 g/L, 58% greater than in the presence of a low biotin concentration. Similarly, the final acetate concentration was 0.57 g/L in the high biotin fermentations, an increase of 66%. The maximum molar ratio of succinate-to-acetate was 2.0 for the low biotin concentration, while this ratio was 1.9 for the high biotin concentration. Intermediate biotin concentrations (0.01 mg/L) showed intermediate succinate and acetate final concentrations, while higher biotin concentrations (1.0 g/L) resulted in reduced succinate and acetate final concentrations (data not shown).

Biotin is a cofactor in many biological carboxylation reactions, and an elevated concentration in these present experiments indeed enhanced the synthesis of succinate. However, since the ratio of succinate-to-acetate remained unchanged by the biotin

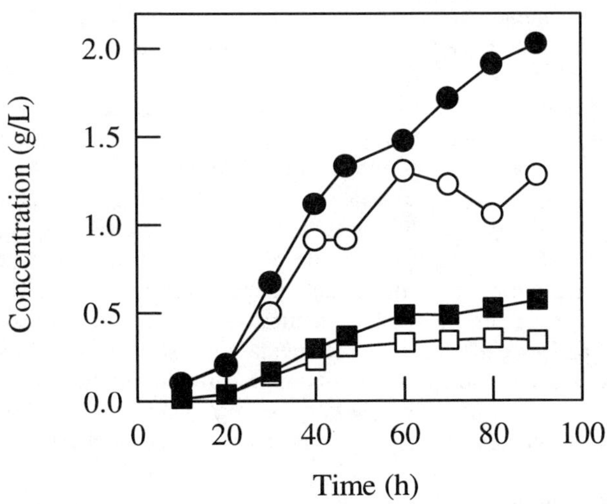

Figure 7 Succinate (○,●) and acetate (□,■) concentration during fermentation by *F. succinogenes* with high (filled symbols) or low (hollow) biotin concentrations.

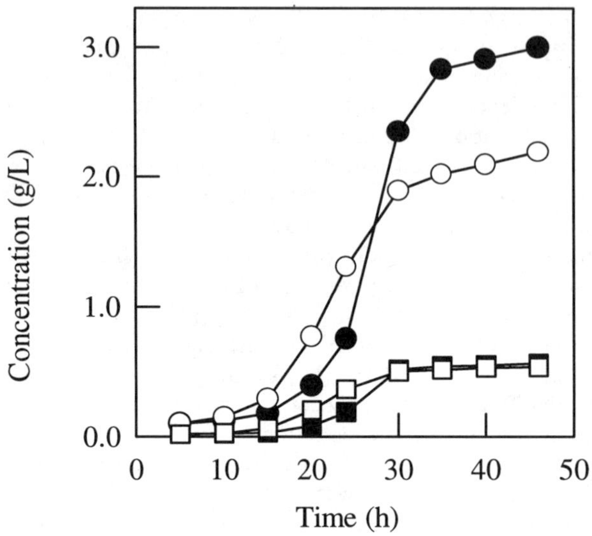

Figure 8 Succinate (○,●) and acetate (□,■) concentrations during fermentation by F. succinogenes with high (filled symbols) or low (hollow) carbonate concentrations.

concentration, these results suggest that biotin does not preferentially enhance the flux of carbon through the succinate branch of the biochemical pathway. Additional studies are required to elucidate the exact role of biotin in succinate production by *F. succinogenes*.

In order to determine the effect of carbon dioxide availability on succinate production, *F. succinogenes* fermentations with glucose as the substrate were conducted in serum bottles at a low initial molar ratio of sodium carbonate to glucose (0.85) and a high initial molar ratio of carbonate to glucose (1.70). The succinate and acetate production under these two conditions is shown in Figure 8 (formate data not shown). In the presence of a low carbon dioxide concentration the final succinate concentration was 2.2 g/L, while the final acetate concentration was 0.54 g/L. Since only 52% of the substrate was consumed, this concentration corresponds to a succinate yield of 0.33 g/g. For the low carbon dioxide fermentation, the maximum succinate-to-acetate molar ratio was 2.1. In the presence of a high carbon dioxide concentration the final succinate concentation was 3.0 g/L, while the final acetate concentration was 0.57 g/L. In this case about 61% of the substrate was utilized, and based on this consumption the succinate yield was 0.40 g/g. For the high carbon dioxide fermentation, the maximum succinate-to-acetate molar ratio was 2.7. The acetate yield was essentially identical for the low and high carbon dioxide concentrations at 0.07 g/g. Additional fermentations were also conducted at other intial sodium carbonate concentrations. For both the initial carbonate-to-glucose molar ratio of 0.34 and for the initial ratio of 3.4, very poor growth of *F. succinogenes* was observed.

The higher relative succinate production resulting from the higher carbonate-to-glucose (i.e., 1.70) indicates that elevated carbon dioxide availability enhances succinate synthesis, and the flux of carbon through the succinate branch of the pathway.

Clostridium coccoides. Since previous studies did not quantify the level of succinate production, several batch glucose fermentations of *C. coccoides* were performed. Figure 9 shows the average glucose concentration and optical density during these fermentations. During the 24 h of the fermentation, the pH decreased from 6.7 to 5.7. Figure 10 shows the production of organic acids during the fermentations. Acetate was the principal fermentation product, achieving a final concentration of 3.0 g/L, corresponding to a yield of 0.57 g acetate/g glucose. Succinate and lactate were produced to similar final concentrations, respectively, of 0.57 g/L and 0.58 g/L, corresponding to yields of approximately 0.11 g/g. After 8 hours of fermentation, the succinate-to-acetate molar ratio remained near 0.10. The maximum acetate productivity was 0.17 g/L·h, while the maximum succinate productivity was 34.6 mg/L·h. The maximum succinate productivity for *C. coccoides* on glucose is about 40% greater than the maximum succinate productivity for *F. succinogenes* on glucose, but just half of the maximum productivity for *F. succinogenes* on pulped paper.

With *C. coccoides* lactate was a significant fermentation end-product, whereas with *F. succinogenes* this product was not detected, and has not been noted in the extensive literature. With *C. coccoides* fumarate was never detected, even in trace quantities.

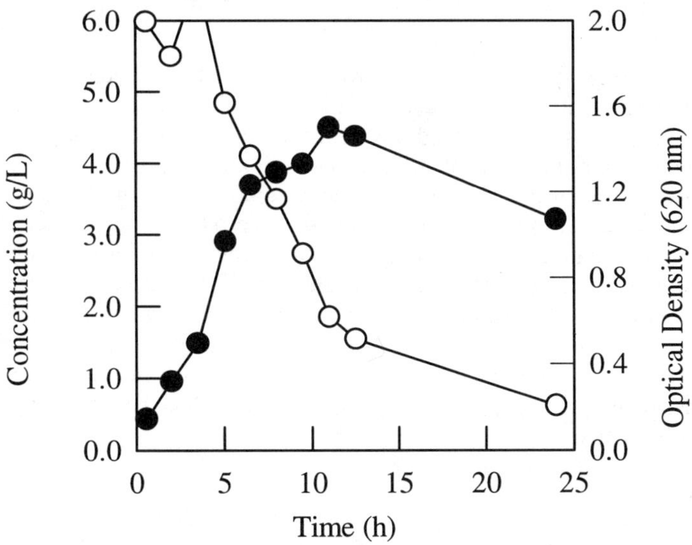

Figure 9 Optical density (●) and glucose concentration (O) during glucose fermentation by *C. coccoides*.

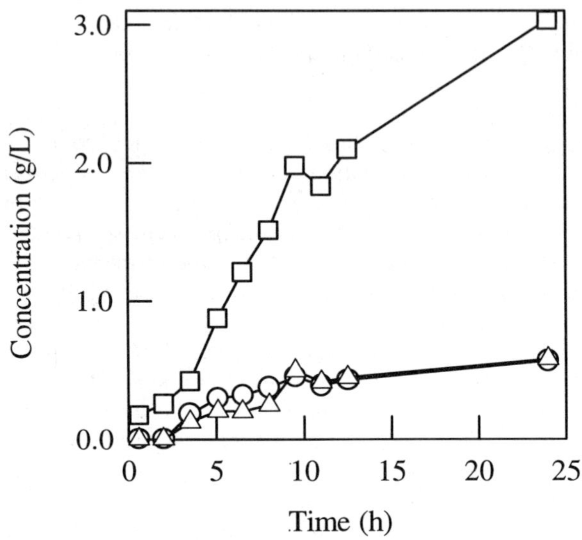

Figure 10 Succinate (○), acetate (□) and lactate (△) concentration during the glucose fermentation by *C. coccoides*.

Conclusions

Fibrobacter succinogenes degrades cellulose or glucose to form succinate, acetate and formate. Trace amounts of pyruvate and fumarate are also detected throughout the fermentation, supporting the proposed biochemical pathways which have these acids as intermediates. Optimal intermediate carbon dioxide availability (1.70 carbon dioxide-to-glucose molar ratio) and biotin concentration (0.1 mg/L) exist for the production of succinate. Interestingly, this organism produces more succinate more quickly with less acetate from paper pulp than from glucose as a substrate.

Clostridium coccoides also degrades glucose to form succinate, acetate and lactate. In contrast to *F. succinogenes*, the principal fermentation product for this organism is acetate, not succinate. However, *C. coccoides* has the greater succinate productivity when grown on glucose. Since fumarate is never detected during the *C. coccoides* fermentation, the pathway to succinate synthesis may be quite different than for the rumen organism.

Acknowledgments

The authors express thanks to Drs. Scott A. Martin and Paul J. Weimer, as well as to Ms. M. Julia Chastain and Ms. Tracey J. Olexa for technical assistance. The financial support of the Georgia Experiment Stations is gratefully acknowledged.

Literature Cited

1. Scheifinger, C. C.; Wolin, M. J. *Appl. Microbiol.* 1973, 26, 789–795.
2. Wolin, M. J.; Miller, T. L. In *The rumen microbial ecosystem*; Hobson, P. N., Elvsevier Science Publishers Ltd.: Essex, England, 1988, pp 343–359.
3. Hungate, R. E. *Bacteriol. Rev.* 1950, 14, 1–49.
4. Montgomery, L.; Macy, J. M. *Appl. Environ. Microbiol.* 1982, 44, 1435–1443.
5. Varel, V. H.; Fryda, S. J.; Robinson, I. M. *Appl. Environ. Microbiol.* 1984, 47, 219–221.
6. Davies, M. E. *J. Appl. Bacteriol.* 1964, 27, 373–378.
7. Brown, J. F.; Balish, E. *Appl. Environ. Microbiol.* 1978, 36, 144–159.
8. Bauchop, T.; Martucci, R. W. *Science* 1968, 161, 698–700.
9. Hungate, R. E. *Science* 1959, 130, 1192–1194.
10. Montgomery, L.; Flesher, B.; Stahl, D. *Int. J. Syst. Bacteriol.* 1988, 38, 430–435.
11. Bryant, M. P.; Doetsch, R. N. *J. Dairy Sci.* 1954, 37, 1176–1183.
12. Wegner, G. H.; Foster, E. M. *J. Bacteriol.* 1963, 85, 53–61.
13. Scott, H. W.; Dehority, B. A. *J. Bacteriol.* 1965, 89, 1169–1175.
14. Bryant, M. P.; Robinson, I. M. *Appl. Microbiol.* 1961, 9, 96–103.
15. Dehority, B. A. *J. Bacteriol.* 1971, 105, 70–76.
16. Caldwell, D. R.; Keeny, M; Van Soest, P. J. *J. Bacteriol.* 1969, 98, 668–676.
17. Halliwell, G.; Bryant, M. P. *J. Gen. Micriobiol.* 1963, 32, 441–448.
18. Groleau, D.; Forsberg, C. W. *Can. J. Microbiol.* 1981, 27, 517–530.
19. Forsberg, C. W.; Beveridge, T. J.; Hellstrom, A. *Appl. Environ. Microbiol.* 1981, 42, 886–896.
20. Groleau, D.; Forsberg, C. W. *Can. J. Microbiol.* 1983, 29, 504–517.
21. Weimer, P. J.; Lopez-Guisa, J. M.; French, A. D. *Appl. Environ. Microbiol.* 1990, 56, 2421–2429.
22. Hiltner, P.; Dehority, B. A. *Appl. Environ. Microbiol.* 1983, 46, 642–648.
23. Matte, A.; Forsberg, C. W. *Appl. Environ. Microbiol.* 1992, 58, 157–168.
24. Morris, E. J.; Van Gylswyk, N. P. *J. Agri. Sci.* 1980, 95, 313–323.
25. Dehority, B. A. *J. Bacteriol.* 1965, 89, 1515–1520.
26. Matte, A.; Forsberg, C. W.; Gibbins, A. M. V. *Can. J. Micrbiol.* 1992, 38, 370–376.
27. Marounek, M.; Wallace, R. J. *J. Gen. Microbiol.* 1984, 130, 223–229.
28. Weimer, P. J. *Arch. Microbiol.* 1993, 160, 288–294.
29. Russell, J. B.; Dombrowski, D. B. *Appl. Enivron. Microbiol.* 1980, 39, 604–610.
30. Chow, J. M.; Russell, J. B. *Appl. Environ. Microbiol.* 1992, 58, 1115–1120.
31. Russell, J. B. *Appl. Environ. Microbiol.* 1985, 49, 572–576.
32. Martin, S. A. *J. Anim. Sci*, 1994, 72, 3019–3031.
33. Joyner, A. E.; Baldwin, R. L. *J . Bacteriol.* 1966, 92, 1321–1330.
34. Miller, T. L. *Arch Microbiol.* 1978, 117, 145–152.
35. Dawson, K. A.; Preziosi, M. C.; Caldwell, D. R. *J. Bacteriol.* 1979, 139, 384–392.

36. Bryant, M. P. *Bacteriol. Rev.* 1959, 23, 125–151.
37. Bryant, M. P.; Small, N.; Bouma, C.; Robinson, I. M. J. Bacteriol. 1958, 76, 529–537.
38. Dehority, B. A.; Scott, H. W. *J. Dairy Sci.* 1967, 50, 1136–1141.
39. Allison, M. J.; Bryant, M. P.; Doetsch, R. N. *Science* 1958, 128, 474–475.
40. Pettipher, G. L.; Latham, M. J. *J. Gen. Microbiol.* 1979, 110, 21–27.
41. Weimer, P. J.; Shi, Y.; Odt, C. L. *Appl. Microbiol. Biotechnol.* 1991, 36, 178–183.
42. Hopgood, M. F.; Walker, D. J. *Aust. J. Biol. Sci.* 1967, 20, 165–182.
43. Hopgood, M. F.; Walker, D. J. *Aust. J. Biol. Sci.* 1967, 20, 183–192.
44. Hopgood, M. F.; Walker, D. J. *Aust. J. Biol. Sci.* 1967, 22, 1413–1424.
45. Hamlin, L. J.; Hungate, R. E. *J. Bacteriol.* 1956, 72, 548–554.
46. Stackbrandt, E.; Hippe, H. *Syst. Appl. Microbiol.* 1986, 8, 204–207.
47. Blackman, T. H.; Hobson, P. N. *J. Gen. Microbiol.* 1962, 29, 69–81.
48. Bryant, M. P.; Robinson, I. M. *J. Bacteriol.* 1962, 84, 605–614.
49. Blackman, T. H. *J. Gen. Microbiol.* 1968, 53, 27–36.
50. Caldwell, D. R., Keeney, M.; Barton, J. S.; Kelley, J. P. *J. Bacteriol.* 1973, 114, 782–789.
51. Anderson, K. L. *Appl. Environ. Microbiol.* 1995, 61, 1488–1491.
52. Wetzstein, H. G.; Gottschalk, G. *Arch. Microbiol.* 1985, 143, 157–162.
53. Bryant, M. P.; Small, N.; Bouma, C.; Chu, H. *J. Bacteriol.* 1958, 76, 15–23.
54. Williams, A. G.; Withers, S. E.; Coleman, G. S. *Curr. Microbiol.* 1984, 10, 287–294.
55. Bryant, M. P.; Small, N. *J. Bacteriol.* 1956, 72, 22–26.
56. Wilson, S. N. *J. Gen. Microbiol.* 1953, 9, 1–2.
57. Gomez-Alarcon, R. A.; O'Dowd, C.; Leedle, J. A. Z.; Bryant, M. P. *Appl. Environ. Microbiol.* 1982, 44, 346–350.
58. Patterson, J. A.; Hespell, R. B. *Appl. Environ. Microbiol.* 1985, 50, 1014–1020.
59. Caldwell, D. R.; Hudson, R. F. *Appl. Microbiol.* 1974. 27, 549–552.
60. O'Herrin, S. M.; Kenealy, W. R. *Appl. Environ. Microbiol.* 1993, 59, 748–755.
61. Shah, H.; Collins, D. M. *Int. J. Syst. Bacteriol.* 1990, 40, 205–208.
62. Fulghum, R. S.; Worthington, J. M. *Appl. Environ. Microbiol.* 1984, 48, 675–677.
63. Howlett, M. R.; Mountfort, D. O.; Turner, K. W.; Roberton, A. M. *Appli. Environ. Microbiol.* 1976, 32, 274–283.
64. Hazlewood, G. P.; Edwards, R. *J. Gen. Microbiol.* 1981, 125, 11–15.
65. Strobel, H. J. *Appl. Environ. Microbiol.* 1992, 58, 2331–2333.
66. Wallnofer, P.; Baldwin, R. L. *J. Bacteriol.* 1967, 93, 504–505.
67. Russell, J. B.; Baldwin, R. L. *Appl. Environ. Microbiol.* 1979, 37, 531–536.
68. Russell, J. B.; Baldwin, R. L. *Appl. Environ. Microbiol.* 1979, 37, 537–543.
69. Russell, J. B. *Appl. Environ. Microbiol.*, 1992, 58, 2040–2045.
70. Russell, J. B.; Sharp, W. M.; Baldwin, R. L. J. Anim. Sci. 1979, 48, 251–255.

71. Mountfort, D. O.; Roberton, A. M. *J. Gen. Microbiol.* 1978, 106, 353–360.
72. Wolin, M. J.; Wolin, E. A.; Jacobs, N. J. *J. Bacteriol.* 1961, 81, 911–917.
73. Tanner, A. C. R.; Badger, S.; Lai, C. H.; Listgarten, M. A.; Visconti, R. A.; Socransky, S. S. *Int. J. Syst. Bacteriol.* 1981, 31, 432–445.
74. Yoshinari, T. *Appl. Environ. Microbiol.* 1980, 39, 81–84.
75. Macy, J. M.; Schroder, I.; Thauer, R. K.; Korger, A. *Arch Micribiol.* 1986, 144, 147–150.
76. Hungate, R. E. The rumen and its microbes, Academic Press, New York, USA, 1966, 1–90.
77. Niederman, R. A.; Wolin, M. J. *J. Bacteriol.* 1972, 109, 546–549.
78. Kroger, A.; Winkler, E. *Arch Micribiol.* 1981, 129, 100–104.
79. Bronder, M.; Mell, H.; Stupperich, E.; Kroger, A. *Arch Microbiol.* 1982, 131, 216–223.
80. Davis, C. P.; Cleven, D.; Brown, C. J.; Balish, E. *Int. J. Syst. Bacteriol.* 1976, 26, 498–504.
81. Samuelov, N. S.; Lamed, R.; Lowe, S.; Zeikus, J. G. *Appl. Environ. Microbiol.* 1991, 57, 3013–3019.
82. Glassner, D. A. *Eur. Patent* 389-103, 1989.
83. Podkovyrov, S. M.; Zeikus, J. G. *J. Gen. Microbiol.* 1993, 139, 223–228.
84. Datta, R.; Glassner, D. A.; Jain, M. K.; Roy, J. R. *Eur. Patent* 405-707, 1990.
85. Glassner, D. A.; Datta, R. *U.S. Patent* 5,143,834, 1992.
86. Lemme, C. J.; Datta, R. *Eur. Patent* 249-773, 1987.
87. Woods, D. R. *The Clostridia and Biotechnology*; II Series: Biotechnology series, Butterworth-Heinemann: Stoneham, MA, 1993; pp 443.
88. Drent, W. J.; Lahpor, G. A.; Wiegant, W. M.; Gottschal, J. C. *Appl. Environ. Microbiol.* 1991, 57, 455–462.
89. Cato, E. P.; George, W. L.; Finegold, S. M. In *Bergey's Manual of Systematic Bacteriology*; Sneath, P. H. A.; Mair, N. S.; Sharpe, M. E.; Holt, J. G., Eds.; The Williams & Wilkins Co.: Baltimore, MD, 1986, Vol. 2; pp 1141–1200.
90. Ollivier, B.; Smiti, N.; Garcia, J. L. *Biotechnol. Lett.* 1985, 7, 847–852.
91. Paster, B. J.; Russell, J. D.; Yang, C. M. J.; Chow, J. M.; Woese, C. R.; Tanner, R.*Int. J. Syst. Bacteriol.* 1993, 43, 107–110.
92. Sharma, V. K.; Hagen, J. C. *Bioresource Tech.* 1995, 51, 61–74.
93. Yang, J. C.; Chynoweth, D. P.; Williams, D. S.; Li, A. *Int. J. Syst. Bacteriol.* 1990, 40(3), 268–272.
94. Crociani, F.; Biavati, B.; Castagnoli, P.; Matteuzi, D. *J. Appl. Bacteriol.* 1984, 57, 83–88.
95. Doelle, H. W. *Bacterial Metabolism*; Academic Press, Inc.: New York, NY, 1975; pp 588–591.
96. Lamanna, C.; Mallett, M. F. *Basic Bacteriology-Its Biological and Chemical Background*; Williams & Wilkins: Baltimore, MD, 1965.
97. Kaneuchi, C.; Benno, Y.; Mitsuoka, T. *Int. J. Syst. Bacteriol.* 1976, 26, 482–486.

98. Benno, Y.; Endo, K.; Namba, Y.; Komori, T. ; Mitsuoka, T. *Appl. Environ. Microbiol.* 1989, 55, 1100–1105.
99. Lee, W. K.; Fujisawa, T.; Kawamura, S.; Itoh, K.; Mitsuoka, T. *Laboratory Animals* 1991, 25, 9–15.
100. Wood, H. G.; Werkman, C. H. *Biochem. J.* 1936, 30, 618–623.
101. Wood, H. G.; Werkman, C. H. *Biochem. J.* 1938, 32, 1262–1271.
102. Wood, H. G.; Werkman, C. H. *Biochem. J.* 1940, 34, 129-138.
103. de Vries, W.; van Wyck-Kapteyn, W. M. C.; Stouthamer, A. H. *J. Gen. Microbiol.* 1972, 71, 515–524.
104. de Vries, W.; van Wyck-Kapteyn, W. M. C.; Stouthamer, A. H. *J. Gen. Microbiol.* 1973, 76, 31–41.
105. Crow, V. L. *Appl. Environ. Microbiol.* 1986, 52, 359–365.
106. Crow,V. L.; Turner, K. W. *New Zealand J. Dairy Sci. Tech.* 1986, 21, 217–227.
107. Wood, H. G. *Current Topics in Cellular Regulation* 1981, 18, 255–287.
108. Crow, V. L.; Martley, F. G.; Delacroix, A. *New Zealand J. Dairy Sci. Tech.* 1988, 23, 75–85.
109. Rosner, B.; Schink, B. *Arch. Microbiol.* 1990, 155, 46–51.
110. Lewis, V. P.; Yang, S. T. *Appl. Microbiol. Biotechnol.* 1992, 37, 437–442.
111. Hsu, S. T.; Yang, S. T. *Biotechnol. Bioeng.* 1991, 38, 571–578.
112. Czelovsky, J.; Wolf, G.; Hammes, W. P. *Appl. Environ. Microbiol.* 1992, 37, 94–97.
113. Kaneuchi, C.; Seki, M.; Komagata, K. *Appl. Environ. Microbiol.* 1988, 54, 3053–3056.
114. Radler, F. In *Lactic Acid Bacteria in Beverages and Food*; Carr, J. G.; Cutting, C. V.; Whiting, G. C., Eds.; Academic Press,Inc.: London, 1975, pp 17–27.
115. Cato, E. P.; Moore, W. E. C.; Johnson, J. L. *Int. J. Syst. Bacteriol.* 1983, 33, 426–428.
116. Whiting, G. C. In *Lactic acid bacteria in beverages and food* ; Carr, J. G.; Cutting, C. V.; Whiting, G. C., Eds.; Academic Press,Inc.: London, 1975, pp 69–85.
117. Lindgren, S. E.; Axelsson, L. T.; McFeeters, R. F. *FEMS Microbiol. Lett.* 1990, 66, 209–214.
118. McFeeters, R. F.; Chen, K. H. *Food Microbiol.* 1986, 3, 73–81.
119. Macy, J. M.; Probst, I. *Ann. Rev. Microbiol.* 1979, 33, 561–594.
120. Caspari, D.; Macy, J. M. *Arch. Microbiol.* 1983, 135, 16–24.
121. Macfarlane, G. T.; Gibson, G. R. *Appl. Environ. Microbiol.* 1991, 57, 1–6.
122. Kotarski, S. F.; Salyers, A. A. *J.Bacteriol.* 1981, 146, 853–860.
123. Jackson, F. L.; Goodman, Y. E. *Int. J. Syst. Bacteriol.* 1978, 28, 197–200.
124. Smibert, R. M.; Holdeman, L. V. *J. Clin. Microbiol.* 1976, 3, 432–437.
125. Wong, J. C.; Dyer, J. K.; Tribble, J. L. *Appl. Environ. Microbiol.* 1977, 33, 69–73.
126. Reddy, C. A.; Bryant, M. P. *Can. J. Microbiol.* 1977, 23, 1252–1256.
127. Loesche, W. J.; Socransky, S. S.; Gibbons, R. J. *J. Bacteriol.* 1964, 88, 1329–1337.

128. Janssen, P. H.; Schink, B. *Arch. Microbiol.* 1995, 163, 188–194.
129. Utkin, I.; Woese, C.; Wiegel, J. *Int. J. Syst. Bacteriol.* 1994, 44, 612–619.
130. Kane, M. D.; Breznak, J. A. *Arch. Microbiol.* 1991, 156, 91-98.
131. Maczulak, A. E.; Wolin, M. J.; Miller, T. L. *Appl. Environ. Microbiol.* 1993, 59, 657–662.
132. Ohta, K.; Beall, D. S.; Mejia, J. P.; Shanmugam, K. T.; Ingram, L. O. *Appl. Environ. Microbiol.* 1991, 57, 2810–2815.
133. Delahaye, E. P.; Bouchet, B.; Andrieux, C.; Raibund, P.; Szylit, O.; Gallant, D. J. *J. Sci. Food. Agric.* 1994, 65, 487–495.
134. Ouattara, A. S.; Traore, A. S.; Garcia, J. L. *Int. J. Syst. Bacteriol.* 1992, 42, 390–397.
135. Prins, R. A.; Lankhorst, A.; van der Meer, P.; Van Nevel, C. J. *Antonie van Leeuwenhoek* 1975, 41, 1–11.
136. Anderson, R. L.; Ordal, E. J. *J. Bacteriol.* 1961, 81, 139–146.
137. Marvin-Sikkema, F. D.; Richardson, A. J.; Stewart, C. S.; Gottschal, J. C.; Prins, R. A. *Appl. Environ. Microbiol.* 1990, 56, 3793–3797.
138. Muratsubaki, H. *J. Biochem.* 1987, 102, 705–714.
139. Kertes, A. S.; King, C. J. *Biotechnol. Bioeng.*, 1986, 28, 269–282.
140. Tamada, J. A.; Kertes, A. S.; King, C. J. *Ind. Eng. Chem. Res.* 1990, 29, 1319–1326.
141. Tamada, J. A.; King, C. J. *Ind. Eng. Chem. Res.* 1990, 29, 1327–1333.
142. Tamada, J. A.; King, C. J. *Ind. Eng. Chem. Res.* 1990, 29, 1333–1338.
143. Poole, L. J.; King, C. J. *Ind. Eng. Chem. Res.* 1991, 30, 923–929.
144. Starr, J. N.; King, C. J. *Ind. Eng. Chem. Res.* 1992, 31, 2572–2579.
145. Tung, L. A.; King, C. J. *Ind. Eng. Chem. Res.* 1994, 33, 3217–3223.
146. Tung, L. A.; King, C. J. *Ind. Eng. Chem. Res.* 1994, 33, 3224–3229.
147. Eyal, A. M.; Bressler, E. *Biotechnol. Bioeng.* 1993, 41, 287–295.
148. Mok, Y. S.; Lee, W. K. *Separ. Sci. Technol.* 1994, 29, 743–764.
149. Berglund, K. A.; Elankovan, P.; Glassner, D. A.; *Eur. Patent* 410-728, 1991.
150. Glassner, D. A.; Elankovan, P.; Beacom, D. R.; Berglund , K. A. *Appl. Biochem. Biotechnol.* 1995, 51/52, 73–82.
151. Hostettler, F.; Borel, E.; Deuel, H. *Helv. Chim. Acta.* 1951, 34, 2133-2139.

Chapter 14

Microbial Conversion of Glycerol to 1,3-Propanediol: Recent Progress

A.-P. Zeng, H. Biebl, and W.-D. Deckwer

Biochemical Engineering Division, Gesellschaft für Biotechnologische Forschung mbH, Mascheroder Weg 1, D–38124 Braunschweig, Germany

The microbial conversion of glycerol to 1,3-propanediol has recently received much attention because of the appealing properties of 1,3-propanediol and the anticipated surplus of glycerol on the market. Our knowledge of the metabolic pathway, the intrinsic metabolic potential and kinetic limitation of this bioconversion process has been substantially augmented. Progress has also been made in improving the process performance and strains both on process engineering and molecular biology levels. These recent advances are summarized in this communication. Further research and development needs are also discussed.

Glycerol is a by-product from the cleavage of natural fats:

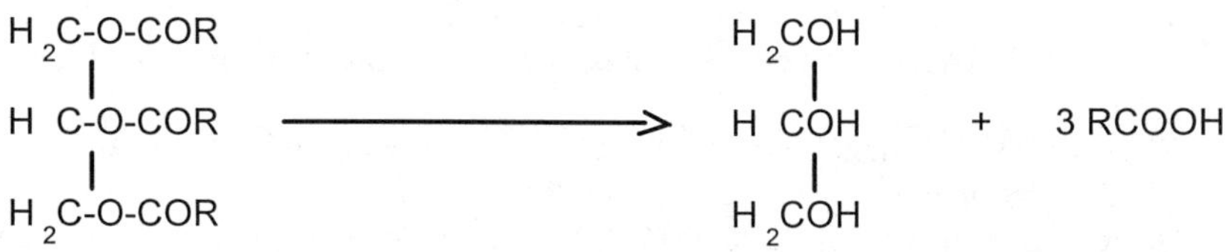

Fat (Triglyceride) — Glycerol — Fatty acid

The fatty acid is used by the oleo-chemical industry as a feedstock for the production of detergents and other chemical intermediates. Glycerol has been traditionally used in the production of pharmaceuticals, cosmetics, resins, food, beverages, tabacco as well as cellophan and explosives. The oleo-chemical industry has considerably grown in the last decade due to the relatively low price increase of natural fats compared to petrochemicals. However, no major new application has been found for glycerol during this period, leading to a surplus of glycerol on the market. This surplus of glycerol is expected to expand further with the envisaged application of rape-seed oil as a diesel substitute. The conversion of rape-seed oil to an appropriate diesel fuel yields about 10% glycerol by weight (*13,15*). The use of glycerol as a stock compound for chemical processing is therefore of industrial interest.

One potential use of surplus glycerol is its conversion to 1,3-propanediol (1,3-PD). Recently, this bioconversion has received world-wide attention. It is mainly driven by two factors. First, 1,3-PD is an appealing product and finds applications in the synthesis of heterocycles and polyesters. Polyesters based on 1,3-PD have special properties such as biodegradability, improved light stability, anti-sliding and re-stretching qualities when used as material for the manufacture of carpet ware in combination with terephthalic acid. Recently, two large chemical companies Shell and Degussa anounced the commercialization of 1,3-PD production based on petrochemical feedstocks. Second, 1,3-PD from biological route represents a rare case for a primary chemical the biological production of which is competitive or even more economical compared to the chemical route (*12*). This fermentation process has been successfully scaled up on a pilot plant scale with batch culture of *Clostridium butyricum* (*17*). The recovery and purification of 1,3-PD has also been intensively studied. Recently, Deckwer (*12*) reviewed the microbial conversion of glycerol to 1,3-PD, covering major work up to about 1993 and including economical aspects. In this communication advances achieved in the last few years are briefly summarized, focusing mainly on pathway, stoichiometric and kinetic analysis, strain improvement and process optimization.

Metabolic Pathways of Glycerol Bioconversion

Glycerol can be converted to 1,3-PD by a number of bacteria including strains of the genera *Klebsiella, Citrobacter*, and *Clostridium* under anaerobic conditions. Among these, *Klebsiella pneumoniae* and *Clostridium butyricum* have been most intensively studied. The pathway of anaerobic dissimilation in these strains is depicted in Fig.1.

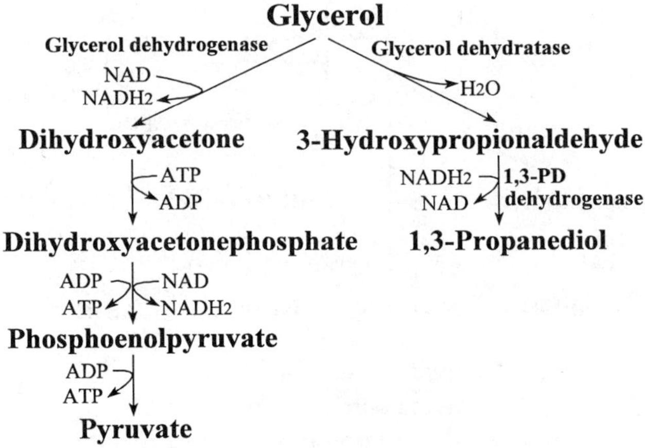

Fig.1 Pathway of anaerobic glycerol dissimilation in *K. pneumoniae* and *C. butyricum.*

Glycerol is fermented by a dismutation process involving two parallel pathways. Through the oxidative pathway, glycerol is dehydrogenated by an NAD^+-linked enzyme, glycerol dehydrogenase, to dihydroxyacetone (DHA) which is then further metabolized to pyruvate. Through the parallel reductive pathway, glycerol is dehydrated by a B_{12}-dependent glycerol dehydratase to form 3-hydroxypropionaldehyde which is then reduced to 1,3-PD by an NADH-linked oxidoreductase, 1,3-PD dehydrogenase. The physiological role of the 1,3-propanediol pathway is to regenerate the reducing equivalents ($NADH_2$) which are released from the formation of DHA and during the further oxidation of dihydroxyacetonephosphate (DHAP) as well as from biosynthesis. The enzymes leading to the formation of 1,3-PD and DHAP have been studied by many researchers with strains of Enterobacteriaceae (*9,19,20,25,29*). Enzymes active in the glycerol metabolism of *C. butyricum* were measured very recently (*1,2*).

The further metabolism of DHAP is essential to provide ATP for cell growth and for the necessary phosphorylation of dihydroxyacetone. In addition, it provides the reducing equivalents for the 1,3-PD pathway, leading to the formation of 1,3-PD. Despite its importance for an optimum production of 1,3-PD the metabolism of DHAP and its subsequent metabolites has received little attention in the past. In fact, the pathways of DHAP (and pyruvate) oxidation have been taken from glucose metabolism without experimental evidence. There is few work concerning the enzymes which catalyze the metabolism of pyruvate in pure glycerol fermentation. Zeng et al. (*33,38*) applied the pathway stoichiometry to analyze the fermentation of glycerol in both *K. pneumoniae* and *C. butyricum*, with particular emphasis on the regulation of pyruvate metabolism and its influences on the 1,3-PD yield and selectivity. Fig. 2 summarizes the pathways of pyruvate metabolism in *K. pneumoniae* and *C. butyricum* grown anaerobically on glycerol, respectively.

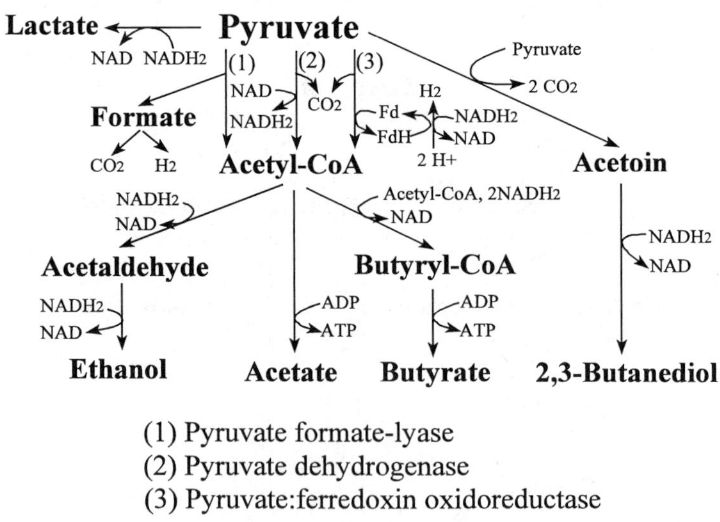

Fig. 2. Pyruvate metabolism during anaerobic fermentation of glycerol.

The cleavage of pyruvate to acetyl-CoA and CO_2 is assumed to be carried out by the enzyme pyruvate: ferredoxin oxidoreductase in *C. butyricum* and by the enzyme pyruvate-formate lyase in *K. pneumoniae* in the literature. Acetic acid and butyric acid are the main fermentation products of pyruvate in *C. butyricum*. *K. pneumoniae* produces no butyric acid, but ethanol as one of the main products. Minor products include 2,3-butanediol, lactic and formic acid. At low pH value a significant amount of 2,3-butanediol is formed in the glycerol fermentation of *K. pneumoniae* (*7*).

Analysis of continuous culture data of *C. butyricum* revealed that the reduced ferredoxin (Fd_{red}) formed during oxidation of pyruvate to acetyl-CoA is not completely cleaved into hydrogen and oxidized ferredoxin under conditions of glycerol excess (*38*). Instead, part of the reducing power from Fd_{red} is transferred to NAD^+ under the formation of $NADH_2$. The enzymes catalysing this reaction had been previously described for *C. acetobutylicum* (*21*) and recently for *C. butyricum* (*2*). In *K. pneumoniae* reducing equivalents released from pyruvate cleavage by the pyruvate formate lyase are trapped in formate and cannot be transferred to NAD. It was therefore surprising to find substantial deviations of the ratio of 1,3-PD to hydrogen from the calculated one based on the action of pyruvate formate lyase in this species. We could recently demonstrate by enzyme assays that pyruvate dehydrogenase, which is normally the enzyme complex for an aerobic pyruvate decarboxylation in Enterobacteriaceae, is simultaneously involved in this anaerobic fermentation process (Menzel et al., GBF, unpublished data). Factors that affect the pyruvate dehydrogenase activity in *K. pneumoniae* are being studied in continuous culture with the goal of further increasing the activity of this enzyme for a high yield and flux of 1,3-PD. Although the activity of this enzyme is desirable its simultaneous involvement in addition to pyruvate formate lyase in the *K. pneumoniae* culture gives rise to unfavorable dynamic behavior of the pathways such as oscillation and hysteresis under a variety of conditions (*37*).

Stoichiometric and Energetic Considerations of Yield Coefficients and Pathway Selectivity under Different Conditions

The yield of 1,3-PD from glycerol depends significantly on the selectivity of pathways involved in the pyruvate metabolism and on the regulation of the reducing equivalent balance (*33,38*). This is illustrated below by the glycerol fermentation of *C. butyricum*. Similar analysis can be done for the glycerol fermentation of *K. pneumoniae*. For the glycerol fermentation of *C. butyricum* we first consider the situation in which the reduced ferredoxin (Fd_{red}) is cleaved into hydrogen and oxidized ferredoxin (Fd_{ox}) (reaction 3 in Fig.2). The main reactions involved in the glycerol fermentation by *C. butyricum* under these conditions can be written as follows:

(1) biomass formation

$$C_3H_8O_3 + 3NH_3 + 3\frac{101}{Y_{ATP}}ATP \rightarrow 3C_4H_7O_2N + 8[H] + 6H_2O \quad (1a)$$

Where $C_4H_7O_2N$ denotes the elemental composition of biomass and corresponds to a molecular biomass weight (M_G) of 101 g/mol (*38*). The same formula for biomass has been used for the analysis of glycerol fermentation by *K. pneumoniae* (*33*). Y_{ATP} is the energetic yield of biomass (g biomass/mol ATP) which was found to be about 8.5 g/mol ATP for *C. butyricum* (*8*). For simplicity Y_{ATP} is taken here as 8.6 g/mol ATP. In Eq. 1 [H] represents one g atom available hydrogen which is equivalent to 1/2 mole $NADH_2$. Thus, Eq. (1a) becomes

$$C_3H_8O_3 + 3NH_3 + 35ATP \rightarrow 3C_4H_7O_2N + 4NADH_2 + 6H_2O \quad (1b)$$

(2) acetate formation

$$C_3H_8O_3 \rightarrow C_2H_4O_2 + 2NADH_2 + 2ATP + CO_2 + H_2 - H_2O \quad (2)$$

(3) butyrate formation

$$2C_3H_8O_3 \rightarrow C_4H_8O_2 + 2NADH_2 + 3ATP + 2CO_2 + 2H_2 \quad (3)$$

(4) 1,3-propanediol formation

$$C_3H_8O_3 + NADH_2 \rightarrow C_3H_8O_2 + 2H_2O \quad (4)$$

From Eqs. 2 and 3 it is obvious that the formation of acetate and/or butyrate is necessary for the generation of energy (ATP) for biosynthesis. These two pathways are also associated with the generation of reducing equivalent which is regenerated by the 1,3-PD pathway. Thus, the overall energetic and product yields depend on the degree of involvement of the acetic acid and butyric acid pathways. Two extreme cases can be considered.

Case I assumes that only the acetate pathway is used for energy generation. Under the prerequisite of ATP and $NADH_2$ balances a fermentation equation can be obtained through the operation 17.5 x Eq.2 + Eq.1b + 39 x Eq.4:

$$60.5C_3H_8O_3 + 3NH_3 \rightarrow 3C_4H_7O_2N + 39C_3H_8O_2 + 17.5C_2H_4O_2 + 17.5CO_2 + 17.5H_2 \quad (5)$$

According to this reaction, the yield coefficients on glycerol can be calculated for biomass, ATP, acetic acid and propanediol which are given in row I of Table I.

Case II assumes that only the butyrate pathway is used for the energy generation. Under the prerequisite of ATP and $NADH_2$ balances it follows from Eqs. (1-4) (11.7 x Eq.3 + Eq.1b + 27.4 x Eq.4):

$$54.8C_3H_8O_3 + 3NH_3 \rightarrow 3C_4H_7O_2N + 27.4C_3H_8O_2 + 11.7C_4H_8O_2 + 23.4CO_2 + 23.4H_2 \quad (6)$$

The yield coefficients on glycerol according to this fermentation equation are calculated and given in row II of Table I. It can be seen that the butyrate pathway renders higher ATP and biomass yields than the acetic acid pathway under these conditions. Thus, for the biosynthesis of cell material the butyrate pathway is more efficient than the acetate pathway. The maximum ATP and biomass yields on glycerol would be 0.64 mol/mol and 5.52 g/mol, respectively. The involvement of the acetic acid pathway results in reduced yields of ATP and biomass. However, acetic acid formation involves a 1,3-PD yield which is about 30% higher than butyric acid formation. Thus, for the production of 1,3-PD the acetic acid pathway is more attractive. According to the above calculations the theoretical maximum 1,3-PD yield would be 0.65 mol PD/mol glycerol. It will be reduced as soon as butyric acid is produced.

Table I. Theoretical maximum yields of growth and metabolism of *C. butyricum* under different conditions

	Conditions	$Y_{PD/S}$ mol/mol	$Y_{HAc/S}$ mol/mol	$Y_{HBu/S}$ mol/mol	$Y_{X/S}$ g/mol	$Y_{ATP/S}$ mol/mol
Case I	No butyrate Maximum H_2	0.65	0.29	0.00	5.05	0.56
Case II	No acetate Maximum H_2	0.50	0.00	0.21	5.53	0.64
Case III	No butyrate No H_2	0.72	0.22	0.00	3.84	0.45
Case IV	No acetate No H_2	0.65	0.00	0.15	3.84	0.45

S: Glycerol; PD: 1,3-propanediol; HBu: butyric acid; HAc: acetic acid; X: biomass

Table II shows typical experimental data of glycerol fermentation by *C. butyricum*. Both the two fermentation cases considered above have been experimentally encountered under low residual glycerol conditions. The steady-state at D = 0.052 h^{-1} under conditions of low residual glycerol in glycerol-limited culture represents the case of no acetate formation (case II of Table I). Under these conditions the experimental butyric acid yield reached exactly the theoretical maximum (0.21 mol/mol) as calculated above. The propanediol and ATP yields are also very close to the theoretical values. The steady-state at D = 0.60 h^{-1} approximately approached the case of no butyric acid formation (case I of Table I). These results confirm the general approach used for the theoretical calculation of yields.

Inspection of data in Table II shows that at similar growth rate (= dilution rate) the energetic yield $Y_{ATP/S}$ and biomass yield $Y_{X/S}$ are generally higher under low residual glycerol conditions than under glycerol-excess conditions. This applies also to the formation of butyric acid. Furthermore, at low residual glycerol the yield of butyric acid decreases with increasing dilution rate. In contrast, the yield of acetic acid increases with growth rate and is lower under low residual glycerol conditions than under glycerol-excess conditions. $Y_{PD/S}$ is less dependent on growth rate. Significant difference is however observed for cultures under glycerol limitation and glycerol excess: $Y_{PD/S}$ is generally higher under glycerol excess than under glycerol limitation. It is noted that the highest $Y_{PD/S}$ value (0.71 mol/mol) achieved under glycerol excess is distinctively higher than the theoretical maximum value (0.65 mol/mol) as calculated above for acetic acid formation alone. Values of $Y_{PD/S}$ in the range 0.70-0.71 mol/mol have been frequently obtained in continuous culture of *C. butyricum* (*18,24*). Similar high values of $Y_{PD/S}$ (0.68 - 0.73 mol/mol) were also obtained for glycerol fermentation by *K. pneumoniae* (*33*; see also Menzel, K.; Zeng, A.-P.; Deckwer, W.-D. *Enzyme Microbiol. Tech-*

nol. in press). It was found that the increase of 1,3-PD yield in both the cultures of *C. butyricum* and *K. pneumoniae* is due to an altered regulation and balance of reducing equivalents released from the oxidation of pyruvate into acetyl-CoA under conditions of glycerol excess. As pointed out above, reducing equivalents of the reduced ferredoxin formed during the oxidation of pyruvate by *C. butyricum* can be either released as H_2 or transferred to NAD^+ under formation of $NADH_2$:

$$H_2 + Fd_{red} \Leftrightarrow Fd_{red} + (2H^+ \text{ or } NAD^+) \Leftrightarrow NADH_2 + Fd_{oxd} \quad (7)$$

Table II. Experimental yields of growth and metabolism of *C. butyricum* under different culture conditions

Culture	D h^{-1}	Residual glycerol mmol/l	$Y_{PD/S}$ mol/mol	$Y_{HAc/S}$ mol/mol	$Y_{HBu/S}$ mol/mol	$Y_{X/S}$ g/mol	$Y_{ATP/S}$ mol/mol
Low	0.052	0.98	0.51	0.00	0.21	4.48	0.62
residual	0.098	1.63	0.56	0.009	0.19	6.01	0.59
glycerol	0.20	1.63	0.55	0.028	0.18	7.63	0.59
	0.29	2.99	0.60	0.099	0.11	6.84	0.54
	0.39	1.21	0.52	0.13	0.077	6.80	0.49
Glycerol	0.06	27	0.71	0.092	0.12	0.92	0.54
excess	0.1	30.5	0.69	0.068	0.14	3.32	0.56
	0.11	34.5	0.70	0.088	0.14	3.20	0.59
	0.23	134	0.61	0.13	0.074	3.91	0.47
	0.35	237	0.66	0.14	0.072	4.63	0.50
	0.45	254	0.60	0.10	0.10	4.94	0.50
	0.56	409	0.65	0.13	0.14	8.70	0.67
	0.60	139.7	0.67	0.22	0.06	4.85	0.62

The distribution of reducing equivalents of Fd_{red} to $NADH_2$ and H_2 has significant influence on the metabolic pathways and maximum yields of the fermentation. It is obvious from reaction (7) that the maximum 1,3-PD yield will be obtained if H_2 is not produced at all. In the following the maximum possible propanediol yields are calculated for the two fermentation pathways under ideal conditions. For the case of the acetate pathway (**Case III**) the following reaction can be written:

$$C_3H_8O_3 \rightarrow C_2H_4O_2 + 3NADH_2 + 2ATP + CO_2 - H_2O \quad (8)$$

Under the prerequisite of ATP and $NADH_2$ balances it follows

$$78C_3H_8O_3 + 3NH_3 \rightarrow 3C_4H_7O_2N + 56.5C_3H_8O_2 + 17.5C_2H_4O_2 + 17.5CO_2 \quad (9)$$

The yield coefficients on glycerol under these conditions are calculated and given in row III of Table I. A propanediol yield as high as 0.72 mol/mol is calculated. This means an increase of propanediol yield of 12.3 % compared with the case of maximal hydrogen evolution. In contrast, the biomass and ATP yields would be reduced by about 22%.

For the case of the butyrate pathway (**Case IV**) the reaction can be written as:

$$2C_3H_8O_3 \rightarrow C_4H_8O_2 + 4NADH_2 + 3ATP + 2CO_2 \quad (11)$$

Under the prerequisite of ATP and $NADH_2$ balances it follows

$$78.2C_3H_8O_3 + 3NH_3 \rightarrow 3C_4H_7O_2N + 50.8C_3H_8O_2 + 11.7C_4H_8O_2 + 23.4CO_2 \quad (12)$$

The yield coefficients on glycerol under these conditions are given in row IV of Table I. The calculations indicate an increase of PD yield of 30 % compared with the case of maximum hydrogen evolution. This is more significant than in the acetate pathway. The biomass and ATP yields would also be reduced by about 30%. It is interesting to note that the ATP and biomass yields of these two pathways under conditions of no hydrogen production are exactly the same. The selectivity of the two pathways under these conditions would be determined by the toxicity of the products. The theoretical maximum propanediol yield (0.72 mol/mol) calculated above for the acetic acid pathway is slightly higher than the experimental maximum value. It can be seen in Table 2 that under conditions of low dilution rate and high residual concentration glycerol fermentation by *C. butyricum* can proceed with nearly theoretical maximum PD yield.

The formation rate of hydrogen in the glycerol fermentation of *C. butyricum* was measured in continuous culture at different dilution rate (*28*). This enabled the calculation of distribution of reducing equivalents produced during glycerol degradation to acetyl-CoA and biomass formation to the $NADH_2$-dependent products (i.e., 1,3-PD, butyric acid and H_2). Figs. 3a and 3b show the distribution of reducing equivalents for substrate-limited and product inhibited cultures, respectively. It can be seen that a proportion of as high as 40-60% of the total reducing equivalents is allocated to the formation of butyric acid and hydrogen under conditions of substrate limitation, whereas it accounts for about only 25-40% of the total reducing equivalents under conditions of substrate excess (product inhibition). The relatively higher level of reducing equivalent allocation to butyric acid under conditions of substrate limitation may be explained by the higher energy and biomass yields of the butyric pathway (see Table I). The energetic consideration gives also a possible explaination for the relatively high allocation of reducing equivalents to H_2 formation under these conditions. Comparing the cases I - IV in Table I it is obvious that an increased formation of H_2 enhances both the energy and biomass yields. However, under conditions of substrate excess the energetics of the cell machinery may no longer have the highest priority due to energy excess, leading to a diminished allocation of reducing equivalents to the formation of hydrogen. Simultaneously, the allocation of reducing equivalents to the formation of butyric acid is also slightly reduced, propabaly due to its strong toxicity to growth (*34*). It should be mentioned that the experimentally measured formation rates of butyric acid and hydrogen under substrate excess are somewhat higher than one would expect from the stoichiometry and the high 1,3-PD yield under these conditions. It is possible that more reducing equivalents than assumed in eq. (1) might have been released during biomass formation. Theoretically, an 1,3-PD yield higher than 0.72 mol/mol would be thus possible if the formation of butyric acid and/or hydrogen can be further reduced. Efforts have been made in obtaining mutants without butryric acid and hydrogen formation (see below).

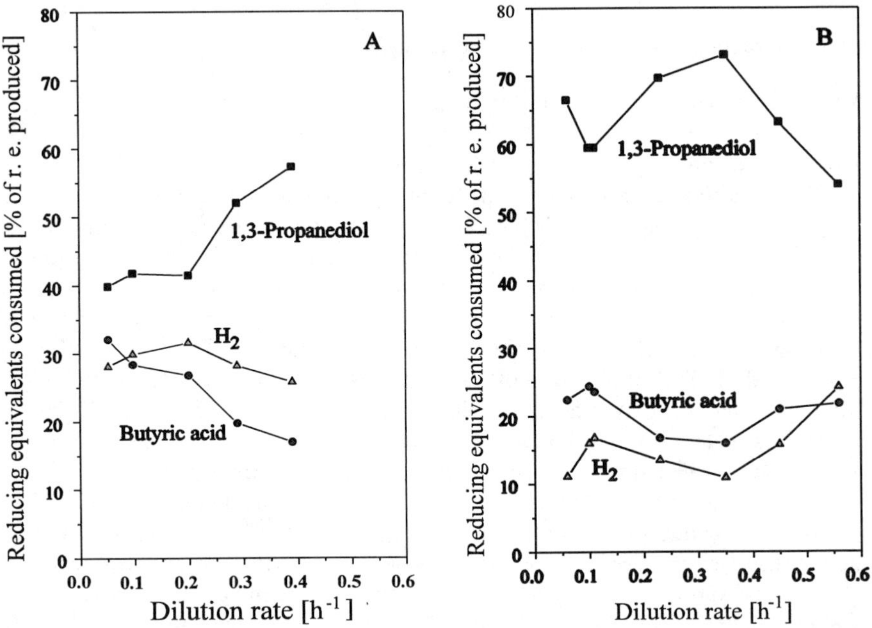

Fig.3. Distribution of reducing equivalents produced during glycerol catabolism to acety-CoA and biomass formation to the NADH-dependent products. Values normalized to $R_H = 1.0$. Data from Solomon et al. (*8*). (A) Substrate limited culture. (B) Product limited culture.

Kinetic Analysis of Cell Growth and Product Formation

Biebl (*4*) studied the inhibition potentials of 1,3-PD, acetic acid, butyric acid and glycerol on the growth of *C. butyricum* in a pH-auxostat culture and found that all these substances are toxic to *C. butyricum*. The critical concentrations of these substances, i.e., concentrations above which cells cease to grow, were found to be: 27 g/l for acetic acid (0.49 g undissociated acetic acid/l); 19 g/l for butyric acid (0.39 g undissociated acid/l), 64.0 g/l for 1,3-PD, and 97.0 g/l for glycerol at pH 6.5 in this study with externally added substances. Zeng et al. (*34*) further examined the inhibition potentials of substrate and products both self-produced and externally added on *C. butyricum* and *K. pneumo-*

niae with the help of a mathematical model. Whereas the inhibition potential of externally added and self-produced 1,3-propanediol reveals to be essentially the same, butyric acid produced by the culture is found to be more toxic than that externally added. The same seems to apply for acetic acid. Furthermore, the inhibitory effect of butyric acid is shown to be due to the total concentration instead of its undissociated form.

The inhibition effects of products and substrate in the glycerol fermentation are essentially irrespective of the strains. Thus, a common growth model could be proposed for the two strains anaerobically grown on glycerol at different pH value:

$$\mu = \frac{\mu^*_{\max}}{1+\frac{H^+}{K_H}+\frac{K_{OH}}{H^+}} \frac{C_{Gly}}{C_{Gly}+K_S}\left(1-\frac{C_{HAc}}{C^*_{HAc}}\right)\left(1-\frac{C_{HBu}}{C^*_{HBu}}\right)\left(1-\frac{C_{EtOH}}{C^*_{EtOH}}\right)\left(1-\frac{C_{PD}}{C^*_{PD}}\right)\left(1-\frac{C_{Gly}}{C^*_{Gly}}\right) \quad (13)$$

where μ is the specific growth rate; H^+ is the hydrogen ion concentration; $\mu^*_{\max}$, K_H, and K_{OH} are constants; K_s is the saturation constant; C^*_{HAc}, C^*_{HBu}, C^*_{EtOH}, C^*_{PD}, C^*_{Gly} are the critical concentrations of acetic acid, butryic acid, ethanol, 1,3-PD and glycerol respectively. In the above model the parameters $\mu^*_{\max}$, K_s, K_H, and K_{OH} are strain-specific and can be estimated from experimental data for each strain (*34*). The critical concentrations of acetic acid, 1,3-PD and glycerol are assumed to be identical for both *C. butyricum* and *K. pneumoniae*. The best estimates of the critical concentrations (C^*_{pi}) are as follows: 0.35 g/l for undissociated acetic acid, 10.1 g/l for total butyric acid, 16.6 g/l for ethanol, 71.4 g/l for 1,3-PD, and 187.6 g/l for glycerol. Eq. (13) describes the product and substrate inhibition of both *C. butyricum* and *K. pneumoniae* in different types of continous cultures over a pH range of 5.3-8.5 satisfactorily.

Zeng and Deckwer (*35*) and Zeng (*36*) studied also the kinetic of substrate consumption and product formation of *K. pneumoniae*. They found that in order to achieve a high substrate uptake rate and a high production rate of 1,3-PD a certain level of glycerol excess in the culture is necessary. The dependence of increase of substrate uptake rate and product formation rate on the excess concentration of glycerol can be described with a saturation function similar to that of Michaelis-Menten kinetics.

The maximum achievable concentration of 1,3-PD in the continuous fermentation of glycerol can be predicted from considerations of growth and product formation stoichiometry and product inhibition (Eq.13). As discussed above, the maximum 1,3-PD production would be obtained in both strains if only 1,3-PD and acetic acid are formed as fermentation products and at the same time there is no formation of hydrogen. Under these conditions the mole ratio of acetic acid to 1,3-PD should be 0.31 mol/mol. At pH = 7 and a residual glycerol concentration of 2-100 mmol/l ($C^*_{Gly} >> C_{Gly} >> K_S$) eq. (13) reduces to

$$\mu = \mu_{\max}\left(1-\frac{C_{HAc}}{C^*_{HAc}}\right)\left(1-\frac{C_{PD}}{C^*_{PD}}\right) \quad (14)$$

Substituting C_{HAc} by 0.31 C_{PD} Eq. (14) can be used to predict the maximum achievalbe 1,3-PD concentration and productivity under different dilution rate. Since all the parameters of Eq. 14 are the same for *C. butyricum* and *K. pneumoniae* at pH 7.0 ($\mu_{\max}$ =

0.67 h^{-1}) the predicted theoretical maximum product concentration and productivity are applied to both strains. Fig. 4 shows the theoretical maximum propanediol concentration and experimental maximum values achieved so far at different dilution rate. At high dilution rate (D > ca. 0.35 h^{-1}) nearly the same level of 1,3-PD has been obtained for *C. butyricum* and *K. pneumoniae*, being about 70 - 80% of the theoretical maximum. Whereas at low dilution rate (< 0.35 h^{-1}) *K. pneumoniae* reached 80 - 96% of the theoretical maximum *C. butyricum* reached only about 40 - 60%. At the lowest dilution rate (0.1 h^{-1}) where experimental data are available for both strains *C. butyricum* reached less than half of the propanediol concentration and productiviy of *K. pneumoniae*. Obviously, the maximum values measured for *C. butyricum* do not necessarily represent the real maximum values achievable at the respective dilution rate. As shown for *K. pneumoniae* (Menzel, K.; Zeng, A.-P.; Deckwer, W.-D. *Enzyme Microbiol Technol.* in press) there exists an optimum residual glycerol concentration for propanediol formation at each dilution rate. So far, no systematical work has been carried out to find the optimum propanediol production at each dilution rate for *C. butyricum*. Another reason for the lower propanediol production by *C. butyricum* seems to be due to the less favorable distribution of the by-products. At lower dilution rates the acetic acid production of *C. butyricum* is far below the acetate level for optimal 13-PD formation. As mentioned above, for an optimal production of 1,3-PD the formation of butyric acid in *C. butyricum* and ethanol in *K. pneumoniae* should be as low as possible. It can be shown that the butyric acid concentration in *C. butyricum* culture is often higher than the ethanol concentration in the *K. pneumoniae* culture. However, butyric acid is nearly twice as toxic as ethanol (*34*). Thus, 1,3-PD concentration is limited in the *C. butyricum* culture.

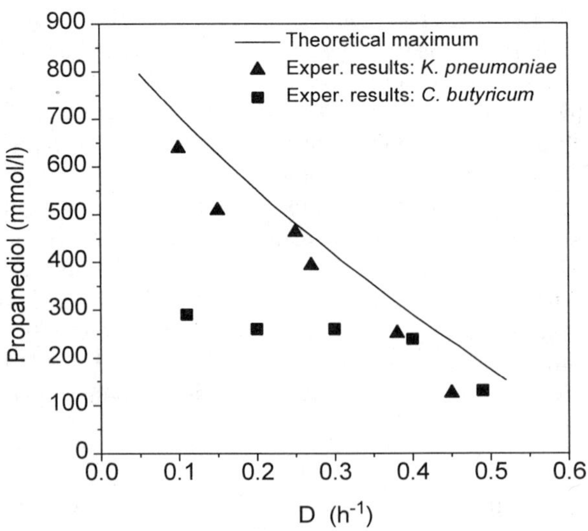

Fig.4. Comparison of maximum 1,3-propanediol concentration experimentally achieved for *C. butyricum* and *K. pneumoniae* with theoretical maximum values under ideal conditions. (Reproduced with permission from reference 38. Copyright 1996 Springer-Verlag.

In addition, the hydrogen production in *C. butyricum* appeared to be higher than in *K. pneumoniae* under substrate excess conditions. This also impedes the production of 1,3-PD in *C. butyricum*. Recently, Barbirato et al. (*3*) found that the accumulation of 3-hydroxypropionalde-hyde strongly inhibits glycerol fermentation of three enterobacterial species including *K. pneumoniae*. Cameron et al. (*10*) showed that glycerol metabolism in *E. coli* is inhibited by the intracellular accumulation of glycerol-3-phosphate. The identification of possible inhibitory effects of intermediate metabolite(s) in *C. butyricum* is desirable.

Process Development and Optimization

It is still an open question if the enterobacteria of the genera *Klebsiella* and *Citrobacter* or the clostridia are more suitable for a 1,3-PD production process. From the viewpoint of an interested company there will probably be a preference for the clostridia as both enterobacterial species are classified as opportunistic pathogens and thus would require costly safety precautions. On the other hand, as shown in the preceding section the strains of *Clostridium butyricum* which are presently in use cannot entirely compete with strains like *Klebsiella pneumoniae* DSM 2026 at least if productivity is concerned (*38*). Several efforts have been made in the last five years to meet this shortcoming. The medium has been improved, better culture techniques have been elaborated, new strains have been isolated from nature or selected from culture collections, and mutants have been obtained which are considerably increased in product tolerance and yield.

Medium and Culture Conditions. For estimations of the performance of *Klebsiella* and *Clostridium* cultures virtually different media had been in use. When the medium used for *Klebsiella* (*14*) was applied to *Clostridium butyricum* DSM 5431 indeed the steady state product concentration in continuous cultures could be increased. Higher iron content and the presence of citrate as a complexing agent were found to be the main cause of this stimulation (*24*). The yeast extract formerly used in a concentration of 1 g/l (*17*) could be replaced by biotin with only a slight loss in productivity.

Continuous culture is a good tool for anaerobic glycerol fermentation particularly when a second stage is used to increase the final product concentration (*8*). However, for maximum propanediol content and simple operation batch cultures appear to be more advantageous. Günzel et al. (*17*) described a fedbatch culture that consumed about 110 g/l of glycerol supplied in three successive additions over a period of 24 h which resulted in a 1,3-propanediol concentration of 56 g/l. Recently this process was automated using the pH decrease as a signal for glycerol addition. The culture was kept at a slight glycerol excess so that substrate limiting intervals were avoided; the residual glycerol was used up towards the end of the fermentation after turning off the glycerol supply. The results were about the same as in discontinuous feeding (*24a*). A similar system was described by Saint-Amans et al. (*27*) for *C. butyricum* VPI 3266 using CO_2 production for control of glycerol supply. Probably due to the properties of the particular strain (see below) the final product concentration was higher (65 g/l), but the fermentation time was about three times as long.

Procedures to increase the productivity by immobilization or cell recycling have been rarely elaborated for the glycerol fermentation. Pflugmacher and Gottschalk (*23*) described a fixed bed loop reactor culture of *Citrobacter freundii* using polyurethane foam as carrier substance. In comparison to a stirred tank reactor culture (*8*) the propanediol productivity was more than doubled, but the propanediol concentration could not be increased beyond 19 g/l. Similar results were obtained with *Clostridium butyricum* using a crossflow filtration module. Although it was possible to increase the productivity up to 3 times of the conventional continuous culture, the steady state propanediol concentration was not higher than 14 g/l (Reimann and Biebl, unpublished results).

Use of Cosubstrates. As pointed out above a 1,3-propanediol yield of 72 mol per 100 mol of glycerol cannot be surpassed. If however another fermentable electron donor substrate could be applied in addition to glycerol, a 100 % product yield is conceivable. In the patent literature procedures are described in which glucose is the cosubstrate and enterobacteria the converting organisms. It is proposed to grow the cultures initially with glycerol alone to induce the 1,3-propanediol forming enzymes. Glucose is added either directly to the growing culture (*32*) or used in a mixture with glycerol in a resting-cell culture without an ammonium source (*16*). 91 to 100 % of the glycerol were converted to 1,3-propanediol by this technique.

If glucose is considered as cosubstrate it has to be kept in mind that hexoses are less reduced substrates than glycerol, so that twice as much glucose than glycerol is needed on a weight basis to provide the same amount of reducing equivalents. This implies that use of glucose in 1,3-propanediol production would only be useful if glucose is available at a considerably lower price than glycerol, which appears to be presently the case, especially in the USA.

Another prerequisite for glucose addition should be that it is converted mainly to acetate in the presence of glycerol and not to ethanol or butyrate. This has been verified for *C. butyricum* (*6*). However, there are other physiological barriers in simultaneous metabolization of glycerol and glucose. As reported by Biebl and Marten (*6*) glucose is fermented much slower than glycerol by glycerol-fermenting clostridia so that a complete conversion of the glycerol cannot be obtained. A fermentable substance other than glucose which is a cheaply available chemical bulk product, ethylenglycol, proved to be unsuitable as a cosubstrate. In the presence of glycerol this diol did not release reducing equivalents in the course of acetate formation as expected, but utilized hydrogen equivalents to yield ethanol as product, while glycerol was oxidized to acetate.

New Strains of Clostridium butyricum. Whereas in the case of *Klebsiella pneumoniae* the type strain of the species is still unsurpassed in glycerol fermentation, in the case of *Clostridium butyricum* new screening efforts have led to strains that are markedly improved in product tolerance (*22*). In fed batch culture the best strain (E5) produced up to 66 g/l of propanediol from 122 g/l of glycerol compared to 55 g/l from 108 g/l of glycerol with strain DSM 5431 (*5,17*). The new strain was distinguished by very low hydrogen evolution (*24a*). However, its maximum growth rate and productivity were distinctly lower. The strain used by Saint-Amans et al. (*26*), *C. butyricum* VPI 3266, is probably closely related.

Genetic Approaches to Strain Improvement in C. butyricum have been undertaken only very recently (*22*). NTG mutations of strain DSM 5431 were selected in the presence of high propanediol concentrations and further on bromide-bromate glucose medium to obtain mutants that sustained substantially higher product concentrations and were strongly reduced in hydrogen evolution (Reimann and Biebl, unpublished data). In fedbatch culture the best of them was able to convert up to 130 g of glycerol to about 70 g of 1,3-propanediol. Thus it resembles the new isolates obtained by the same group as well as the VPI strain used by Saint-Amans et al. (*27,28*). If these strains will be subjected to genetic improvement further increase in product concentration can be expected.

Attempts to generate a strain from the product tolerant, hydrogen reduced mutants that was defect or affected in butyrate formation failed with several methods. In contrast it was possible, to obtain butyrate reduced mutants from the isolate E5 by selection on allylalcohol (*22*). These mutants were not much changed in propanediol production but exhibited a hydrogen production that was near the physiological maximum whereas the parent strain was very low in hydrogen evolution. This result in addition to fermentation data and experiments that showed stimulation of 1,3-propanediol production in the presence of an aldehyde prompted the authors to assume that glycerol dehydration the product of which is the toxic 3-hydroxypropionaldehyde is the rate limiting step. Consequently, a certain amount of reducing equivalents has to be disposed via butyrate and/or hydrogen, as the 1,3-propanediol pathway can be only varied within a narrow range. A possibility to claim all reducing equivalents produced in glycolysis for 1,3-propanediol formation would be to overexpress the glycerol dehydratase along with the 1,3-propandiol dehydrogenase using recombinant DNA techniques. Such experiments are presently in progress.

Genetic Improvement with Enterobacterial Strains. From *Citrobacter* and *Klebsiella* strains over-producing mutants have not been isolated. On the other hand basic genetic research on the key enzymes of the process is more advanced than in clostridia. In both species the genes of the *dha* regulon which encode for the glycerol dehydrogenase, the dihydroxyacetone kinase, the gycerol dehydratase and the 1,3-propanediol dehydrogenase have been cloned and expressed in *E. coli* (*10,11,30,31*). A new construction of the *dha* regulon has been achieved by Cameron (personal communication). The amounts of 1,3-propanediol wich have been reached in the recombinant *E. coli* strains do not yet approach that of the original strains (i.e. 9 g/l for strains with *Citrobacter* and 6 g/l for strains with the *Klebsiella* genes), but in the long term highly productive recombinant *Klebsiella* strains can be expected as soon as the optimized genes have been reintroduced into the donor strains.

Conclusion and Outlook

Recent theoretical and experimental work has advanced our understanding of the metabolic pathways and elucidated the intrinsic metabolic potential and kinetic limitations of microbial conversion of glycerol to 1,3-propanediol by the major strains isolated. The maximum product yield conceivable is about 0.72 mol/mol in fermentation with glycerol as a sole carbon and 1.0 mol/mol in fermentation with cosubstrates. However, these high yields are normally not achieved in fermentations with high final product concentration which is kinetically constrained to about 65 -70 g/l. Further work is needed to optimize the distribution of reducing power released from the metabolism of glycerol and/or cosubstrate and to reduce the formation of toxic by-products for simultaneously achieving a high PD yield and concentration. This may be best achieved by re-designing the metabolic pathway and properly controling the cultivation conditions. Progress has already been made in these directions. The use of recombinant techniques also opens new ways for producing PD from other cheap carbon sources. It will become more evident that the microbial production of 1,3-propanediol is technically more flexible and economically more competitive than the chemical route.

Literature

1. Abbad-Andaloussi, S.; Guedon, E.; Spiesser, E; Petitdemange, H. *Lett. Appl. Microbiol.* **1996**, 22, 311-314

2. Abbad-Andaloussi, S.; Dürr, C.; Raval, G.; Petitdemange, H. *Microbiology*, **1996**, 142, 1149-1158.

3. Barbirato, F.; Grivet, J.P.; Bories, A.: Paper (#MEP 55) presented at the 7th European Congress on Biotechnology, Nice, France, **1995**.

4. Biebl, H. *Appl. Microbiol. Biotechnol.* **1991**, 35, 701-705.

5. Biebl, H.; Marten, S.; Hippe, H.; Deckwer, W.-D. *Appl. Microbiol. Biotechnol.* **1992**, 36, 592-597.

6. Biebl, H.; Marten, S. *Appl. Microbiol. Biotechnol.* **1995**, 44, 15-19.

7. Biebl, H,; Zeng, A.-P. Proceedings of Ninth Forum for Applied Biotechnology. Gent, Belgium, **1995**, pp 1927-1932.

8. Boenigk, R.; Bowien, S.; Gottschalk, G. *Appl. Microbiol. Biotechnol.* **1993**, 38, 453-457.

9. Bouvet, O.M.M.; Lenormand, P.; Ageron, E.; Grimont, P.A.D. *Res. Microbiol.* **1995**, 146, 279-290.

10. Cameron, D.C.; Skraly, F.A.; Wilard, B. L.; Zhu, M.-Y.; Held, A.M.; Shaw, A.J.: Paper presented at "Biochemical Engineering IX", Davos, Switzland, **1995**.

11. Daniel, R.; Gottschalk, G. *FEMS Microbiol. Lett.* **1992**, 100: 281-286

12. Deckwer, W.-D. *FEMS Microbiol. Rew.* **1995**, 16, 143-149.

13. Eggersdorfer, M.; Meyer, J.; Eckes, P. *FEMS Microbiol. Rew.* ***1992**, 103, 355-364.*

14. Evans, C.G.T.; Herbert. D.; Tempest, D.W. In *Methods in microbiology*, vol. 2; Editor, J.P., Norris; D.W., Ribbons (eds.); Academic Press: London, New York, **1970**; pp 413-419.

15. Gottlieb, K.; Neitsch, H.; Wessendorf, R. *Chem.-Ing.-Tech.* **1994**, 66: 64-66

16. Gottschalk, G.; Averhoff , B. European patent no 0 373 230 A1, **1988**.

17. Günzel, B.; Yonsel, S.; Deckwer, W.-D. *Appl. Microbiol. Biotechnol.* **1991**, 36, 289-294.

18. Günzel, B. Dissertation, Technische Universität Braunschweig, Germany, **1991**.

19. Lin, E.C.C.: *Ann. Rev. Microbiol.* **1976**, 30, 535-578.

20. Neijssel, O. M.; Hueting, S.; Crabbendam, K. J.; Tempest, D. W. *Arch. Microbiol.* **1975**, 104, 83-87.

21. Petitdemange, H.; Cherrier, C.; Raval, G.; Gay, R. *Biochim. Biophys. Acta* **1976**, 421, 334-347.

22. Petitdemange, H.; Dürr, C.; Abbad-Andaloussi, S.; Raval, G. *J. Ind. Microbiol.* **1995**, 15, 498-502.
23. Pflugmacher, U.; Gottschalk, G. *Appl. Microbiol. Biotechnol.* **1994**, 41, 313-316.
24. Reimann, A.; Biebl, H.; Deckwer, W.-D. *Appl. Microbiol. Biotechnol.* **1996**, 54, 47-50
24a.Reimann, A.; Biebl, H. *Biotechnol. Lett.* **1996**, 18, 827-823.
25. Ruch, F.E.; Lengeler J.; Lin, E.C.C. *J. Bacteriol.* **1974**, 119, 50-56.
26. Saint-Amans, S.; Soucaille, P. *Biotechnol. Lett.* **1995**, 17, 211-216.
27. Saint-Amans, S.; Perlot, P; Goma,G; Soucaille, P. *Biotechnol. Lett.* **1994**, 16, 831-836.
28. Solomon, B.O.; Zeng, A.-P.; Biebl, H.; Posten, C.; Deckwer, W.-D. *J. Biotechnol.* **1995**, 39, 107-117.
29. Streekstra, H.; Teixeira de Mattos, M.J.; Neijssel, O.M.; Tempest, D.W. *Arch Microbiol.* **1987**,147, 268-275.
30. Tong, I-T.; Liao, H.; Cameron, D.C. *Appl. Environ. Microbiol.* **1991**, 57, 3541-3546.
31. Tong, I.T.; Cameron, D.C. *Appl. Biochem. Biotechnol.* **1992**, 34/35, 149-159.
32. Tran-Dinh, K.; Hill, H.H. German patent no. DE 37 34 376 A1, **1987**.
33. Zeng, A.-P.; Biebl, H.; Schlieker, H.; Deckwer, W.-D. *Enzyme Microb. Technol.* **1993**, 15, 770-779.
34. Zeng, A.-P.; Ross, A.; Biebl, H.; Tag, C.; Günzel, B.; Deckwer, W.-D. *Biotechnol. Bioeng.* **1994**,44, 902-911.
35. Zeng, A.-P; Deckwer, W.-D. *Biotechnol. Prog.* **1995**, 11, 71-79.
36. Zeng, A.-P. *Biotechnol. Bioeng.* **1995**, 46, 314-324.
37. Zeng, A.-P.; Menzel, K.; Biebl, H.; Deckwer, W.-D. In *Biochemical Engineering 3*; Schmid, R.D., Ed.; E.Kurz & Co, Druckerei, Stuttgart, **1995**; pp 174-177.
38. Zeng, A.-P. *Bioprocess Eng.* **1996**, 14, 169-175.

Chapter 15

Biological Production of 2,3-Butanediol from Renewable Biomass

C. S. Gong, Ningjun Cao, and G. T. Tsao

Laboratory of Renewable Resources Engineering, Purdue University, West Lafayette, IN 47907

A simple and effective method of treatment of lignocellulosic material was used for the preparation of poplar wood chip and corn cob for the production of 2,3-butanediol by *Klebsiella oxytoca* ATCC 8724 in a simultaneous saccharification and fermentation (SSF) process. During the treatments, lignin and hemicellulose fractions of lignocellulosic materials were sequentially removed by aqueous ammonia (10-20%) steeping at 24°C for 24 h followed by dilute hydrochloric acid (1%, w/v) hydrolysis at 100-108°C for 1 h. The cellulose fractions (80 g/L) were then converted to butanediol by *K. oxytoca* in the presence of a fungal cellulase (8.5 g IFPU per g cellulose). The butanediol concentrations of 24 and 25.5 g/L, and ethanol concentrations of 6 and 7 g/L were produced by *K. oxytoca* from wood chip and corn cob, respectively. The average butanediol volumetric productivity was 0.26 g/L/h from wood chip and 0.35 g/L/h from corn cob.

2,3-Butanediol (2,3-butylene glycol) is a metabolic product of simple carbohydrates produced by many species of enterobacteria through a fermentative metabolic pathway (*1*). It is a colorless and odorless liquid that has a high boiling point of 180-184° C and a low freezing point of -60°C. The heating value of butanediol (27,198 J/g) is very similar to that of ethanol (29,055 J/g) and methanol (22,081 J/g) which makes it a potentially valuable liquid fuel and fuel additive (*2*). Butanediol can be dehydrated to methyl ethyl ketone (MEK) and used as an octane booster for gasoline or as high-grade aviation fuel (*2*). MEK can also be further dehydrated to 1,3-butadiene and dimerized to styrene (*3*). Therefore, butanediol has a diverse industrial usage, particularly as a polymeric feedstock, in addition to its use for manufacturing butadiene or antifreeze. Currently, butanediol is enjoying an annual growth rate of 4 to 7 percent, buoyed by the increased demand for polybutylene terephthalate resins, γ-butyrolactone, Spandex, and its precursors (*4*).

2,3-Butanediol is the only isomer, among many, that can be produced by microorganisms. Bacterial species, particularly those belonging to Klebsielleae, are known to metabolize carbohydrates to produce neutral compounds such as 2,3-butanediol, acetoin, and ethanol as metabolic products. Other groups of enterobacteria, such as *Erwinia*, produced mixed acid and 2,3-butanediol. In contrast, *E. coli* produces mainly mixed acid and ethanol under similar fermentation conditions (*1*). Research on the production of butanediol and its immediate precursor, acetoin, has centered around the microorganisms *Klebsiella oxytoca* (*K. pneumonia*) and *Enterobacter aerogenes (Aerobacter aerogenes*). Although other bacterial species, such as *Bacillus polymyxa,* are also known to produce butanediol (*5*).

Butanediol-producing bacteria utilize all the major sugars (hexoses and pentoses) to produce butanediol in high yield and high concentration (up to 10%, w/v) under optimum conditions e. g. dissolve oxygen, pH, etc., (*6,7*). Three enzymes, α-acetolactate synthetase (ALS), α-acetolactate decarboxylase (ALDC), and acetoin reductase (AR), are directly responsible for the accumulation of 1 mole of butanediol by bacterial cultures from 2 moles of pyruvate (*8*). In *K. terrigena* and *E. aerogenes*, they are encoded by the gene *bud*B (ALS), *bud*A (ALDC), and *bud*C (AR), respectively. The three genes are organized in an operon (*9*) and activated by transcriptional products of gene *budR* (*10*). The expression of the operon is optimal under anaerobic conditions at low pHs, and in the presence of low concentrations of acetate (*9*). The metabolic pathway leading to butanediol formation is shown in Figure 1.

Dissolved oxygen (DO) has a profound effect on butanediol and acetoin production and end-products distribution. Oxygen supply is important because it determines the flow of carbon source via the respiratory pathway versus the butanediol-producing fermentative pathway. The role of oxygen supply on biomass and butanediol accumulations has been studied extensively (*7,11-15*). Butanediol yield may be optimized by minimizing the available oxygen. However, with a small oxygen transfer rate, the cell yield will be low. Therefore, the total reaction rate will be slow. In the presence of excessive oxygen, the metabolism leads to the oxidation of NADH to NAD^+, and a high NAD^+/NADH ratio leads to excessive biomass accumulation. There is a general agreement among investigators that the oxygen supply rate is perhaps the most critical criterion for the performance of butanediol fermentation [see review articles (*6,16*) for detail].

In addition to DO, pH environment also exerts great influence on the product distribution and yield. In enterobacteria, the types and the ratio of fermentation products formed depend on two biochemical constraints: the need to maintain redox balance, and the requirement for maintaining the pH of the medium in a physiological range. Therefore, in a different pH environment, different relative amounts of end products are formed, with the proportion of neutral compounds increasing while the pH of the medium decreases (*17*). Under a neutral or high pH environment, bacteria tend to utilize the carbon source for cell growth and organic acid production. It has been shown that when the pH is lower than 6, over 50% of the pyruvate were

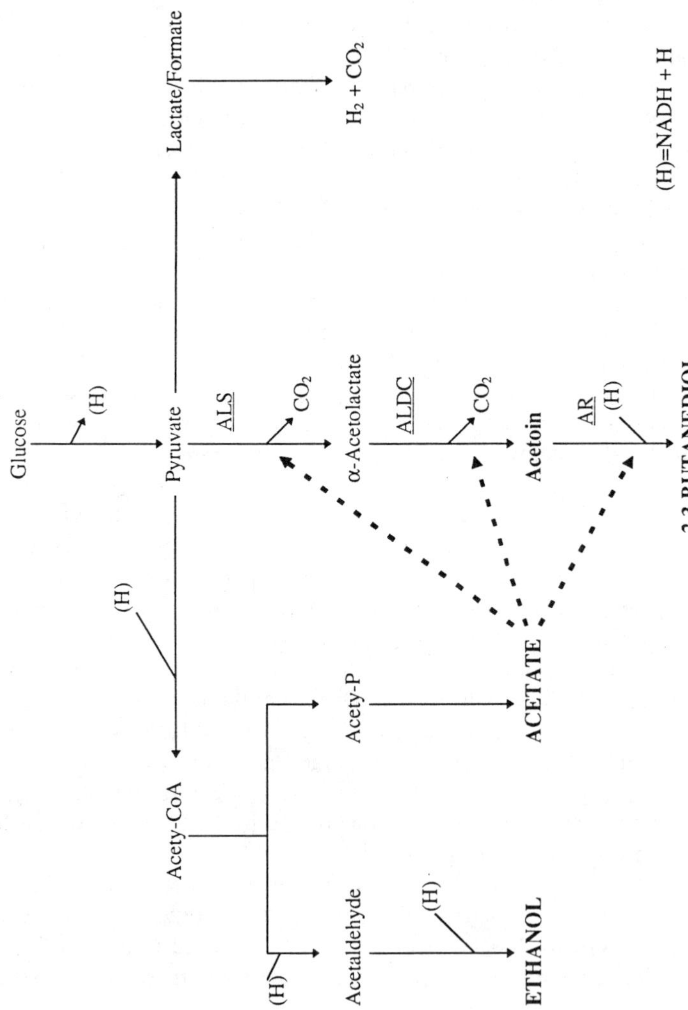

Figure 1. The Metabolic Pathway Leading To The Production Of 2,3-Butanediol And The Role Of Acetic Acid. ALS:α-acetolactate synthetase; ALDC:α-acetolactate decarboxylase; AR:acetoin reductase.

channeled into butanediol production. The enzymes of the butanediol pathway can constitute as much as 2.5% of the total protein in *E. aerogenes* (*18*).

Zeng et al. (*19*) investigated the effect of pH on growth and product formation of glucose by *E. aerogenes* in a continuous culture operation and found the optimal pH range of 5.5-6.5 for butanediol and acetoin production. Similar pH optimum was also observed in *K. oxytoca* (*7*). In general, the biomass concentration increases steadily with increased pH. At high pH, butanediol concentration decreases with the increase of acetic acid production. Acetic acid has a dual role in the regulation of butanediol formation. It serves as the activator for butanediol accumulation at low concentrations, and at a concentration of 10 g/L or higher, it inhibits butanediol production (*19-21*). The strength of acetic acid inhibition depends on the concentration of its undissociated form, HAc. The concentration of HAc was in turn determined by the pHs.

The production of butanediol from lignocellulosic materials has been considered as an alternative approach in the conversion of biomass substrates to liquid fuels and chemical feedstocks (*3,22*). Over the years, there have been many studies conducted utilizing agricultural residues and wastes for butanediol production. The materials studied include: citrus waste (*23*), water hyacinth (*24*), wheat and barley straws (*25*), corn stover (*25*), and hard wood hemicellulose fraction (*25-30*).

For the biological production of 2,3-butanediol to be economically competitive with petrochemical-based processes, the substrate source must be inexpensive, while reactor yields and productivity should be high. Lignocellulosic materials from forestry and agriculture residues, such as wood chips and corn crop residues, are inexpensive and abundant and can be used in many bioprocesses for the production of products of high economical value. An effective utilization of xylose, arabinose, and other minor sugars in addition to glucose is important in the process economics. *K. oxytoca* ATCC 8724 is capable of producing butanediol from both hexoses and pentoses with good yield (*7*). In this study, we used *K. oxytoca* to produce butanediol from pretreated poplar wood chip and ground corn cob cellulose fraction in the presence of a fungal cellulase.

Materials and Methods

Organism and Medium. *Klebsiella oxytoca* ATCC 8724 was purchased from American Type Culture Collection, Rockville, MD and was maintained on YMA (Difco) slants. The medium (YMP) for cell growth contained the following: yeast extract (Difco), 3 g; malt extract (Difco), 3 g; peptone (Difco), 5 g; glucose, 10 g; and distilled water, 1 liter. Sterilization was accomplished by autoclaving at 15 lb/in^2 for 15 min. Bacterial cells were grown aerobically in Erlenmeyer flasks (250-ml) containing 100 ml YMP medium at 25°C on a rotary shaker at 200 rpm for 24 h.

Materials. Hybrid poplar wood chip (9% moisture) with an average size of 1/4 in^2 was obtained from National Renewable Energy Laboratory (NREL), Golden, CO. Corn cob (8% moisture) with an average size of 1/8 in^2 was purchased from Andersons Inc., Maumee, OH. Wood chip and corn cob were composed of 42.9 and

44.88% cellulose, 22.5 and 32.68% xylan, and 25 and 7.41% lignin, respectively. The cellulase preparation in liquid form was provided by Iogen Co., Ottawa, Canada. The specific activity of the enzyme as determined by Iogen Co. was 170 IFPU/ml. Aqueous ammonia (30%) and hydrochloric acid (37%) were purchased from Mallinckrodt Chemical, Inc., Paris, KY.

Pretreatment.

Ammonia Steeping. Dry wood chip was mixed with 20% (w/v) aqueous ammonia with a solid to liquid ratio of 1 to 5 in a 250-ml flask and was incubated on a shaker at 24°C for 24 h. The mixture was then filtered to separate the solid from the ammonia solution. The partial delignined wood chip was collected by filtration, washed, and vacuum dried. The same procedures were also used for grounded corn cob, except the ammonia concentration was 10% (w/v). The lignin fraction was recovered as a precipitate after the vacuum removal of ammonia and the evaporated ammonia was collected by cooling trap.

Dilute Acid Hydrolysis. The partial delignined materials were subjected to further treatment by 1% hydrochloric acid solution at 100-108°C for 1 h with a solid to liquid ratio of 1 to 5. The solid cellulose fractions were separated from hemicellulose hydrolysates by filtration and washed in deionized water to remove residual acid.

Enzymatic Hydrolysis of Cellulose Fraction. Cellulose fractions of corn cob (20 g) were mixed with 0.05 M phosphate buffer (pH 5.0) and 1.0 ml cellulase enzyme (equivalent to 8.5 IFPU/gm corn cob) to a total volume of 50 ml in a 250-ml flask. The saccharification was carried out at 50°C for 48 h.

Simultaneous Saccharification And Fermentation Of Cellulose Fraction. Wood chip and corn cob cellulose fractions obtained after pretreatment were autoclaved for 15 min at 121°C in a 250-ml Erlenmeyer flask. After cooling, sterile nutrient solution (YMP) was added to a final volume of 80 ml. Initial pH was adjusted to pH 5.5 by potassium phosphate buffer. Concentrated bacterial cells (2 g/L dry equivalent) and cellulase (8.5 IFPU/gm original wood chip or corn cob) were added, and the flasks were incubated in a Controlled Environment Incubator Shaker (New Brunswick Scientific Inc.) at 200 rpm at a constant temperature of 32°C. Dissolved oxygen and pH were not monitored, nor were they controlled during fermentation. Samples (0.2 ml) were taken at 12 hour intervals for 96 h. Samples were then centrifuged and analyzed.

Analysis. Glucose, xylose, acetoin, 2,3-butanediol, ethanol, and acetic acid were analyzed by HPLC method (column: HPX-87H, 300x7.8 mm, Bio-Rad, Richmond, CA). Cellulose and hemicellulose were determined according to the procedures described (*31*). Lignin was determined as Klason lignin by weight method (*32*). The residual ammonium ion in solution was determined by 05800-05 Solution

Analyzer equipped with ammonium electrode (Cole-Parmer Instrument Co., Niles, IL).

Results and Discussion

One method of producing butanediol from biomass is the simultaneous saccharification and fermentation (SSF) process. In this process, cellulose and hemicellulose are hydrolyzed by cellulase to glucose, xylose and a mixture of minor sugars such as arabinose, and simultaneously conversion of sugars by a butanediol producing microorganism to butanediol. The result is improved hydrolysis rates and yields of product when compared to those involved with separate hydrolysis and fermentation steps (*25,33*). Similar to ethanol production in the SSF process, the cost of cellulase enzyme accounts for a large portion of the overall cost of biomass conversion to butanediol. Therefore, a reduction in the enzyme usage would make this process more economically attractive. One way of reducing enzyme usage is to remove lignin prior to SSF. Lignin is known as a physical barrier that impedes the contact of enzyme to cellulose. In addition, it will adsorb cellulase and deactivate cellulase activity (*34-37*). Likewise, the prior removal of hemicellulose will enhance the reactivity of cellulose fraction as well (*31,37*).

The procedure for the preparation of substrate for SSF from wood chip and corn cob for the production of butanediol is summarized in Figure 2. The initial step of substrate preparation involves steeping in aqueous ammonia (10-20%, w/v) at 24°C to remove and separate lignin fraction from carbohydrate polymers since alkaline treatment of lignocellulose is known to dissolve lignin (*38*).

The change in the composition of wood chip and corn cob after ammonia steeping is shown in Table I. The original lignin content of wood chip was reduced by approximately 70%, from 0.25 to 0.07 g per g dry wood chip, after steeping. In addition, other extractives such as acetate group and alkaline extractable materials were removed or reduced. Hemicellulose content, measured as xylan, remained unchanged. Based on the composition of the treated wood chip dry weight, xylan content was increased from 22.5% to 30.37% after lignin removal. Correspondingly, cellulose content was increased from 42.9% to about 58%. Similar results were obtained from grounded corn cob (see Table I).

The steeping process was followed by dilute acid hydrolysis. In this step, treated materials were incubated in the dilute acid solution (1%, w/v) at 100-108°C for 1 h to hydrolyze the hemicellulose. Hemicellulose solubilized by this treatment is in the range of 80 to 85% of the original hemicellulose content. Wood chip hemicellulose was reduced to 0.045 g per g dry from 0.255g. Similarly, corn cob hemicellulose was reduced to 0.05 g per g dry from 0.37 g. The clean, light-amber colored hemicellulose hydrolysates with a carbohydrate concentration of approximately 80 g/L were obtained. Xylose comprised over 85% of the total carbohydrates of hydrolyzed hemicellulose. The hemicellulose hydrolysate has no acetate and alkali extractives present. This is important for the utilization of hemicellulosic carbohydrates since alkali extractives, even at a very low concentration of 0.5-1 g/L, are known to inhibit butanediol fermentation (*26*). Hemicellulose

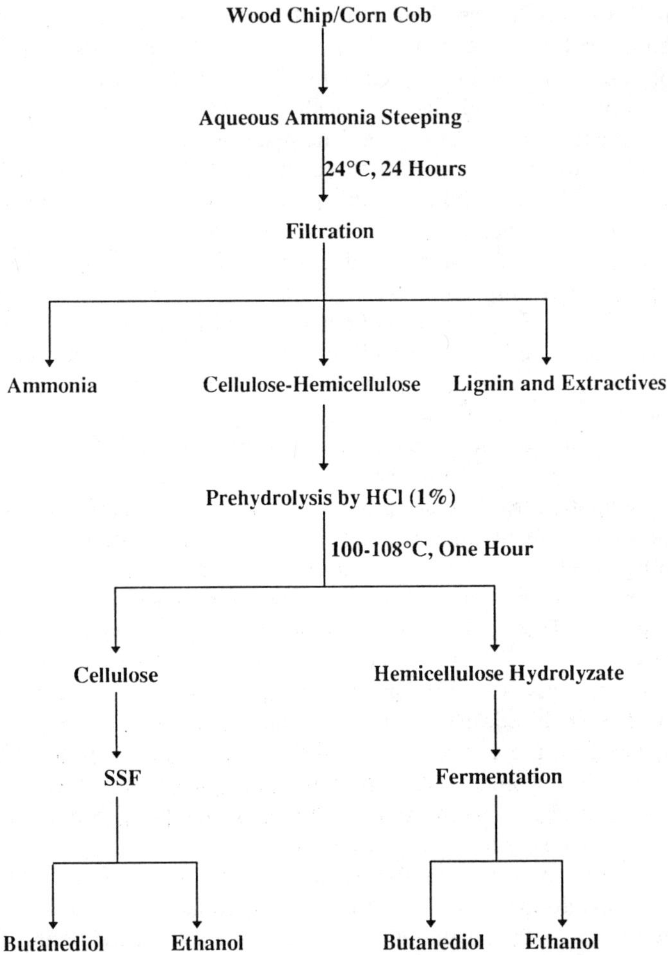

Figure 2. The Schematic Representation Of The Procedures Used For The Preparation Of Substrate From Poplar Wood Chip And Corn Cob.

Table I. Composition of Different Materials after Differet Treatments

Material		Composition (%)					Residual Yield(w/w, %)				
		Cellu-lose[1]	Xylane[2]	Lignin	Acetate	Other	Cellu-lose[1]	Xylane[2]	Lignin	Acetate	Other
Wood Chip	Original	42.9	22.5	25.0	3.2	6.4	100	100	100	100	100
	After NH_3 Steeping	58.09	30.37	7.11	0.0	4.33	100	100	20.6	0.0	50.0
	After HCl hydrolysis	85.37	4.48	6.97	0.0	3.18	100	9.98	14.01	0.0	24.6
Corn Cob	Original	44.88	32.68	7.41	2.51	12.53	100	100	100	100	100
	After NH_3 Steeping	56.2	38.5	0.85	0.0	4.45	100	98.6	9.6	0.0	29.7
	After HCl hydrolysis	90.4	5.29	0.91	0.0	3.4	100	8.01	6.10	0.0	14.68

1. Measured as glucose
2. Measured as xylose

hydrolysate has been used as a substrate for ethanol production by a xylose-fermenting yeast strain (*39*). The same hydrolysate can also be used as the substrate for the production of high-value sweetener, xylitol (*40*), or serves as raw material for xylose production.

Since most of the lignin, acetate, alkali extractives, and hemicellulose have been removed by prior process steps, one would expect the remaining cellulose fractions to be more reactive to cellulase hydrolysis. This is supported by the cellulose hydrolysis results. Figure 3 shows the effect of different stages of treatment on cellulase hydrolysis of corn cob when a cellulase dosage of 8.5 IFPU per g corn cob was used. Results indicated that the combination of ammonia steeping and dilute acid hydrolysis gives the highest glucose yield (91.8%) based on dry cellulose after 48 h of incubation. The ammonia treated or dilute acid treated samples showed similar reactivity toward the cellulase. They are more susceptible to cellulase hydrolysis than the sample without any treatment. The rate of cellulose hydrolysis after ammonia and dilute acid treatments is considerably higher than other pretreatment procedures. Comparable results were also obtained using wood chip cellulose as the substrate (Figure 4).

The effects of ammonia steeping and dilute acid treatment are: exposing cellulose to cellulase and increasing the surface area of cellulose available for the enzymatic hydrolysis due to the swelling of cellulose (*37,41*); and increase the porosity, thereby, enhanced the accessibility of cellulose by cellulase (*42*). Upon drying of the treated cellulose fraction, the effect of alkaline swelling of cellulose was reduced. This effect is shown in figure 5, in that air-dried cellulose is not as susceptible to cellulase as when it had not been dried. The solid material obtained after lignin and hemicellulose removal has a cellulose content of over 85%, based on dry weight. This is more than a 100% increase of cellulose content from original untreated materials (see Table I).

To demonstrate the effectiveness of the current pretreatment process, SSF experiments were performed using bacterial culture *K. oxytoca* and a treated wood chip sample. The time course of product formation during SSF with an ammonia and acid treated sample is shown in Figure 6. The combination of ammonia steeping and dilute acid hydrolysis gives the best results. The amount of butanediol produced from this combination is at least two folds higher than the experiment using ammonia but without an acid treated sample. Compared to the untreated sample, a four-fold increase in butanediol accumulation was obtained with the ammonia-acid treated sample.

The time course of product formation during SSF with ammonia-acid treated wood chip is shown in Figure 7. During the first 24 h of incubation, glucose was accumulated and reached about 5 g/L and then declined, as would be expected, due to the presence of bacterial cells. Cellobiose was also produced and reached a level of about 5 g/L while ethanol concentration was increased to 6 g/L. Other products such as acetoin and acetate were detected in minute quantities. Butanediol concentration increased almost linearly and reached a level of 24 g/L after 96 h of incubation. This is a 60% yield based on initial dry cellulose (80 g/L) from ammonia and acid treated wood chip. Similar results were also obtained with corn cob cellulose as the substrate (Figure 8). The amount of butanediol produced was 25.5 g/L after 72 h and declined

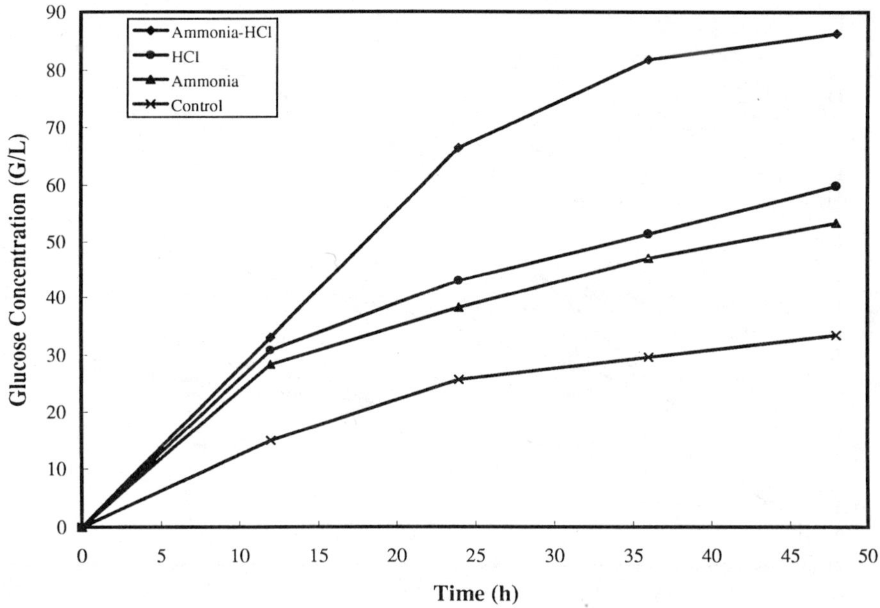

Figure 3. Hydrolysis Of Corn Cob By Cellulase After Different Treatments.

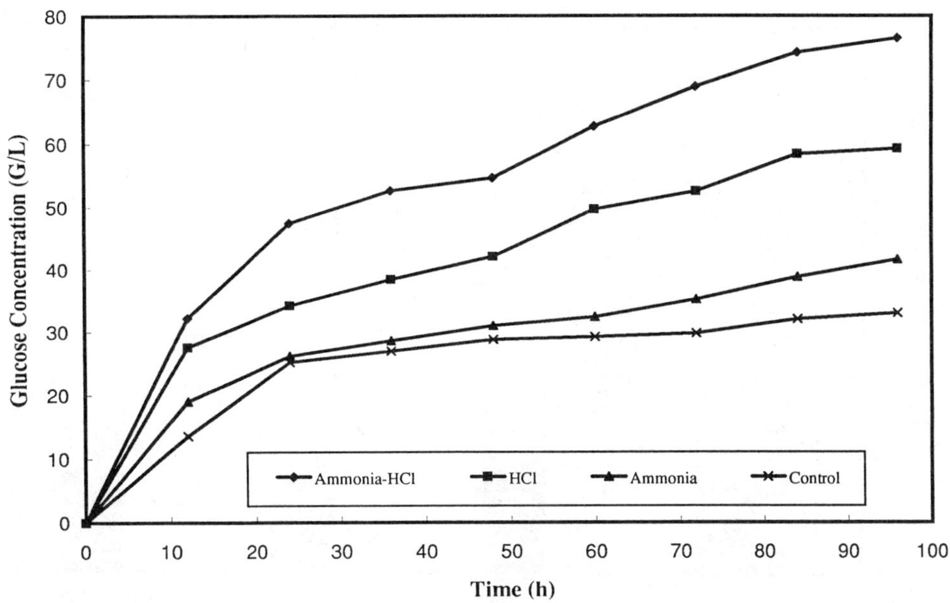

Figure 4. Hydrolysis Of Wood Chip By Cellulase After Different Treatments.

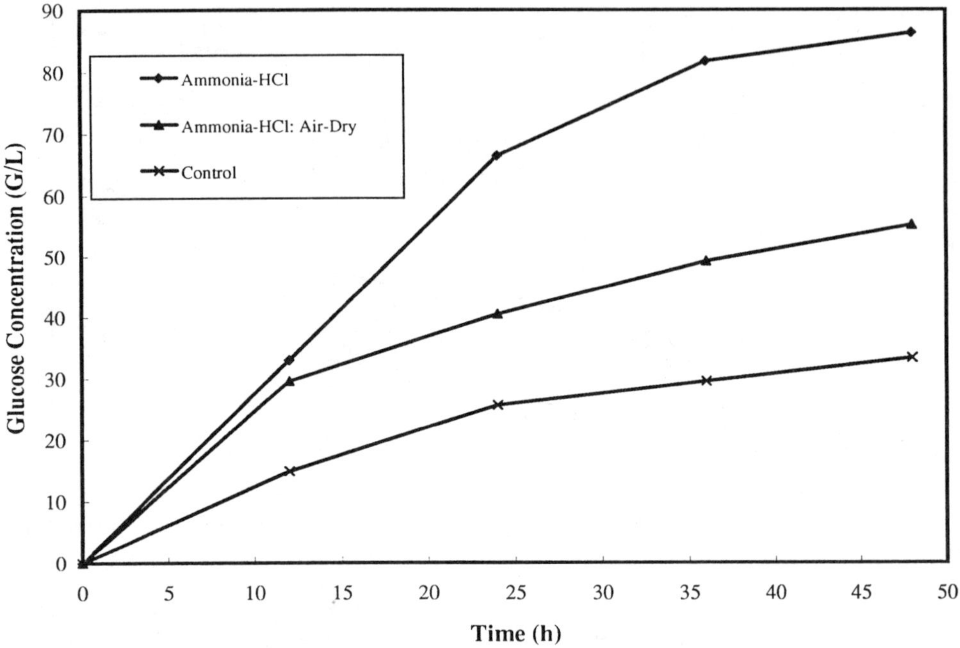

Figure 5. Effect Of Drying On The Reactivity Of Corn Cob Cellulose Toward Cellulase.

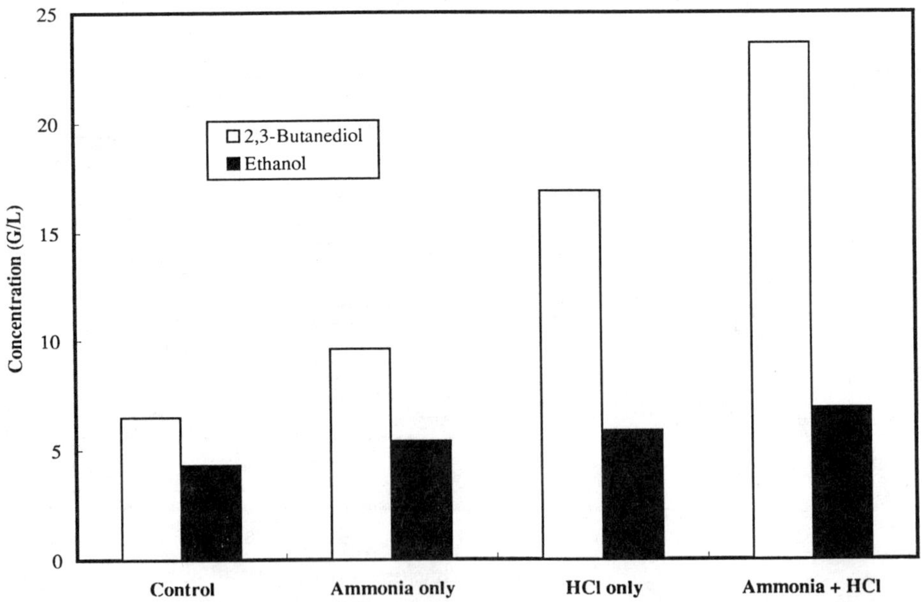

Figure 6. Effect Of Different Treatments On Butanediol And Ethanol Production From Wood Chip Using SSF. pH=5.5; Temperature =32°C; Incubation=96 h.

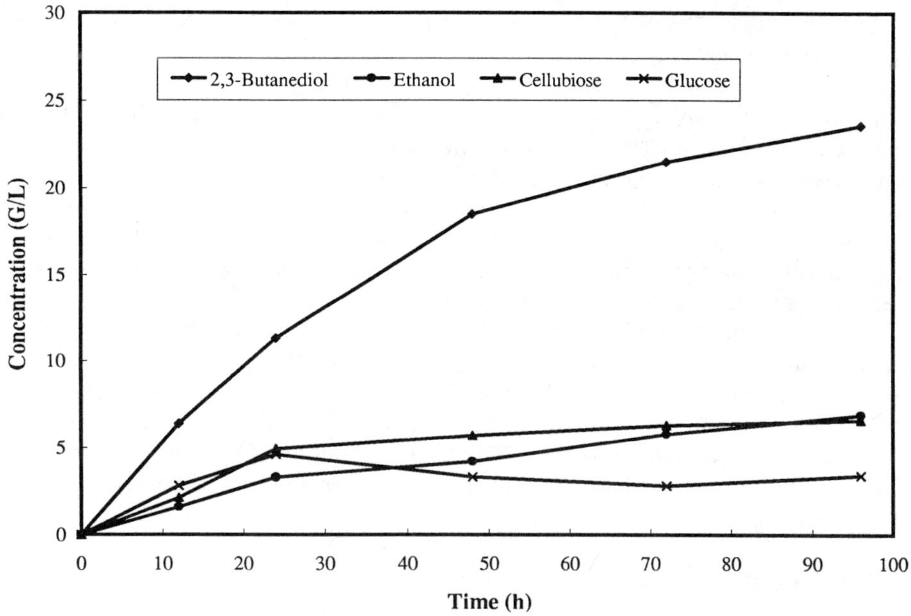

Figure 7. Time Course Of SSF Of Wood Chip Cellulose. Cellulose=80 g/L; pH=5.5; Temperature =32°C.

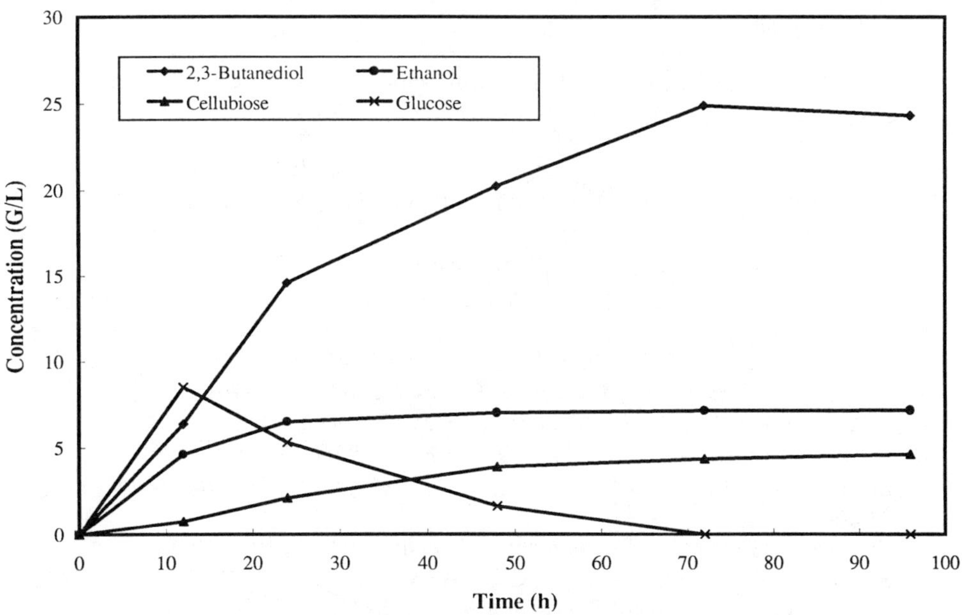

Figure 8. Time Course Of SSF Of Corn Cob Cellulose. Cellulose=80 g/L; pH=5.5; Temperature =32°C.

slightly after 96 h. The yield of butanediol (63.7%) from corn cob is slightly higher than that from wood chip cellulose.

The stepwise removal of lignin and hemicellulose fractions from lignocellulose appears to be effective in the utilization of cellulose fraction for butanediol production. The beneficial effects of ammonia steeping include: the solubilization of lignin, the removal of lignin from cellulose fraction, the chemical swelling of cellulose, the disruption of the crystalline structure of cellulose, and the increase in accessible surface area for cellulase. Consequently, cellulase dosage required for cellulose hydrolysis is much lower than that reported in literature. Another advantage to be gained by the current procedure is the obtaining of lignin and acetate free hemicellulose hydrolysate. This fraction can be converted to products in separate streams. This is significant in view of the fact that pentoses utilization by bacteria is often subjected to inhibition by many extraneous materials produced during harsher hydrolysis process (*26,27*). The potential inhibitors include: acetic acid (*11*) and furfural and phenolic compounds such as vanillin and syringealdehyde (*26*). The utilization of hemicellulose hydrolysates for butanediol production is currently under way.

Acknowledgments

This study was supported in part, through The Consortium for Plant Biotechnology Research, Inc. by DOE cooperative agreement No. DE-FCO5-92OR22072. We also thank Cathy Loayza and Susan Smith for their invaluable help.

Literature Cited

1. Cowan, S. T., In *Bergey's Manual of Determinative Bacteriology*, Buchanan, R. E., & Gibbons, N. E., Eds., The Williams Wilkins Co., Baltimore, **1974**, pp. 290-383.
2. Tran, A. V.; Chamber, R. P., *Biotechnol. Bioeng.*, **1987**, 29, 343-351.
3. Palsson, B. O.; Fathi-Afshar, S., Rudd, D. F., Lightfoot, E. N., *Science*, **1981**, 213, 513-517.
4. Chemical Marketing Reporter (July 24, **1995**).
5. Laube V. M.; Groleau, D., Martin, S. M., *Biotechnol. Lett.* **1984,** 6, 257-262.
6. Jansen, N. B.; Tsao, G. T., *Adv. Biochem. Eng. Biotechnol.* **1983,** 27, 85-99.
7. Jansen, N. B.; Tsao, G. T., *Biotechnol. Bioeng.* **1984,** 26, 362-369.
8. Magee, R. J.; Kosaric, N., *Adv. Appl. Microbiol.* **1987,** 32, 89-161.
9. Blomqvist, K.; Nikkola, M., Lehtovaara, P., *J. Bacteriol.* **1993,** 175:1392-1404.
10. Mayer, D.; Schlensog, V., Bock, A., *J. Bacteriol.* **1995,** 177, 5261-5269.
11. Zeng, A. P.; Biebl, H., Deckwer, W. D., *Appl. Microbiol. Biotechnol.* **1990,** 33,264-268.
12. Zeng, A. P.; Decker, W. D., *Biotechnol. Bioeng.* **1992,** 40, 1078-1084.
13. Silveira, M. M.; Schmidell, W., Berbert, M. A., *J. Biotechnol.* **1993,** 31, 92-102.
14. Beronio, P. B.; G. T. Tsao, *Biotechnol. Bioeng.* **1993,** 42, 1263-1269.
15. Zeng, A. P.; Byun, T. G., Posten, C., Decker, W. D., *Biotechnol. Bioeng.* **1994,** 44, 1107-1114.

16. Garg, S. K.; Jain, A., *Biosource Technol.* **1995**, 51, 103-109.
17. Booth, I. R., *Microbiol. Rev.* **1985,** 49, 356-378.
18. Johansen, L.; Bryn, K., Stormer, F. C., *J. Bacteriol.* **1975,** 123, 1124-1130.
19. Zeng, A. P.; Biebl, H., Deckwer, W. D., *Appl. Microbiol. Biotechnol.* **1990,** 33, 485-489.
20. Johansen, L.; Bryn, K., Stormer, F. C., *J. Bacteriol.* **1975,** 123, 1124-1130.
21. Yu, E. K. C.; Saddler, J. N., *Appl. Environ. Microbiol.* **1982,** 44, 777-784.
22. Flickinger, M. C., *Biotechnol. Bioeng.* **1980,** 22, 27-48.
23. Long, S. K.; Patrick, R., *Appl. Microbiol.* **1961,** 9, 244-248.
24. Motwani, M.; Seth, R., Daginawala, H. F., Khanna, P., *Biores. Technol.* **1993,** 44, 187-195.
25. Yu, E. K. C.; Deschatelets, L., Saddler, J. N., *Biotechnol. Bioeng. Symp.* **1984,** 14, 341-352.
26. Frazer, F. R; McCaskey, T. A., *Enzyme Micro. Technol.,* **1991,** 13, 110-115.
27. Mes-Hartree, M.; Saddler, J. N., *Biotechnol. Lett.* **1983,** 5, 531-536.
28. Saddler, J. N.; Yu, E. K. C., Mes-Hartree, M., Levitin, N., Brownnell, H. H., *Appl. Environ. Microbiol.* **1983,** 45, 153-160.
29. Yu, E. K. C.; Deschatelets, L., Levitin, N., Saddler, J. N., *Biotechnol. Lett.* **1984,** 6, 611-614.
30. Grover, B. P.; Garg, S. K., Verma, J., *World J. Microbiol. Biotechnol.* **1990,** 6, 328-332.
31. McMillan, J. D., In: *Enzymic Conversion of Biomass for Fuels Production.* Himmel M. E., Baker J. O., Overend, R. P., Eds. ACS, Washington, DC., **1994,** pp. 292-324.
32. Ghosh, S., Biomass Handbook, p. 395, Gordon and Breach Science Publishers, New York, **1989.**
33. Yu, E. K. C.; Saddler, J. N., *Trends Biotechnol.* **1985,** 3, 100-104.
34. Sutcliffe, R.; Saddler, J. N., *Biotechnol. Bioeng. Symp.* **1986,** 17,749-762.
35. Ooshima, H.; Burns, D. S., Converse, A. O., *Biotecnnol. Bioeng.* **1990,** 36,446-452.
36. Elander, R. T.; Hsu, T. *Appl. Biochem. Biotechnol.* **1995,** 51/52,463-478.
37. Philippidis, G. P.; Smith, T. K. *Appl. Biochem. Biotechnol.* **1995,** 51/52,117-124.
38. Meller, A.; Britt, K. W., Singh, R. P., in *Handbook of Pulp and Paper Technology*, 2nd ed., Britt, K. W., Ed., Van Nostrand Reinhold Co., N.Y., **1969,** pp. 263-267.
39. Cao, N. J.; Krishnan, M. S., Du, J. X., Gong, C. S., Ho, N. W. Y., Chen Z. D., Tsao, G. T., **1996,** *Biotechnol Lett.* 18, *In Press.*
40. Chen, L. F.; Gong, C. S., *J. Food Sci.* **1985,** 50, 226-228.
41. Yoon, H. H.; Wu, Z. W., Lee, Y. Y., *Appl. Biochem. Biotechnol.* **1995,** 51/52, 5-19.
42. Burn, D. S.; Ooshima, H., Converse, A. O., *Appl. Biochem. Biotechnol.* **1989,** 20/21, 79-94.

Chapter 16

Microbial Production of Biodegradable Plastics from Carbon Dioxide and Agricultural Waste Material

Ayaaki Ishizaki, Naohiko Taga, Toshihiro Takeshita, Toshikazu Sugimoto, Takeharu Tsuge, and Kenji Tanaka

Department of Food Science and Technology, Faculty of Agriculture, Kyushu University, Hakozaki, Higashi-ku, Fukuoka 812, Japan

Fermentative production of biodegradable plastic material, poly-D-3-hydroxybutyrate, P(3HB) from CO_2 or agricultural wastes is expected to contribute to the solution of global environmental pollution problems. The practical cultivation systems to produce P(3HB) from CO_2, H_2 and O_2 by hydrogen-oxidizing bacterium were developed by maintaining O_2 concentration in gas phase below the lower limit for explosion. P(3HB) productivity was increased by improving gas mass transfer with the use of air-lift fermentor and the addition of 0.05 %CMC to the culture medium. P(3HB) was also produced from xylose via L-lactate by two-stage culture method using *Alcaligenes eutrophus* and *Lactococcus lactis* IO-1. P(3HB) productivity was increased by a pH-stat fed-batch culture method with feeding the substrate solution so as to control L-lactate concentration at very low level.

Polyhydroxyalkanoates, PHAs are potential raw materials for manufacturing biodegradable plastics(1). *Alcaligenes eutrophus* is a hydrogen-oxidizing bacterium that is able to grow autotrophically using H_2, O_2 and CO_2 and heterotrophically using organic acids as substrate with the accumulation of poly-D-3-hydroxybutyric acid, P(3HB) in the cell under nutrient-limited conditions(Fig.1). The growth rate of this hydrogen-oxidizing bacterium is much higher than that of other autotrophs such as photosynthetic organisms and hence the bacterium has the potential to be used in industrial processes. Production of P(3HB) from CO_2 or organic acids derived from agricultural wastes by *A.eutrophus*, could contribute to the solution of two environmental pollution problems of increased CO_2 levels in the atmosphere and that of the disposals of non-biodegradable plastic waste. Here, we describe a strategy and system set-up for fermentative production of P(3HB) from CO_2 and agricultural waste materials.

1. Strategy and system-set up for P(3HB) production from CO_2

For practical application of autotrophic production of P(3HB) from CO_2 by *A.eutrophus*, the loss of substrate gas utilization efficiency concomitant with the exhaustion of the gas from fermenter, and the potential of explosion of the substrate gas mixture are very serious problems to be solved. A recycled-gas closed-circuit culture system attains high gas utilization efficiency by reusing the exhausted gas. Figure 2 shows the gas-recycling culture system in bench scale. This system could also eliminate the potential for the gas mixture to explode by maintaining the oxygen concentration in the gaseous phase below the lower limit for explosion(6.9 %(v/v)) and introducing several other safety measures. Oxygen consumption by the cells in autotrophic synthesis of P(3HB) is very large as shown in the following equations, the decrease in the driving force for oxygen from the gas phase into the liquid phase then results in the serious decrease in P(3HB) productivity.

Exponential cell growth; $21.36\ H_2 + 6.21\ O_2 + 4.09\ CO_2 + 0.76\ NH_3 \rightarrow C_{4.09}H_{7.13}O_{1.89}N_{0.76}$

P(3HB) formation; $33\ H_2O + 12\ O_2 + 4\ CO_2 \rightarrow C_4H_6O_2 + 30\ H_2O$

Hence, a doughnut-shaped agitation system to attain a *KLa* of 2,970 h^{-1} was used in the bench plant to compensate for the decrease in oxygen transfer. As a result, cell and P(3HB) concentrations increased to 91.3 $g \cdot dm^{-3}$ and 61.9 $g \cdot dm^{-3}$ respectively, under O_2-limited condition after 40 h of cultivation (Fig.3). While the O_2 concentration in the gas phase was maintained at very low level, the overall productivities of biomass and P(3HB) obtained in this cultivation were 2.28 $g \cdot dm^{-3} \cdot h^{-1}$ and 1.55 $g \cdot dm^{-3} \cdot h^{-1}$, respectively, which were much higher than those reported for other autotrophic cultivation of hydrogen-oxidizing bacteria (Table I).

2. Two-stage culture method for P(3HB) production under low O_2 condition

The use of a fermenter with very high *KLa* is not practical in industrial-scale fermentation process because power consumption for agitation will be uneconomically large. Hence, we developed a new culture method using a conventional-type fermenter under low O_2 conditions. In this method, heterotrophic cultivation using fructose as carbon source was first carried out for exponential cell growth. After the fructose in the medium was exhausted, the culture broth was centrifuged. The harvested cells were suspended in sterilized mineral medium and autotrophic cultivation for P(3HB) accumulation was performed by feeding a substrate gas mixture in which the O_2 concentration was below 6.9%(second culture stage). This method is referred as a two-stage culture method. The cell and P(3HB) concentrations were 26.3 $g \cdot dm^{-3}$ and 21.6 $g \cdot dm^{-3}$, respectively after 40 h of autotrophic cultivation. The average productivity of P(3HB) in the autotrophic stage was about 0.56 $g \cdot dm^{-3} \cdot h^{-1}$. According to previous reports (9-20) on the fermentative production of P(3HB), the productivity of P(3HB) was 0.08 - 4.00 $g \cdot dm^{-3} \cdot h^{-1}$ in heterotrophic cultures and 0.04 -1.54 $g \cdot dm^{-3} \cdot h^{-1}$ in autotrophic cultures. Although the *KLa* of the fermenter used in this experiment was 340 h^{-1} and the O_2

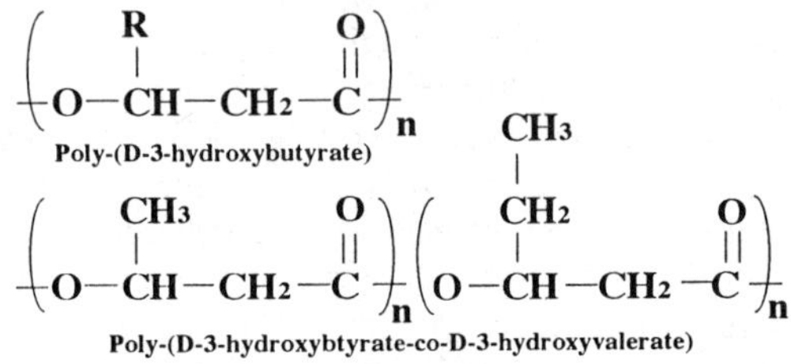

Figure 1 Chemical structure of poly-D-hydroxybutyric acid.

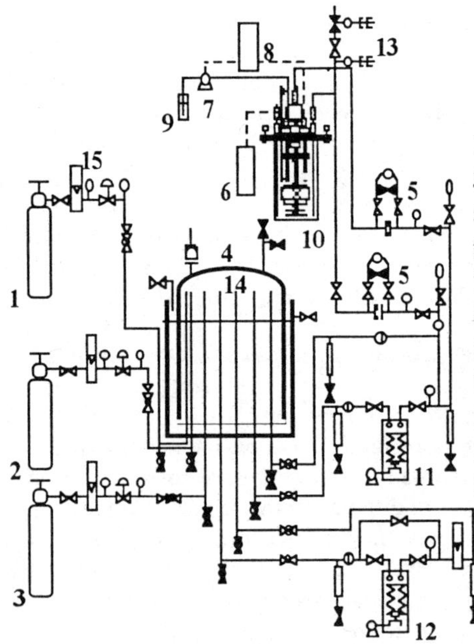

1, O_2 cylinder; 2, H_2 cylinder; 3, CO_2 cylinder; 4, gas reservoir; 5, differential-pressure type flowmeter; 6, DO controller; 7, alkali feeding pump; 8, pH controller; 9, NH_3 solution; 10, fermentor; 11, circulating pump; 12, mixing pump; 13, internal pressure sensor; 14, saline replacement; 15, flowmeter

Figure 2 Schematic diagram of explosion-proof type recycled-gas closed-circuit culture system.

concentration in the gas mixture was very low, a relatively high P(3HB) production rate was obtained. In this culture method, CO_2 is evolved from fructose during the heterotrophic stage. However, the two-stage culture works as a CO_2 absorption system because the amount of CO_2 consumed during the autotrophic stage is about twice that of the heterotrophic stage.

3. Development of pH-stat batch culture with continuous feeding substrate solution to obtain protein-rich cell mass for two-stage culture

Fructose is an expensive carbon source in the fermentation industry, hence we next investigated the application of other economical carbon sources for the heterotrophic culture in the two stage method. We used acetic acid as the carbon source and developed a pH-stat continuos substrate-feeding method for the culture. Flask culture experiments showed the optimum concentration of acetic acid for cell growth was very low (ca.1.0 g•dm^{-3}) and the growth was seriously inhibited by a slight increase in acetic acid concentration. It was, therefore, necessary to control the acetate concentration around this level in high cell density cultivation. The ratio of consumption of acetic acid to that of ammonium by *A.eutrophus* cells was determined by a standard-type batch culture experiment to be about 10 (mol-acetic acid/mol-ammonium). It was therefore expected that the acid-base equilibrium in culture system would be balanced by feeding the substrate solution in which the C/N ratio was 10 (mol/mol) so as to maintain the culture pH at a constant level, in order for acetate concentration in the fermentor to be also controlled at low level. However, in batch culture with such a feeding, acetate concentration of the culture liquid increased after cell concentration reached approximately 5 g•dm^{-3}. The increase in acetate concentration was thought to be due to the depletion of mineral nutrients. Hence, the mineral concentrations in the medium was increased 5 times as that of the basal medium (this was referred as 5-fold medium). As a result, acetate concentration was controlled around 1 g•dm^{-3} and cell concentration reached about 25 g•dm^{-3} after 18 h. Acetate concentration increased after that due to the depletion of phosphate. A pH-stat batch culture was hence carried out by feeding a solution in which the C/P ratio was 118.4 (mol-C/mol-P). Acetate concentration was maintained around 1 g•dm^{-3}, and cell and protein concentrations increased to 48.6 g•dm^{-3} and 35.0 g•dm^{-3}, respectively after 21 h of cultivation (Fig.4). Autotrophic cultivation for P(3HB) accumulation was then performed using the protein-rich cell mass obtained from a pH-stat batch culture into which the modified substrate solution was fed. When the cell concentration reached to about 5 g•dm^{-3}, feeding of the substrate solution was stopped and autotrophic cultivation was performed by feeding a substrate gas mixture into the fermentor. P(3HB) accumulated in the cells up to about 60 % by dry cell weight. This feeding method can therefore be used in fermentation process where the cell growth or P(3HB) accumulation is inhibited by high concentrations of the substrate such as propionate, formate, or lactate, etc.

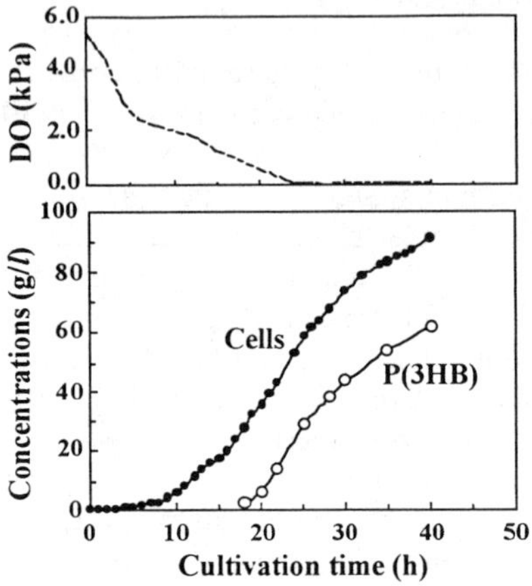

Figure 3 Time course of autotrophic culture of *A.eutrophus* in bench-plant scale culture system.

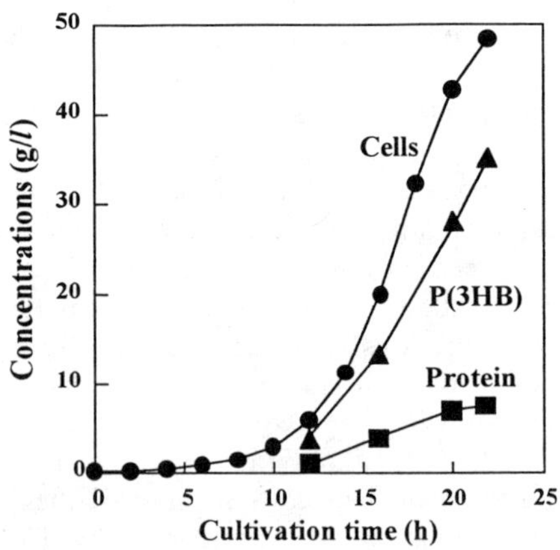

Figure 4 Time course of pH-stat batch culture with feeding acetic acid and inorganic nutrients solution. The C/N and C/P ratios in the feed solution were 10 and 118.4(mol/mol), respectively.

Table I. List of autotrophic cultivations of hydrogen-oxidizing bacteria and fermentaive production of P(3HB) using various microorgansims and substrates

Strains	Substrate	Culture method	Cultivation time (h)	Cell concentration ($g \cdot dm^{-3}$)	Cell productivity ($g \cdot dm^{-3} \cdot h^{-1}$)	P(3HB) concentration ($g \cdot dm^{-3}$)	P(3HB) productivity ($g \cdot dm^{-3} \cdot h^{-1}$)	Ref.
Alcaligenes eutrophus	$H_2/O_2/CO_2$	Batch	25	25.0	1.00	--	--	9
Alcaligenes eutrophus	$H_2/O_2/CO_2$	Batch	40	91.3	2.28	61.9	1.55	5
Pseudomonas hydrogenovora	$H_2/O_2/CO_2$	Batch	48	24.0	0.50	--	--	10
Alcaligenes eutrophus	$H_2/O_2/CO_2$	Continuous			0.40	--	--	11
Alcaligenes eutrophus H16	$H_2/O_2/CO_2$	Batch	70	18.0	0.26	16.0	0.23	12
Alcaligenes hydrogenophilus	$H_2/O_2/CO_2$	Continuous			0.33	--	--	13
Pseudomonas hydrogenothermophila	$H_2/O_2/CO_2$	Continuous			3.00	--	--	14
Alcaligenes eutrophus	$H_2/O_2/CO_2$	Batch	60	60.0	1.00	36.0	0.60	3
Recombinant *E.coli*	Glucose	Fed-batch	42	117.0	2.79	89.0	2.11	15
Protomonas extroquens	Methanol	Fed-batch	121	223.0	1.84	136.0	1.12	16
Alcaligenes eutrophus	Glucose	Fed-batch	50	164.0	3.28	121.0	2.42	17
Alcaligenes latus	Sucrose	Fed-batch[1)]	18	142.0	7.89	68.4	4.0	18
Methylo-bacterium organophilum	Methanol	Fed-batch	70	250.0	5.57	130	1.85	19
Alcaligenes eutrophus	Glucose+ valerate	Fed-batch	50	--	--	90.4[2)]	1.80	20

1)The inoculum cell size was 13.7 $g \cdot dm^{-3}$.
2)The product was poly(hydroxybutyrate-co-hydroxyvalerate).

4. Application of air-lift type fermenter and improvement of gas mass transfer by changing medium rheology

Air-lift fermentors have often been used instead of the traditional stirred-tank fermentor for production of penicillin(21), α-amylase(22), xanthan(23) and single cell protein(24). As the air-lift fermentor does not require mechanical agitation, the energy consumption is lower than that of a stirred-tank fermentor. The effect of the change in medium viscosity by adding carboxymethylcellulose (CMC) in an air-lift type fermenter on gas-hold up, bubble formation, flow pattern and mass transfer of oxygen has been reported by many researchers(25-27). We investigated the application of an air-lift type fermentor and the effect of change in medium viscosity by addition of CMC. Figure 5 shows the air-lift type fermentor used in this study, which was assembled as described by Okabe *et al.*(28). To obtain high mass transfer achieved through the formation of small bubbles, a sintered stainless steel sparger (pore size, 10 mm; diameter, 12 mm; length, 20 mm) was installed at the bottom of the reactor. The feeding rate of the substrate gas mixture in the air-lift fermentor was 2 $dm^3 \cdot min^{-1}$, which is equivalent to a superficial gas velocity of 2.62 $cm \cdot s^{-1}$. Figure 6 shows the changes in medium viscosity and gas hold-up at various concentrations of CMC in the air-lift fermentor. Gas hold-up increased in proportion to the increase in CMC concentration up to 0.1% (w/v) but the gas hold-up decreased above 0.1% (w/v) of CMC. Deducing from this result, addition of CMC up to 0.1% (w/v) was expected to increase oxygen transfer rate with resultant increase in P(3HB) productivity. Figure 7 shows the time courses of autotrophic culture of *A. eutrophus* in the air-lift fermentor with addition of various concentrations of CMC. The productivity of P(3HB) in the culture with addition of 0.05 %(w/v) CMC (shown in Fig.7b) was increased to twice as that of the control culture with no addition of CMC(shown in Fig.7a). In the culture with addition of 0.1 % CMC, P(3HB) productivity was about 1.5 times higher than that of the control culture(Fig.7c). However, there was no apparent effect of the addition of CMC on the productivity of P(3HB) in the cultivation using the stirred-tank fermentor. A comparison was made for the effect of CMC addition on the mass transfer of oxygen in the air-lift and stirred-tank fermenters. When measurements were done by the static method, maximum *KLa* value for the air-lift fermenter was obtained at 0.05% of CMC concentration. The values of *KLa* for the air-lift fermentor measured by the sulfite oxidation method was observed to decrease with an increase in CMC concentration. For stirred-tank fermentor on the other hand, there was no increase in *KLa* values by addition of CMC into the culture medium. In the measurement by the sulfite oxidation method, the *KLa* of the stirred-tank fermenter was lowest at 0.1 % CMC. It is generally known that some kinds of surfactants, such as CMC, affect the sulfite-oxidation reaction. P(3HB) production rate observed in the fermentation experiments using the air-lift fermentor, correlated to the *KLa* measured by the static method but did not correlated to the *KLa* measured by the sulfite oxidation method. The *KLa* value measured by sulfite oxidation method was larger than those measured by the static method. These results mean that the static method is more reliable for the measurement of *KLa* than the sulfite oxidation method in autotrophic culture for P(3HB) production using air-lift fermentor. However, the relationship between the

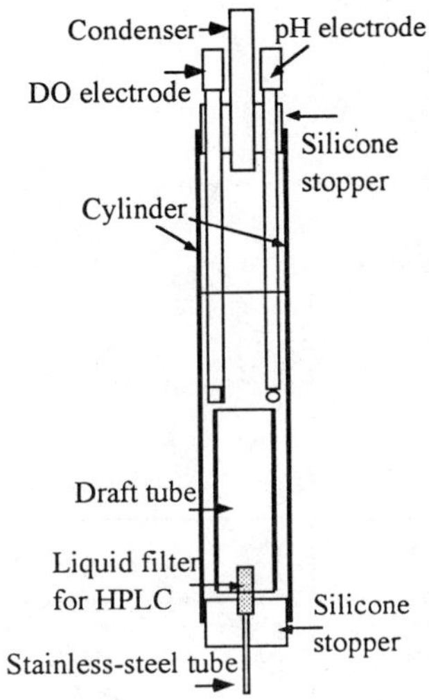

Dimensions of the air-lift fermentor.

Reactor:	
total volume	350 ml
total length	350 mm
Cylindrical part:	
length	300 mm
inner diameter	40 mm
diameter of draft tube	30 mm
length of draft tube	100 mm
Sparger:	
pore size	10 μm
length	20 mm
diameter	10 mm

Dimensions of the stirred-tank fermentor.

Reactor:	
total volume	200 ml
length	112 mm
inner diameter	50 mm
Magnetic stirrer-bar:	
length	40 mm
diameter	5 mm

Figure 5 Schematic diagram of the air-lift type fermenter.

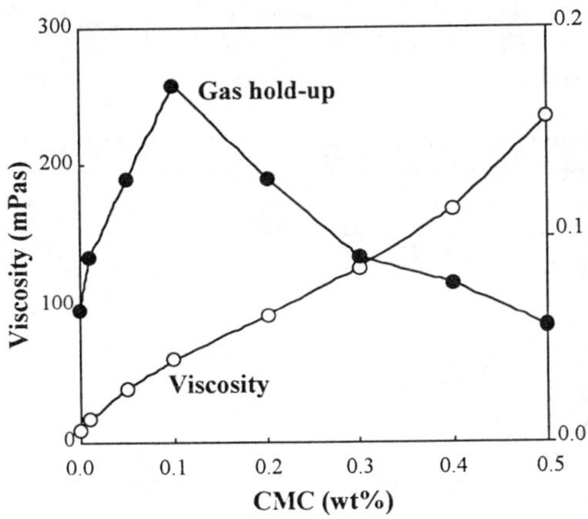

Figure 6 The effect of CMC concentration on viscosity and gas-hold up of culture medium.

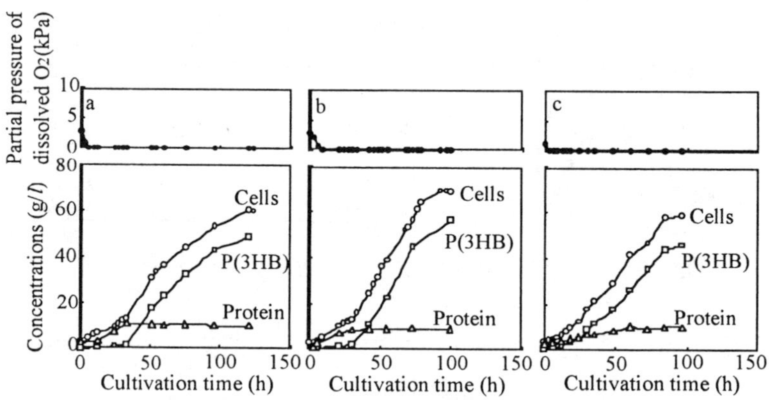

Figure 7 Time course of autotrophic cultivation of *A.eutrophus* in air-lift type fermenter under various concentrations of CMC.
(a): no addition of CMC, (b): 0.05 %(w/v) CMC,
(c): 0.1 % (w/v) CMC.

KLa measured by the static method and the gas hold-up was not close, especially for the case where 0.1% CMC was used. This cannot be easily explained at present.

5. P(3HB) production from xylose via L-lactate fermentation

For practical application of PHAs on commercial scale, consideration has to be given to the origin of carbon source used as well as the reducing the cost of production through the use of economical carbon sources. Carbon source such as 4-hydroxyvaleric acid derived from fossil fuel is often used as raw material for producing PHAs. However, it is known that the consumption of a large amount of fossil fuel results in the increase in CO_2 concentration in atmosphere. Waste materials such as lignocellulose is desirable from the viewpoints of utilization of unexplored resources and solution of the green house effect. Xylose is one of main components of hemicellulose contained in wood waste. At present, few types of commercial products have been produced from xylose. Young *et. al.* reported that *Pseudomonas cepacia* could produce P(3HB) from xylose and lactose, but the productivity was very low(29). *Lactococcus lactis* IO-1 isolated in our laboratory is able to produce L-lactic acid and acetic acid at a high production rate from xylose (30,31). L-Lactate also has the potential to be used as raw material in the manufacture of a biodegradable plastic, poly(L-lactate). *A.eutrophus* cannot utilize xylose but utilize lactate as carbon source. We therefore developed a culture method for the production of PHA from xylose employing these two bacteria. This culture method consisted of an initial fermentative production of L-lactate from xylose employing *L.lactis* IO-1 and a conversion of L-lactate into PHA by *A.eutrophus*. Flask culture experiment showed that the growth rate of *A.eutrophus* decreased according to the increase in L-lactate concentration in the medium and the cells could not grow above 30 $g{\bullet}dm^{-3}$ of L-lactate. In pH-controlled batch fermentations, a maximum specific growth rate of 0.6 h^{-1} was obtained when 5 $g{\bullet}dm^{-3}$ of L-lactate was used. The growth of microorganisms is generally inhibited by the presence of lactate, however, the specific growth rate of *A.eutrophus* when using L-lactate was higher than when other types of carbon source were used. According to our study for *A.eutrophus*, the maximum specific growth rate with using fructose was about 0.2 h^{-1} and the maximum specific growth rate in autotrophic condition was 0.42 h^{-1}. Such growth characteristic of *A.eutrophus* on L-lactate is favorable for production of P(3HB) by the two-stage method. The accumulation of P(3HB) by *A.eutrophus* was next investigated using the culture supernatant containing L-lactate and acetate converted from xylose by *L.lactis* IO-1. A pH-controlled batch culture of *L.lactis* IO-1 was first anaerobically carried out using 30 $g{\bullet}dm^{-3}$ of xylose as carbon source. When xylose in the culture was completely consumed, the culture broth was aseptically centrifuged and the supernatant was returned to the fermenter. *A.eutrophus* cells was then inoculated and the second stage cultivation for P(3HB) accumulation was aerobically carried out. The initial L-lactate concentration in the second stage culture was adjusted to 10 $g{\bullet}dm^{-3}$. After 24 hours of cultivation, 8.5 $g{\bullet}dm^{-3}$ of cells were produced. The final percentage of P(3HB) in the cell reached 55 %(w/w) without nitrogen source on medium being limited (32). The growth of *A.eutrophus* is inhibited by high lactate concentrations, therefore high cell density cultivation can be achieved by pH-stat batch culture with substrate feeding to control

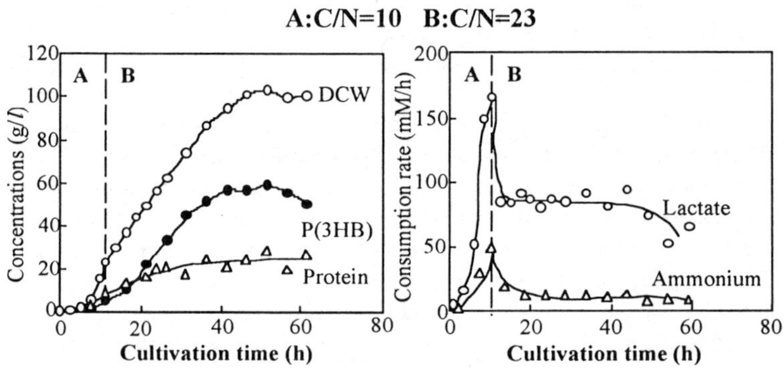

Figure 8 Time course of pH-stat batch culture with feeding L-lactic acid and inorganic nutrients solution. The C/N ratio in the feed solution was changed from 10 to 23.3 (mol/mol) after 12 of cultivation.

L-lactate concentration in medium at low level. As the C/N ratio for the consumption of L-lactate and ammonium by the cells was determined to be 10 (mol/mol) by a standard-type batch culture, the feed solution in which the C/N ratio was prepared to 10, was first used in the pH-stat batch culture as feed substrate. However, it was impossible to control L-lactate concentration at a constant level by using this feed solution. It was observed that the microorganism accumulated P(3HB) in the cell even during exponential growth phase and excreted a small amount of an unknown organic acid, then the acid-base equilibrium was not balanced in the culture system. The C/N ratio in the feed solution was, therefore, changed from 10 to 23.3(mol/mol) after 12 h of cultivation and phosphate and other organic nutrients were also supplied. As a result, cell concentration increased to 102 $g{\bullet}dm^{-3}$ (Fig.8). The P(3HB) content in the cells reached about 60 %(w/w) although nitrogen source in culture medium was not limiting. We are now investigating substrate feeding strategy to increase P(3HB) accumulation.

6. Conclusion

The practical cultivation systems for hydrogen-oxidizing bacterium, *A.eutrophus* to produce a biodegradable plastic, P(3HB) from CO_2 and xylose were developed and P(3HB) accumulation was improved by incorporating new strategies. The application of such culture systems should contribute to the solution of the global environmental pollution problems caused by increased CO_2 level in atmosphere, disposal of non-degradable plastics and utilization of industrial waste materials. For practical application of biodegradable plastics, obviously considerable technological challenges must be overcome, especially in the reduction in production cost and improvement in extraction and refining process of the product. We are tackling this difficult problem and also investigating the conversion of various types of industrial waste materials to other useful compounds.

Literature Cited

1. Anderson, D.; Dawes, E.A. *Microbiol. Rev.*, **1990**, 54, 450-472.
2. Ishizaki, A.; Tanaka, K. *J. Ferment. Bioeng.*, **1990**, 69, 170-174.
3. Ishizaki, A.; Tanaka, K. *J. Ferment. Bioeng.*, **1991**, 71, 254-257.
4. Tanaka, K.; Ishizaki, A.; Stanbury, P.F. *J. Ferment. Bioeng.*, **1992**, 74, 288-291.
5. Tanaka, K.; Ishizaki., A.; Kanemaru., T.; Kawani, T.; *Biotechnol. Bioeng.*, **1995**, 45, 268-275.
6. Takeshita, T.; Tanaka, K.; Ishizaki, A.; Stanbury, P.F. *J. Fac. Agric., Kyushu Univ.*, **1993**, 38, 55-64.
7. Takeshita, T.; Tanaka, K.; Ishizaki, A.; Stanbury, P.F. *J. Ferment. Bioeng.*, **1993**, 76, 148-150.
8. Tanaka, K.; Ishizaki, A. *J. Ferment. Bioeng.*, **1994**, 77, 425-427.
9. Repask, R.; Mayer, R. *Appl. Environ. Biotechnol.*, **1976**, 32, 592-597.
10. Goto, E.; Suzuki, K.; Kodama, T.; Minoda, T. *Agric., Biol. Chem.*, **1977**, 41, 521-525.

11. Morinaga, Y.; Yamanaka, S.; Ishizaki, A.; Hirose, Y. *Agric. Biol. Chem.*, **1978**, 42, 439-444.
12. Sonnleitner, B.; Heinzle, E.; Braubegg, G.; Lafferty, R.M. *Eur. J. Appl. Microbiol. Biotechnol.*, **1976**, 7, 1-10.
13. Miura, Y.; Okazaki, M.; Ohi, K.; Nishimura, T.; Komemushi, S. *Biotechnol. Bioeng.* , **1982**, 24, 1173-1182.
14. Igarashi, Y. *Nippon Nougeikagaku kaishi* (in Japanpsase) **1986**, 60, 935-941.
15. Kim, B. S.; Lee, S. Y.; Chang, H. N. *Bietechnol. lett.*, **1992**, 14, 811-816.
16. Suzuki, T.; Yamane, T.; Shimizu, S. *Appl. Microbiol. Biotechnol.*, **1986**, 23, 322-329.
17. Kim, B. S.; Lee, S.C.; Lee, S.Y.; Chang, H.N.; Chang, Y.K.; Woo, S.I. *Bietechnol. Bioeng.*, **1993,** 43, 892-898.
18. Yamane, T.; Fukunaga, M.; Lee, Y.W. *Biotechnol. Bioeng.*, **1996**, 50, 197-202.
19. Kim, S.W., Kim, P., Lee, H.S., Kim, J. H. *Biotechnol. lett.*, **1996**, 18, 25-30.
20. Lee, I.Y.; Kim, M.K.; Kim, G.J.; Chang, H.N.; Park, Y.H. *Biotechnol. lett.*, **1995**, 17, 571-574.
21. Möller, J.; Niehoff, J.; Hotop, S.; Dors, M.; Schügerl, K. *Appl. Microbiol. Biotechnol.*, **1992**, 37, 157-163.
22. Chevalier, P.; Noüe, J. *Enzyme Microbiol. Technol.*, **1988**, 10, 19-23.
23. Suh, I.S.; Schumpe, A.; Deckwer, W.D. *Biotechnol. Bioeng.*, **1992**, 39, 85-94.
24. Westlake, R. *Chem. Ing. Technol.*, **1986**, 58, 934-937.
25. Deckwer, W.D.; Schumpe, A. *Chem. Eng. Sci.*, **1993**, 48, 889-911.
26. Kennard, M.; Janekeh, M. *Biotechnol. Bioeng.*, **1991**, 38, 1261-1270.
27. Schumpe, A.; Deckwer, W.D. *Ind. Eng. Chem. Process Des. Dev.*, **1982**, 21, 706-711.
28. Okabe, M.; Ohta, N.; Park, Y.S. *J. Ferment. Bioeng.*, **1993**, 76, 117-122.
29. Young, F.K.; Kastner, J, Y.; May, S. W. *Appl. Environment. Microbiol.*, **1994**, 60, 4195-4198.
30. Ishizaki, A.; Ueda, T.; Tanaka, K.; Stanbury, P.F. *Biotechnol. lett.*, **1992**, 14, 599-604.
31. Ishizaki, A.; Ueda, T.; Tanaka, K.; Stanbury, P.F. *Biotechnol. lett.*, **1993**, 15, 489-494.
32. Tanaka, K.; Katamune, K.; Ishizaki, A. *Can. J. Microbiol.*, **1995**, 41, 257-261.

Chapter 17

Microbial Production of Xylitol

Badal C. Saha and Rodney J. Bothast

Fermentation Biochemistry Research Unit, National Center for Agricultural Utilization Research, Agricultural Research Service, U.S. Department of Agriculture, 1815 North University Street, Peoria, IL 61604

Xylitol, a five-carbon polyalcohol, has attracted much attention because of its potential use as a natural food sweetener, as a dental caries reducer and as a sugar substitute in diets for diabetics. Currently, it is produced chemically by catalytic reduction of xylose. Various microorganisms can convert xylose to xylitol. The present review describes microbial production of xylitol from xylose and xylose rich hemicellulose fractions present in various lignocellulosic biomass.

Xylitol, a pentitol of xylose, has attracted much attention because of its potential use as a natural food sweetener, as a dental caries reducer and as a sugar substitute for treatment of diabetics (*1*). It is a normal intermediary product of carbohydrate metabolism in humans and animals. The human body produces 5-15 g of xylitol a day during a normal metabolism (*2*). Xylitol is widely distributed in the plant kingdom, especially, in certain fruits and vegetables (*1, 3, 4*). However, extracting it from these sources is impractical because it is generally present in small quantities. Xylitol is currently produced chemically by catalytic reduction of xylose present in hemicellulose (xylan) hydrolyzate in alkaline conditions (*5, 6*). The recovery of xylitol from the xylan fraction reaches about 50-60% (*4*). Drawbacks of the chemical process are the requirements of high pressure and temperature, use of an expensive catalyst and use of extensive separation and purification steps to remove the by-products mainly derived from hemicellulose hydrolyzate (*7*). The bulk of xylitol produced is consumed in various food products such as chewing gum, candy, soft drinks and ice cream (*2*).

Microorganisms for Xylitol Production

Xylitol is produced from D-xylose as a metabolic intermediate in many xylose utilizing microorganisms in two ways: D-xylose is directly converted to xylitol by NADPH-dependent aldehyde reductase (EC 1.1.1.21), or D-xylose is first isomerized to D-xylulose by D-xylose isomerase (EC 5.3.1.5) and then reduced to xylitol by NADH-dependent xylitol dehydrogenase (EC 1.1.1.9) (Fig. 1) (*8*). Many yeasts and mycelial fungi possess the enzyme xylose reductase which catalyzes the reduction of xylose to xylitol as a first step in xylose metabolism (*9*). Xylitol production is a relatively common feature among xylose-utilizing yeasts (*10*). In xylose fermenting yeasts, the initial reactions of xylose metabolism are the major limiting steps (*11*). This results in the accumulation of xylitol in culture medium, the degree varying with the culture conditions and the yeast strain used (*12*).

Onishi and Suzuki (*13*) examined 58 yeast strains belonging to the genera *Saccharomyces, Debaryomyces, Pichia, Hansenula, Candida, Torulopsis, Kloeckera, Trichosporon, Cryptococcus, Rhodotorula, Monilia* and *Torula* for polyalcohol production from pentose sugars such as D-xylose, L-arabinose and D-ribose. *Candida polymorpha* dissimilated aerobically these three pentoses and produced xylitol from xylose, L-arabitol from L-arabinose and ribitol from D-ribose at the yield of 30-40% of sugar consumed. Gong et al. (*10*) screened 20 strains of *Candida* belonging to 11 different species, 21 strains of *Saccharomyces* belonging to 8 species and 8 strains of *Schizosaccharomyces pombe* for their ability to convert xylose to xylitol. Significant quantities of xylitol were produced by all these yeast strains. Barbosa et al. (*14*) screened 44 yeasts from five genera (*Candida, Hansenula, Kluyveromyces, Pichia* and *Pachysolen*) for conversion of xylose to xylitol. All but two of the strains produced some xylitol with varying rates and yields. The best xylitol producers were localized largely in the species *C. guilliermondii* and *C. tropicalis*. Seven strains of *C. guilliermondii* from diverse isolation sources produced xylitol efficiently when grown in a simple medium containing 5.0% xylose within 24 h (*15*). However, xylitol essentially disappeared from all the cultures within 72 h. Sirisansaneeyakul et al. (*16*) selected *C. mogii* ATCC 18364 as an efficient xylitol producer (Yp/s = 0.62 g/g) from 11 strains of D-xylose utilizing yeasts. *Debaryomyces hansenii* was an efficient xylitol producer exhibiting a xylitol/ethanol ratio above 4 and a carbon conversion of 54% for xylitol (*17*). *C. entomaea* and *Pichia guilliermondii* produced 0.51 and 0.43 g xylitol/g xylose at pH 5.0 and pH 4.0, respectively and 34°C (*18*). *Ambrosiozyma monospora* NRRL Y-1484 produced about 22 g xylitol and 18 g ethanol from 100 g xylose per L when grown at 25°C under moderate aeration (*19*). A strain of *C. tropicalis* converted xylose to xylitol and did not produce ethanol (*20*). Significant quantity of xylitol was produced during ethanol fermentation by *Pachysolen tannophilus* (*21,22*) and *Kluyveromyces cellobiovorus* (*23*). Various thermo-tolerant yeasts have also been evaluated for the bioconversion of xylose into xylitol (*24*). Xylitol production ranged from 0.83 to 4.69 g from 10 g xylose.

A fungal strain of *Petromyces albertensis* produced xylitol when grown in a medium containing D-xylose (*25*). A large amount (36.8 g/L) of xylitol was obtained from a D-xylose (100 g/L) medium containing ammonium acetate and yeast extract at an initial pH of 7.0. The production of xylitol from xylose has been studied with bacteria such as *Enterobacter liquefaciens* (*26, 27*), *Corynebacterium* sp. (*28, 29*), and *Mycobacterium smegmatis* (*30*).

Onishi and Suzuki (*31*) screened 128 yeast strains for their ability to produce xylitol from glucose. They reported a sequential fermentation process of xylitol production from glucose (glucose⇒ D-arabitol⇒D-xylulose⇒xylitol) without isolation

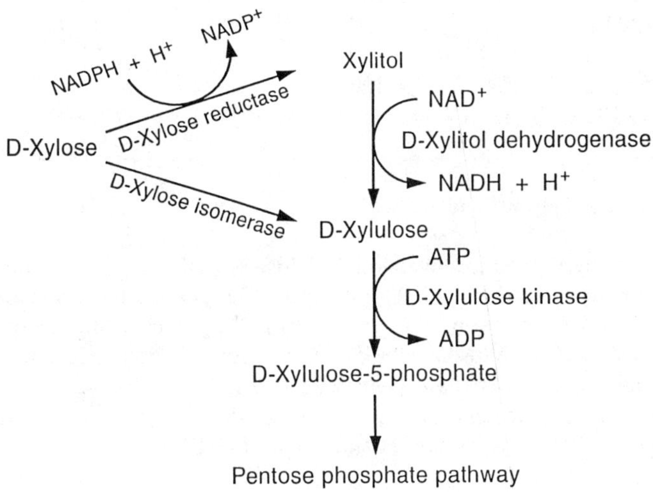

Figure 1. Pathway for xylose utilization by microorganisms.

and purification of the intermediates, and the yield of xylitol was 11% from glucose. *D. hansenii* converted glucose to D-arabitol, *Acetobacter suboxydans* oxidized D-arabitol almost quantitatively to D-xylulose and *C. guilliermondii* var. *soya* reduced D-xylulose to xylitol. Table I summarizes production of xylitol from xylose by some *Candida* species.

Table I. Production of xylitol from xylose by some *Candida* species

Yeast	Fermentation Time (h)	Xylose (g/L)	Xylitol (g/L)	Xylitol Yield (g/g)
Candida sp. B-22 (*32*)	167	249	210	0.84
C. boidinni 2201 (*33*)	120	100	40	0.40
C. guilliermondii FTI-20037 (*14*)	80	104	77.2	0.74
C. guilliermondii NRC 5578 (*34*)	406	300	221	0.75
Candida sp. L-102 (*35*)	65	114	100	0.88

Factors Affecting Xylitol Production

Medium Components. The conversion of xylose to xylitol by *C. guilliermondii* was affected by the nutrient source (*14*). Horitsu et al. (*36*) studied the influence of culture conditions on xylitol formation by *C. tropicalis* and optimized the volumetric xylitol production rate by the Box-Wilson method. In this respect, initial xylose concentration, yeast extract concentration and k_La were chosen as independent factors in 2^3-factorial design. Optimal product formation ($r_{xylitol}$ =2.67 g/L/h, $C_{xylitol}$ = 110 g/L) was obtained at 172 g/L xylose, 21 g/L yeast extract and a k_La of 451.5 L/h.

Xylose Concentration. Initial xylose concentration is an important factor to obtain high xylitol production. Meyrial et al. (*34*) reported that an increase in the initial xylose concentration from 10 g/L to 300 g/L led to activation of xylitol production by *C. guilliermondii*. The xylitol yield increased gradually with substrate, the highest xylitol yield (0.75 g/g xylose) was obtained at a substrate concentration of 300 g/L. However, the growth of the yeast was gradually inhibited by an increase in initial xylose concentration in the medium. Both the yield and specific rate of cells production declined when xylose concentration initially present in the culture increased. Chen and Gong (*32*) reported a xylitol yield of 84.5% of theoretical and a maximum production rate of 0.269 g/g/h from 249 g/L xylose by *Candida* sp. B-22. *C. tropicalis* HXP2 (*37*) and *C. boidinii* (*33*) produced the highest amounts of xylitol (144 g/L and 39 g/L, respectively) at respective values of substrate concentration of 200 g/L and 100 g/L. Dahiya (*25*) reported maximum xylitol production by *P. albertensis* was 36.8 g/L at the initial xylose concentration of 100 g/L. Xylitol production declined when the initial xylose concentration was increased to 150 g/L. This might be due to an osmotic effect on cells of *P. albertensis* or to substrate repression of xylose metabolizing enzymes.

When *C. mogii* was grown under oxygen-limited conditions in synthetic medium containing different concentrations of xylose (5-53 g/L), the xylitol formation rates showed a hyperbolic dependency on the initial substrate concentration (*16*).

Vandeska et al. (*38*) reported that an increase in initial xylose concentration induced xylitol production in *C. boidinii* but simultaneously acted as a growth inhibitory substrate leading to a long fermentation time. To overcome these problems, fed batch cultures were then used in which higher xylitol yields (0.57-0.68 g/g) and production rates (0.32-0.46 g/L/h) were obtained as compared with a batch process (*39*). A fed batch process with highest initial xylose concentration (100 g/L) and lowest level of aeration in the first phase, resulted in the highest yield of xylitol (75% of theoretical). A potentiometric biosensor for xylose to monitor fermentative conversion of xylose to xylitol was devised (*40*).

Presence of Other Sugars. Yahashi et al. (*41*) investigated the effect of glucose feeding on the production of xylitol from xylose by *C. tropicalis*. In the bench-scale fermenter (3 L scale) experiment, xylitol was produced at up to 104.5 g/L at 32 h cultivation and a yield of 0.82 (g/g xylose consumed) which is 1.3 times higher than that without glucose feeding. Meyrial et al. (*34*) evaluated the ability of *C. guilliermondii* to ferment non-xylose sugars such as glucose, mannose, galactose and L-arabinose commonly found in hemicellulose hydrolyzate. The strain did not convert glucose, mannose and galactose into their corresponding polyalcohol but only to ethanol and cell mass. Arabinose was converted to arabitol. Silva et al. (*42*) studied batch fermentation of xylose for xylitol production in stirred tank bioreactor. The efficiency of substrate conversion to xylitol was 66% in a medium containing xylose but decreased to 45% in a medium containing xylose and glucose. Vandeska et al. (*39*) investigated xylitol production by *C. boidinii* in fed batch fermentations with xylose (50, 100 g/L) and a mixture of glucose (25 g/L) and xylose (25 g/L). All fermentations were initially batch processes with high levels of aeration and rapid production of biomass. Faster growth occurred when a mixture of glucose and xylose was used instead of xylose. Glucose was assimilated first and maximal xylitol production was 39-41 g/L, compared with 46.5 and 59.3 g/L with xylose alone.

Nitrogen Sources and Organic Nutrients. Dahiya (*25*) studied the effect of 8 ammonium salts and 4 organic nitrogen sources on the production of xylitol from xylose by *P. albertensis*. Ammonium acetate was most effective for xylitol production. Yeast extract was the most suitable organic nutrient for enhancement of xylitol production. Lu et al. (*35*) investigated the effect of nitrogen sources [asparagine, casein hydrolyzate, glycine, Traders protein, yeast extract, urea, $NaNO_3$, NH_4NO_3, $(NH_4)_2SO_4$, NH_4Cl, $NH_4H_2PO_4$] on xylitol production from xylose in shake flasks by an efficient xylitol producing yeast, *Candida* sp. L-102. Different nitrogen sources influenced xylitol production rate, average specific productivity, and xylitol yield. Maximum xylitol production (100 g/L of xylitol from 114 g/L of xylose) was obtained with urea (3 g/L) as the nitrogen source. Silva et al. (*43*) evaluated the xylose conversion into xylitol by *C. guilliermondii* in semi-synthetic media supplemented with different nitrogen sources [urea, NH_4Cl, $(NH_4)_2SO_4$] in a ratio C/N equal 25.6. The type of nitrogen source did not influence this bioconversion and the xylitol yield was around 80%. On the other hand,

Barbosa et al. (*14*) reported that the use of urea led to higher xylitol productivity by *C. guilliermondii* than with ammonium sulfate, and supplementation of urea with casamino acids improved performance over urea alone only slightly. Yeast extract improved yields, but only slightly.

Magnesium and Biotin. Mahler and Guebel (*44*) studied the influence of Mg^{+2} concentration on growth, ethanol and xylitol production from xylose by *Pichia stipitis* NRRL Y-7124. Under constant oxygen uptake rate, biomass/xylose and biomass/Mg^{+2} yields increased with Mg^{+2} concentration with a maximum value at 4 mm. Ethanol was the main product formed. At low Mg^{+2} levels (1 mM), 49% of carbon flux to ethanol was redirected to xylitol production, and was correlated with intracellular accumulation of NADH.

Lee et al. (*45*) reported that the relative amount of ethanol and xylitol accumulated in xylose fed aerobic batch cultures of *P. tannophilus* and *C. guilliermondii* depended on the limitation by biotin. In high biotin containing media (2 μg/L) *P. tannophilus* favored ethanol production over that of xylitol while *C. guilliermondii* favored xylitol formation.

Methanol Supplementation. Dahiya (*25*) reported that addition of 1% methanol to the medium with 100 g/L xylose increased the xylitol production from 36.8 g/L to 39.8 g/L by *P. albertensis*. No significant difference in fungal biomass and xylulose accumulation was observed and only 0.015% methanol was consumed. This could be due to the oxidation of methanol to yield NADH which would enhance the reduction of xylose and xylulose to xylitol. Vongsuvanlert and Tani (*33*) reported about 18 and 26% increase in xylitol production from xylose in presence of 1 and 2% methanol, respectively by *C. boidinii*. This is also the case with the production of sorbitol from glucose and iditol from L-sorbose by *C. boidinii* (*46*).

Initial Cell Density. Cao et al. (*47*) investigated the effect of cell density on the production of xylitol from xylose by *Candida* sp. B-22. The rate of xylitol production from xylose increased with increasing yeast cell density. At high initial yeast cell concentration of 26 mg/ml, 210 g/L of xylitol was produced from 260 g/L of xylose after 96 h of incubation with a yield of 81% of the theoretical value. Vandeska et al. (*38*) reported that high initial cell densities improved xylitol yields and specific production rates of xylitol by *C. boidinii*. The susceptibility of wood hydrolyzate to fermentation by *D. hansenii* NRRL Y-7426 was strongly dependent on the initial cell concentration (*48*).

Oxygen Supply. A variety of yeasts such as *Candida*, *Hanensula*, *Klyveryomyces*, and *Pichia* require oxygen for sugar uptake (*49*) and availability of oxygen has significant influence on xylose fermentation by these yeasts (*10*). However, oxygen limitation is the main factor stimulating the formation of xylitol (*50*). Roseiro et al (*17*) reported that xylitol production by *D. hansenii* required semianaerobic conditions. The presence of oxygen enhanced NADH oxidation and a high NAD+/NADH ratio led to xylitol oxidation to xylulose; therefore, less xylitol was accumulated. Thus the yield of xylitol depended strongly on the oxygen transfer rate (*51*). Horitsu et al. (*36*) reported that higher level of dissolved oxygen is required only at the earlier phase of cultivation and afterwards it should be decreased to the lower level of respiration by the yeast. Barbosa et al. (*14*) reported that increasing oxygen limitation led to increased xylitol productivity and decreased ethanol production with *C. guilliermondii*. Nolleau et al. (*11*) evaluated the ability of *C. guilliermondii* and *C. parapsilosis* to ferment xylose to xylitol under different oxygen transfer rates. In *C. guilliermondii*, a maximal xylitol yield of 0.66 g/g was obtained when oxygen transfer rate was 2.2 mmol/l· h. Optimal conditions to produce xylitol by *C. parapsilosis* (0.75 g/g) arose from cultures at pH 4.75 with 0.4 mmols of oxygen/l· h. The oxygen is not only an important factor to optimize the xylitol production but it is also an essential component for xylose assimilation. When aerobic batch cultures of *C. guilliermondii* and *C. parapsilosis* provided with xylose, were shifted to anaerobic conditions, the xylose concentration remained at a constant level and all metabolic activities stopped immediately. *C. mogii* produced xylitol from xylose under aerobic and oxygen-limiting conditions, but not without oxygen (*16*). Xylose conversion into xylitol by *C. guilliermondii* FTI 20037 was investigated in a stirred tank bioreactor at different stirring rates (*42*). Maximal xylitol production (22.2 g/L) was obtained at 30°C, with an aeration rate of 0.46 vvm using a stirring rate of 300 per min (k_La = 10.6 h^{-1}). An increase of k_La caused an increase in the consumption of xylose in detriment to xylitol formation. Winkelhausen et al. (*52*) investigated xylitol formation by *C. boidinni* in oxygen limited chemostat culture. The production of xylitol by the yeast occurred under conditions of an oxygen limitation at specific oxygen uptake rates lower than 0.91 mmol/gh. The efffect of aeration on xylitol production from xylose by some yeasts is summarized in Table II.

Xylitol Production by Recombinant *Saccharomyces cerevisiae*

The yeast *Saccharomyces cerevisiae* is not able to use xylose or xylitol as a carbon source for growth or fermentation (*54*). Hallborn et al. (*55*) obtained efficient conversion of xylose to xylitol by transforming *S. cerevisiae* with the gene encoding the xylose reductase (XR) gene (*XYL1*) of *Pichia stipitis*. Due to lack of xylitol dehydrogenase (XDH), the recombinant *S. cerevisiae* needs a co-carbon substrate to regenerate the cofactors and to gain maintenance energy. Hallborn et al. (*56*) studied the influence of cosubstrate and aeration on xylitol formation by the recombinant *S. cerevisiae*. With glucose and ethanol, the conversion yields were close to 1 g xylitol/ g consumed xylose. Decreased aeration increased the xylitol yield based on consumed cosubstrate, while the rate of xylitol formation decreased. Xylitol yields close to 100% could be obtained from a medium with a total xylose concentration corresponding to that of an industrial hemicellulose hydrolyzate by fed-batch cultivation of recombinant *XYL1* expressing *S. cerevisiae* using ethanol as co-substrate (*57*). Recently, Roca et al. (*58*) investigated the effect of hydraulic residence time (1.3-11.3 h), substrate/cosubstrate ratio (0.5 and 1), recycling ratio (0.5 and 10), and aeration (anaerobic and oxygen limited conditions) on xylitol production by immobilized recombinant *S. cerevisiae* in a continuous packed-bed bioreactor.

Enzymatic Production of Xylitol from Xylose

The enzymatic production of xylitol from xylose using xylose reductase of *C. pelliculosa*

Table II. Effect of aeration on xylitol production from xylose by some yeasts

Yeast	Xylose (g/L)	Aeration	Xylitol yield (g/g)
Candida tropicalis (*36*)	100	100 ml/min	0.49
	100	400 ml/min	0.57
	100	500 ml/min	0.45
	100	700 ml/min	0.38
C. guilliermondii (*51*)	100	Microaerobiosis	0.50
	100	Semiaerobiosis	0.49
	100	Aerobiosis	0.56
C. parapsilosis (*51*)	100	Microaerobiosis	0.74
	100	Semiaerobiosis	0.61
	100	Aerobiosis	0.50
C. parapsilosis (*53*) (continuous culture)	10	0.15 vvm	0.31
	10	0.30 vvm	0.27
	10	0.60 vvm	0.08
	10	1.00 vvm	0.04
	10	1.50 vvm	0.02
	10	2.00 vvm	0.04

coupling with the oxidoreductase system of *Methanobacterium* sp. capable of recycling NADP (H) has been demonstrated by Kitpreechavanich et al. (*59*). A sulfonated polysulfone membrane reactor for *in situ* regeneration and retention of coenzymes NADP (H) using the xylose reductase of *C. pelliculosa* coupled with oxidoreductase system of *Methanobacterium* sp. in the reduction of xylose to xylitol with hydrogen gas was also used (*60*). The membrane rejected the permeation of NADP (H) (92 and 97%) F_{420} (97%) and the required enzymes (100%) almost completely, but did not reject for the permeation of xylitol. Nishio et al. (*61*) reported the enzymatic conversion of xylose into xylitol by the immobilized cells of *C. pelliculosa* (NADP+ dependent xylose reductase) coupled with the immobilized cells of *Methanobacterium* sp. HU (hydrogenase and F_{420}-NADP+ oxidoreductase) using hydrogen as an electron donor. The continuous production of xylitol in a column reactor packed with the co-immobilized cells could operate stably for 2 weeks. Xylitol was produced from xylose using commercial immobilized xylose isomerase from *Bacillus coagulans* and immobilized cells of *M. smegmatis* (*30*). From 10 g xylose, 4 g of xylitol was produced and 5 g xylose remained in the reaction mixture; no xylulose was detected. The washed cells of *M. smegmatis* converted xylulose to xylitol under aerobic and anaerobic conditions. The washed cells of a gluconate-utilizing *Corynebacterium* strain grown in a gluconate-xylose medium produced xylitol from xylose in the presence of gluconate (*29*). Xylose was reduced to xylitol by coupling the xylose reductase activity to the 6-phosphogluconate dehydrogenase activity with NADP as a cofactor using cell-free extract and the fractionated enzymes of *Corynebacterium* strain.

Production of Xylitol from Hemicellulose Hydrolyzate

Hemicellulose is one major component of plant cell wall materials, comprising up to 40% of agricultural residues and hardwood. It can be hydrolyzed by using dilute acids under mild hydrolysis conditions to yield a mixture of sugars (glucose, xylose, L-arabinose, mannose) of which xylose is the major component. These xylose containing hemicellulose hydrolyzates can serve as potential substrates for xylitol production. However, during acid hydrolysis, many potentially toxic compounds such as acetic acid, furfural, phenolic compounds, or lignin-degradation products are formed which inhibit growth of yeast.

Chen and Gong (*32*) studied the fermentation of sugarcane bagasse hemicellulose hydrolyzate to xylitol by a hydrolyzate-acclimatized yeast strain *Candida* sp. B-22. With this strain, a final xylitol concentration of 94.74 g/L was obtained from 105.35 g/L xylose in hemicellulose hydrolyzate after 96 h of incubation. *C. guilliermondii* FTI 20037 was able to ferment a sugar cane bagasse hydrolyzate producing 18.4 g/L xylitol from 29.5 g/L of xylose, at a production rate of 0.38 g/L/h (*62*). This lower value, compared to that (0.66 g/L/h) of the synthetic medium, may be attributed to the various toxic substances that interfere with microbial metabolism (e.g., acetic acid). Dominguez et al. (*63*) studied different treatments (neutralization, activated charcoal and neutralization, cation-exchange resins and neutralization) of sugar cane bagasse hemicellulose hydrolyzate to overcome the inhibitory effect on xylitol production by *Candida* sp. 11-2. The highest xylitol productivity (0.205 g/L/h), corresponding to 10.54 g/L, was obtained from hydrolyzates treated with activated charcoal (initial xylose, 42.96 g/L). To obtain higher xylitol productivity, treated hydrolyzates were concentrated by vacuum evaporation in rotavator to provide higher initial xylose concentration. The rate of xylitol production increased with increasing initial xylose concentration from 30 to 50 g/L, reaching a maximum of 28.9 g/L after 48 h fermentation. The decrease in xylitol production was dramatic with further increases in the initial xylose concentration. Parajo et al. (*48*) later reported a xylitol production of 39-41 g/L from concentrated *Eucalyptus globulus* wood acid hydrolyzate containing 58-78 g xylose/L by *Debaryomyces hansenii* NRRL Y-7426 using an initial cell concentration of 50-80 g/L.

Roberto et al. (*64, 65*) tested hydrolyzed hemicellulosic fractions of sugar cane bagasse and rice straw for xylitol production in batch fermentation by *C. guilliermondii* under semi-aerobic condition and compared these with synthetic medium containing xylose. For all media tested, simultaneous utilization of hemicellulosic sugars (glucose and xylose) was observed and the highest substrate uptake rate was attained in sugar cane bagasse medium. Increased xylitol concentration (40 g/L) was achieved in synthetic and rice straw media, although the highest xylitol production rate was obtained in sugar cane bagasse hydrolyzate. They concluded that both hydrolyzates can be converted into xylitol with satisfactory yields and productivities. Roberto et al (*66, 67*) evaluated xylitol production by *C. guilliermondii* in a rice straw hemicellulose hydrolyzate under different conditions of initial pH, nitrogen sources and inoculum level. The xylitol yields were 0.68 g/g for the medium containing ammonium sulfate at pH 5.3 and 0.66 g/g with urea at pH 4.5. Under appropriate inoculum conditions rice straw hemicellulose hydrolyzate was converted into xylitol by the yeast with efficiency values as high as 77% of the theoretical maximum. The production of xylitol from various hemicellulosic hydrolyzates is presented in Table III.

Gurgel et al. (*68*) studied xylitol recovery from fermented sugarcane bagasse hydrolyzate. The best clarifying treatment was found by adding 25 g activated carbon to 100 ml fermented broth at 80°C for 1 h at pH 6.0. The clarified medium was treated with ion-exchange resins after which xylitol crystallization was attempted. The ion exchange resins were not efficient but the crystallization technique showed good performance, although the crystals were involved in a viscous, colored solution.

Table III. Fermentative production of xylitol from hemicellulose hydrolyzates

Yeast	Substrate source	Fermentation Time (h)	Xylose (g/L)	Xylitol (g/L)	Xylitol (g/g)
Candida sp. B-22 (*32*)	Sugar cane	96	105.4	96.8	0.89
Candida sp. 11-2(*63*)	Sugar cane	48	42.96	10.54	-
C. guilliermondii FTI 20037 (*62*)	Sugar cane	-	29.50	18.40	-
C. guilliermondii FTI 20037 (*66*)	Rice straw	72	64	37.6	0.62
Debaryomyces hansenii NRRL Y-7426(*48*)	Wood	78	78	41	0.73

Concluding Remarks

The demand for xylitol in the food and pharmaceutical industries as an alternative sweetener has created a strong market for the development of low cost xylitol production process. Various xylose rich hemicellulosic materials can serve as abundant and cheap feedstocks for production of xylitol by fermentation. The cellulosic fraction can be converted to glucose, which is then fermented to fuel ethanol by *S. cerevisiae*. Much research needs to be done to select a suitable microorganism that can convert xylose into xylitol efficiently in presence of other hemicellulosic sugars and to understand the regulation and optimization of xylitol production by fermentation.

Literature Cited

1. Emodi, A. *Food Technol.* **1978**, 32, 20-32.
2. Pepper, T.: Olinger, P. M. *Food Technol.* **1988**, 10, 98-106.
3. Wang, Y.-M.; van Eys, *Ann. Rev. Nutr.* **1981**, 1, 437-475.
4. Washuttle, J.; Riederer, P.; Banchen, E. J. Food Sci. **1973**, 38, 1262-1263.
5. Hyvonen, L.; Koivistoinen, P. In: *Advances in Food Research;* O. Chichester Ed. Academic Press, New York, **1983**, Vol. 28, pp 373-403.
6. Malaja, A.; Hamalainen, L. *US Patent* 4,008,285, **1977**.
7. Meinander, N.; Hahn-Hagerdal, B.; Linko, M.; Linko, P.; Ojamo, H. *Appl. Microbiol. Biotechnol.* **1994**, 42, 334-339.

8. Hofer, M.; Betz, A.; Kotyk, A. *Biochem. Biophys. Acta* **1971**, 252, 1-12.
9. Chiang, C.; Knight, S. G. *Nature (London)*, **1960**, 188, 79-81.
10. Gong, C. S.; Claypool, L. D.; McCracken, C. M.; Maun, P. P.; Tsao, G. T. *Biotechnol. Bioeng.* **1983**, 85-102.
11. Nolleau, V.; Preziosi-Belloy, L.; Navarro, J. M. *Biotechnol. Letts.* **1995**, 17, 417-422
12. Van Dijken, J. P.; Scheffers, W. A. *FEMS Microbiol. Rev.* **1986**, 32, 199-224.
13. Onishi, H.: Suzuki, T. *Agr. Biol. Chem.* **1966**, 30, 1139-1144.
14. Barbosa, M. F. S.; de Medeiros, M. B.; de Mancilha, I. M.; Schneider, H.; Lee, H. *J. Industrial. Microbiol.* **1988**, 3, 241-251.
15. Leathers, T. D.; Gupta, S. C. *Abstract of Annual Meeting, Soc. Ind. Microbiol., Toronto, Ontario*, **1993**, p. 49.
16. Sirisansaneeyakul, S.; Staniszewski, M.; Rizzi, M. *J. Ferment. Biotechnol.* **1995**, 80, 565-570.
17. Roseiro, J. C.; Peito, M. A.; Girio, F. M.; Amaral-Collaco, M. T. *Arch. Microbiol.* **1991**, 484-490.
18. Saha, B. C.; Bothast, R. J. *Appl. Microbiol. Biotechnol.* **1996**, 45, 299-306.
19. Dien, B. D.; Kurtzman, C. P.; Saha, B. C.; Bothast, R. J. *Appl. Biochem. Biotechnol.* **1996**, 57/58, 233-242.
20. McMillan, J. D.; Boynton, B. L. *Appl. Biochem. Biotechnol.* **1994**, 45/46, 569-584.
21. Schvester, P.; Robinson, C. W.; Moo-Young, M. *Biotechnol. Bioeng. Symp.* **1983**, 13, 131-152.
22. Dekker, R. F. H. *Biotechnol. Letts.* **1982**, 4, 411-416.
23. Morikawa, Y.; Takasawa, S.; Masunaga, I.; Takayama, K. *Biotechnol. Bioeng.* **1985**, 27, 509-513.
24. McCracken, L. D.; Gong, C-S. *Biotechnol. Bioeng. Symp.* **1982**, 12, 91-102.
25. Dahiya, J. S. Can J. Microbiol. *Can J. Microbiol.* **1991**, 37, 14-18.
26. Yoshitake, J.; Ishizaki, H.; Shimamura, M.; Imai, T. *Agric. Biol. Chem.* **1973**, 37, 2261-2267.
27. Yoshitake, J.; Shimamura, M.; Ishizaki, H.; Irie, Y. *Agric. Biol. Chem.* **1976**, 40, 1493-1503.
28. Yoshitake, J.; Ohiwa, H.; Shimamura, M. *Agric. Biol. Chem.* **1971**, 35, 905-911.
29. Yoshitake, J.; Shimamura, M.; Imai, T. *Agr. Biol. Chem.* **1973**, 37, 2251-2259.
30. Izumori, K.; Tuzaki, K. *J. Ferment. Technol.* **1969**, 66, 33-36.
31. Onishi, H.; Suzuki, T. *Appl. Microbiol.* **1969**, 18, 1033-1035.
32. Chen, L.-F.; Gong, C-S. *J. Food Sci.* **1985**, 50, 226-228.
33. Vongsuvanlert, V.; Tani, Y. *J. Ferment. Bioeng.* **1989**, 67, 35-39.
34. Meyrial, V.; Delgenes, J. P.; Moletta, R.; Navarro, J. M. *Biotechnol. Letts.* **1991**, 13, 281-286.
35. Lu, J.; Tsai, L. B.; Gong, C. S.; Tsao, G. T. *Biotechnol. Letts.* **1995**, 17, 167-170.
36. Horitsu, H.; Yahashi, Y.; Takamizawa, K.; Kawai, K.; Suzuki, T.; Watanabe, N. *Biotechnol. Bioeng.* **1992**, 40, 1085-1091.
37. Gong, C. S.; Chen, L. F.; Tsao, G. T. *Biotechnol. Letts.* **1981**, 3, 125-130.
38. Vandeska, E.; Amartey, S.; Kuzmanova, S.; Jeffries, T. W. *World J. Microbiol. Biotechnol.* **1995**, 11, 3-8.

39. Vandeska, E.; Amartey, S.; Kuzmanova, S.; Jeffries, T. W. *Process Biochem.* **1996**, 31, 265-270.
40. Reshetilov, A. N.; Donova, M. V.; Dovbnya, D. V.; Boronin, A. M.; Leathers, T. D.; Greene, R. V. *Biosensors & Bioelectronics*, **1996**, 11, 401-408.
41. Yahashi, Y.; Horitsu, H.; Kawai, K.; Suzuki, T.; Takamizawa, K. *J. Ferment. Bioeng.* **1996**, 81, 148-152.
42. Silva, S. S.; Roberto, I. C.; Felipe, M. G. A.; Mancilha, I. M. *Process Biochem.* **1996**, 31, 549-553.
43. Silva, S. S.; Mancilha, I. M.; Queiroz, M. A.; Felpe, M. G. A.; Roberto, I. C.; Vitolo, M. *J. Basic. Microbiol.* **1994**, 34, 205-208.
44. Mahler, G. F.; Guebel, D. V. *Biotechnol. Letts.* **1994**, 16, 407-412.
45. Lee, H.; Atkin, A. L.; Barbosa, M. F. S.; Dorscheid, D. R.; Schneider, H. *Enzyme Microb. Technol.* **1988**, 10, 81-84.
46. Vongsuvanlert, V.; Tani, Y. *J. Ferment. Technol.* **1988**, 66, 517-523.
47. Cao, N.-J.; Tang, R.; Gong, C. S.; Chen, L. F. *Appl. Biochem. Biotechnol.* **1994**, 45/46, 515-519.
48. Parajo, J. C.; Dominguez, H., Dominguez, J. M. *Biotechnol. Letts.* **1996**, 18, 593-598.
49. Sims, A. P.; Barnett, J. A. *J. Gen. Microbiol.* **1978**, 106, 277-282.
50. Furlan, S. A.; Bouilloud, P.; de Castro, H. F. *Process Biochem.* **1994**, 29, 657-662.
51. Nolleau, V.; Preziosi-Belloy, L.; Delgenes, J. P.; Navarro, J. M. *Curr. Microbiol.* **1993**, 27, 191-197.
52. Winkelhausen, E.; Pittman, P.; Kuzmanova, S.; Jeffries, T. W. *Biotechnol. Letts.* **1996**, 18, 753-758.
53. Furlan, A.; Bouilloud, P.; Strehaiano, P.; Riba, J. P. *Biotechnol. Letts.* **1991**, 13, 203-206.
54. Gong, C.-S.; Chen, L. -F.; Flickinger, M. C.; Chiang, L.-C.; Tsao, G. T. *Appl. Environ. Microbiol.* **1981**, 41, 430-436.
55. Hallborn, J.;l Walfridsson, M.; Airaksinen, U.; Ojamo, H.; Hahn-Hagerdal, B.; Penttila, M.; Keranen, S. *Biotechnology* **1991**, 9, 1090-1095.
56. Hallborn, J.; Gorwa, M.-F.; Meinander, N.; Penttila, M.; Keranen, S.; Hahn-Hagerdahl, B. *Appl. Microbiol. Biotechnol.* **1994**, 42, 326-333.
57. Meinander, N.; Hahn-Hagerdal, B.; Linko, M.; Linko, P.; Ojamo, H. *Appl. Microbiol. Biotechnol.* **1994**, 42, 334-339.
58. Roca, E.; Meinander, N.; Hahn-Hagerdal, B. *Biotechnol. Bioeng.* **1996**, 51, 317-326.
59. Kitpreechavanich, V.; Hayashi, M.; Nishio, N.; Nagai, S. *Biotechnol. Letts.* **1984**, 6, 651-656.
60. Kitpreechavanich, V.; Nishio, N.; Hayashi, M.; Nagai, S. *Biotechnol. Letts.* **1985**, 7, 657-662.
61. Nishio, N.; Sugawa, K.; Hayase, N.; Nagai, S. *J. Ferment. Bioeng.* **1989**, 67, 356-360.
62. Pfeifer, M. J.; Silva, S. S.; Felipe, M. G. A.; Roberto, I. C.; Mancilha, I. M. *Appl. Biochem. Biotechnol.* **1996**, 57/58, 423-430.
63. Dominguez, J. M.; Gong, C. S.; Tsao, G. T. *Appl. Biochem. Biotechnol.* **1996**, 57/58, 49-56.

64. Roberto, I. C.; Felipe, M. G. A.; Mancilha, I. M.; Vitola, M.; Sato, S.; Silva, S. S. *Bioresour. Technol.* **1995**, 51, 255-257.
65. Roberto, I. C.; Mancilha, I. M.; Souza, C. A.D.; Felipe, M. G. A.; Sato, S.; Castro, H. F. D. *Biotechnol. Letts.* **1994,** 16, 1211-1216.
66. Roberto, I. C.; Silva, S. S.; Felipe, M. G. A.; Mancilha, I. M.; Vitola, M. *Appl. Biochem. Biotechnol.* **1996**, 57/58, 339-347.
67. Roberto, I. C.; Sato, S., de Mancilha, I. M. *J. Ind. Microbiol.* **1996**, 16, 348-350.
68. Gurgel, P. V.; Mancilha, I. M.; Pecanha, R. P.; Siqueira, J. F. M. *Bioresour. Technol.* **1995**, 52, 219-223.

Chapter 18

Engineering Issues in Synthesis-Gas Fermentations

R. M. Worden[1], M. D. Bredwell[1], and A. J. Grethlein[2]

[1]Department of Chemical Engineering, Michigan State University, East Lansing, MI 48824
[2]Athena Neurosciences, 800 Gateway Boulevard, South San Francisco, CA 94080

Biomass-derived synthesis gas can be readily converted into fuels and chemicals by anaerobic microorganisms. However, synthesis-gas fermentations typically exhibit low volumetric productivities due, in part, to low cell densities, production of unwanted by-products, and slow transfer of the synthesis gas into the liquid phase. Engineering approaches to improve bioreactor productivities are discussed, and recent advances in this area are summarized. Particular emphasis is placed on the use of bioreactor design to increase biocatalyst concentrations, development of metabolic models to study pathway regulation and the use of microbubble dispersions to enhance synthesis-gas mass transfer.

Synthesis gas, which consists primarily of carbon monoxide (CO), and hydrogen (H_2), is produced by the partial oxidation of an organic feedstock at high temperature in the presence of steam. Although coal and petroleum have historically been the most commonly used feedstocks for synthesis-gas production, several new gasification plants have recently been based on biomass (*1*). Biomass offers several advantages over the traditional feedstocks. First, it has a much lower sulfur content than many coals. Synthesis gas produced from wood chips at the GE gasification plant in Schenectady, NY contained 28 ppm H_2S (*2*), compared to 1-2% for coal-derived synthesis gases (*3*). Purification steps to remove sulfur from coal-derived synthesis gas are energy-intensive and add significantly to the product costs (*4*). Second, biomass materials are more reactive and thus require lower gasification temperatures and/or residence times. Fluidized-bed coal gasifiers are typically run at 1000°C using residence times of 0.5-3.0 h. By comparison, a temperature of 850°C was sufficient to gasify biomass using a residence time of 30 s to 5.0 min (*5*). Third, gasification of wood waste has the potential to solve a disposal problem while producing a valuable product. When wood waste is in short supply, short-rotation forestry can serve as a steady source of feedstock.

Synthesis gas can be catalytically converted into chemical products (e.g., methanol) in reactors operated at high temperatures and pressures. The status of such catalytic processes has recently been reviewed (*6*). Production of higher molecular weight alcohols (e.g., butanol) from synthesis gas is problematic, because existing catalysts yield a broad mix of alcohols. Anaerobic bioconversion of synthesis gas into fuels and chemicals represents an alternative approach that offers the advantages of lower temperatures and pressures, higher reaction specificity of the biological catalysts, and higher tolerance to sulfur compounds. A variety of anaerobes are able to convert synthesis-gas components into fuels and chemicals, including ethanol, butanol, acetic acid, butyric acid, and methane. The pathways of most of these microbes involve the conversion of CO or CO_2 and H_2 to the intermediate acetyl-CoA, which serves as a branch point for production of cell mass and two- and four-carbon alcohols and acids. Several reviews of microbial CO metabolism have been published (*7,8,9,10*). Whole-cell biocatalysts capable of converting synthesis gas to fuels and chemicals can tolerate orders of magnitude higher H_2S concentrations (greater than 1%) than iron and nickel catalysts (1-10 ppm) used for Fischer Tropsch conversion of synthesis gas (*11,12,13*).

Synthesis-gas fermentations have potential for commercial development, and some of the engineering issues have been addressed (*14,15,16*). Low bioreactor productivity is a major obstacle to commercialization. Several factors contribute to the low productivity, including low cell density, an inability to regulate branched pathways to obtain only the most desirable product, inhibition of the biocatalysts by the reactants and products, and low rates of transfer of CO and H_2 from the gas to the liquid phase. Several engineering approaches have been used in the past few years to overcome these limitations. The purpose of this paper is to summarize these approaches and discuss their impact on the feasibility of producing fuels and chemicals via synthesis gas fermentations.

Bioreactor Experiments

Cell Immobilization Studies. Cell immobilization can greatly increase the cell density in bioreactors. Immobilized-cell biocatalysts are well suited for energy-efficient bioreactor configurations, such as airlift and fluidized-beds. The suitability of several support matrices for immobilization of *Butyribacterium methylotrophicum* has been evaluated (*34*). The mass of cell protein immobilized was measured as a function of time for celite, wood powder, activated carbon, ion exchange resin, molecular sieves, and alumina during batch growth on CO in the presence of the support. The fluidization properties (bed expansion as a function of superficial liquid velocity), rate of product formation, and proportion of two- and four-carbon products were also measured for celite, ion-exchange resin and molecular sieves. All of these latter three supports were deemed satisfactory for immobilized-cell culture of *B. methylotrophicum.*

Continuous, Cell-Recycle Fermentations. High cell densities can also be achieved in bioreactors using cell recycle. This approach was evaluated during long-term, continuous CO fermentations by *B. methylotrophicum,* in which a 0.3 μm pore size cellulosic membrane was used in a cross-flow mode to achieve total cell recycle (*17*). Runs were conducted at different pH values, because previous experiments had shown that pH strongly regulates this fermentation (*18,19*), as evidenced by changes in relative proportions of the products. The experiments were performed with the same dilution rate used during previous continuous runs done without cell recycle (19) to study the effect of cell density on reactor productivity. At pH values of 7.2, 6.8, 6.4, and 6.0, steady-state was achieved. At pH values of 5.75 and 5.5, oscillations in the concentrations of several fermentation products were observed. A viable culture could not be maintained at pH values below 5.5.

Butyribacterium methylotrophicum has perhaps the most versatile metabolic capabilities of known microbes capable of anaerobic CO metabolism. It grows on a wide range of carbon and energy substrates, including 100% CO, H_2/CO_2, methanol, formate, and glucose (*20*). When grown on CO, it produces acetate, butyrate, ethanol, and butanol as catabolic products. The direct pathway from CO to butanol is apparently unique to *B. methylotrophicum (21).*

The reaction stoichiometries for the steady-state runs, balanced for carbon and electrons, are given in Table I. As was evident in similar chemostat experiments done without cell recycle (*19*), a reduction in pH resulted in the production of less acetate and more butyrate and alcohols. This trend is also evident in the Tables II and III, which show the partitioning of carbon and available electrons in the fermentation products. The fractions of carbon and electrons going into alcohols approximately double between a pH of 6.4 and 6.0. Production of 4-carbon compounds (i.e., butyrate and butanol) accounts for over 50% of the total electrons from CO at a fermentation pH of 6.0. Between a pH of 7.2 and 6.0, the partitioning of carbon and electrons to acetate decreases by approximately 35%. However, even at a pH of 6.0, on a molar basis, acetate remained the predominant product. Previously, butyrate was found to be the major product during batch culture with a pH shift from 6.8 to 6.0 at the onset of the stationary phase (*18*).

At pH values of 5.75 and 5.5, the cultures initially exhibited transient primary butanol production, followed by prolonged oscillations in acetate and butyrate formation. Table IV shows the initial fermentation stoichiometries for these two oscillatory fermentations during the period of primary butanol production. Average product concentrations over the time period of primary butanol production were used to obtain these balances. These initially high butanol production levels, up to 2.7 g/L, were significant in that they demonstrated that *B. methylotrophicum* is capable of producing butanol as the major product from CO metabolism (21). Butanol accounts for as much as 44% of the total available electrons from the CO feed at a pH of 5.5. The $\Delta G^{\circ\prime}$ of butanol production is sufficiently exergonic to drive ATP synthesis (*15*). An obvious research challenge is how to control the pathway fluxes so as to sustain high butanol yields. Interpretation of the experimental data with the metabolic model described below can help elucidate the trends in pathway regulations.

Table I. Effect of pH on steady-state fermentation stoichiometries

pH	Fermentation Stoichiometry
7.2	$4CO \dashrightarrow 2.21CO_2 + 0.410CH_3COOH + 0.105C_3H_7COOH + 0.019C_2H_5OH + 0.032C_4H_9OH + 0.387$ CELLS
6.8	$4CO \dashrightarrow 2.25CO_2 + 0.334CH_3COOH + 0.124C_3H_7COOH + 0.025C_2H_5OH + 0.040C_4H_9OH + 0.377$ CELLS
6.4	$4CO \dashrightarrow 2.26CO_2 + 0.316CH_3COOH + 0.152C_3H_7COOH + 0.018C_2H_5OH + 0.032C_4H_9OH + 0.279$ CELLS
6.0	$4CO \dashrightarrow 2.32CO_2 + 0.260CH_3COOH + 0.142C_3H_7COOH + 0.050C_2H_5OH + 0.055C_4H_9OH + 0.279$ CELLS

Table II. Effect of pH on Carbon Partitioning During Steady-State Fermentations

	Carbon Partitioning (%Carbon)					
pH	CO_2	CH_3COOH	C_3H_7COOH	C_2H_5OH	C_4H_9OH	Cells
7.2	55.2	20.5	10.5	1.0	3.2	9.7
6.8	56.2	16.7	12.4	1.2	4.0	9.4
6.4	56.5	15.8	15.2	0.9	3.2	7.0
6.0	58.0	13.0	14.2	2.5	5.5	7.0

Table III. Effect of pH on Available-Electron Partitioning During Steady-State Fermentations

	Electron Partitioning (% Available Electrons)				
pH	CH_3COOH	C_3H_7COOH	C_2H_5OH	C_4H_9OH	Cells
7.2	41.0	26.2	2.8	9.6	20.4
6.8	33.4	31.0	3.8	12.0	19.8
6.4	31.6	38.0	2.7	9.6	14.7
6.0	26.0	35.5	7.5	16.5	14.7

Table IV. Fermentation Stoichiometries During the Initial Period of Oscillatory Fermentations

pH	Fermentation Stoichiometry
5.75	4CO ---> 2.36CO2 + 0.126CH3COOH + 0.074C3H7COOH + 0.021C2H5OH + 0.115C4H9OH + 0.595 CELLS
5.5	4CO ---> 2.40CO2 + 0.112CH3COOH + 0.049C3H7COOH + 0.029C2H5OH + 0.149C4H9OH + 0.533 CELLS

Metabolic Model for CO Bioconversion

The complexity of the catabolic pathway of *B. methylotrophicum* makes it difficult to extract patterns of metabolic regulation from the experimental data. Metabolic models that calculate the fluxes of carbon, electron, and ATP fluxes through the various branches from the experimental data can aid in this process. However, development of such models requires that the stoichiometry of the individual reactions of the pathway be known. Sufficient information about the CO metabolism by *B. methylotrophicum* is available (*17,14,21*) for development of a metabolic model.

Model Development. Following the approach developed by Papoutsakis (*22*), an equation was written for cell mass production that is balanced for carbon, electrons and ATP. In acetogenic anaerobes, such as *B. methylotrophicum*, acetyl-CoA is a precursor for cell mass production and links catabolism with anabolism (*7*). Consequently, cell mass was assumed to be produced from acetyl-CoA, as shown in Equation 1:

$$\text{Acetyl-CoA} + y_1\ NADH_2 + y_2\ ATP \rightarrow 2\ \text{Cell Mass} \quad (1)$$

Applying an electron balance to Equation 1, the coefficient y_1 can be estimated to be 0.2 mol $NADH_2$/mol Acetyl-CoA. In an alternative approach (17), the value of y_1 was estimated to be 1.5 mol $NADH_2$/mol Acetyl-CoA from the following stoichiometric equation that was determined from batch growth of *B. methylotrophicum* on CO (*20*):

$$4\ CO \rightarrow 2.17\ CO_2 + 0.74\ CH_3COOH + 0.45\ \text{Cell mass} \quad (2)$$

This equation balances to within 3% for both carbon and electrons. The elemental composition measured for *B. methylotrophicum* cells closely matched the average cell formula ($CH_{1.8}O_{0.5}N_{0.5}$) suggested by Roels (*23*).

Estimation of y_2 required the assumption of a second mechanism of ATP production. When only substrate-level phosphorylation (SLP) is considered, conversion of CO to acetate is an ATP-neutral process (one ATP is consumed by formyl-THF synthase for each ATP produced by acetate kinase), and production of butyrate, ethanol, and butanol result in net consumption of ATP by SLP. Zeikus et al. proposed that an electron-transport phosphorylation (ETP) mechanism contributes the remaining ATP needed for cell maintenance and growth (*7*). In this mechanism, electrons generated by CO dehydrogenase are shuttled through two membrane-bound electron carriers. One of these carries both a proton and an electron, and the other carries only electrons. The net result is the ejection of protons from the cell, generating a transmembrane proton gradient. The protons reenter the cell via a proton-translocating ATP synthase that generates ATP. Six moles of electrons are produced by CO dehydrogenase per mole acetyl-CoA produced, so a theoretical maximum of 6 protons could be ejected per acetyl-CoA produced. A conservative value of 2 moles of protons ejected per mole of acetyl-CoA produced was used, along with a standard ratio of 1 mole ATP produced per 3 moles of protons translocated (*24*), to calculate ATP yields by both SLP and ETP for the growth data given in Equation 2. The net ATP yield for production of 0.74 mol of acetate and the acetyl-CoA used to produce the cell mass was calculated to be 2 mol ATP/mol acetyl-CoA used for cell mass. This value equals the amount of ATP available to convert one mole of acetyl-CoA into cell mass (i.e., y_2). This y_2 value translates into a $Y_{X/ATP}$ value of 26 g cells/mole ATP. By comparison, the accepted value for cell growth on glucose is 10.5 g cells/mole ATP (25).

Reaction equations were written that capture the stoichiometry and structure of the branched pathway. Consecutive reactions that did not involve branch points reactions were lumped together.

$$CO \rightarrow CO_2 + NADH_2 \quad (3)$$
$$CO_2 + 3\ NADH_2 + ATP \rightarrow [CH_3OH] \quad (4)$$
$$CO + [CH_3OH] \rightarrow \text{Acetyl-CoA} \quad (5)$$
$$\text{Acetyl-CoA} \rightarrow CH_3COOH + ATP \quad (6)$$
$$\text{Acetyl-CoA} + 2\ NADH_2 \rightarrow C_2H_5OH \quad (7)$$
$$\text{Acetyl-CoA} \rightarrow 0.5\ \text{Acetoacetyl-CoA} \quad (8)$$
$$\text{Acetoacetyl-CoA} + 2\ NADH_2 \rightarrow \text{Butyryl-CoA} \quad (9)$$
$$\text{Butyryl-CoA} \rightarrow C_3H_7COOH + ATP \quad (10)$$
$$\text{Butyryl-CoA} + 2\ NADH_2 \rightarrow C_4H_9OH \quad (11)$$

Equations 1 and 3-11 were each assigned an unknown rate (or flux) coefficient. An expression for the rate of production of each species (r_i) was then written from these equations in terms of the unknown flux coefficients and the reaction stoichiometries. The reaction rates for non-secreted intermediates $NADH_2$, $[CH_3OH]$, acetyl-CoA, butyryl-CoA, and acetoacetyl-CoA were set equal to zero, based the pseudo-steady-state assumption (22). The r_i terms for other compounds (acetate, butyrate, ethanol, butanol) were calculated from experimental measurements of their liquid-phase concentrations (C_i) , using the following, unsteady-state conservation equation:

$$r_i = \frac{dC_i}{dt} - DC_i \quad (12)$$

where D is the dilution rate. The CO_2 production rate was determined from the flow rate and CO_2 concentration of the effluent gas stream. For the steady-state experiments, the time derivative was set equal to zero. The resulting set of 10 equations with 10 unknowns was solved using Gaussian Elimination to calculate the flux coefficients.

Application of the Metabolic Model to Experimental Data. Calculation of the pathway fluxes allowed the relative ATP contributions from ETP and SLP to be calculated (*17*). Even though ATP is a non-secreted intermediate, the pseudo-steady-state assumption does not apply, because ATP can be consumed in a variety of unknown reactions in the cell, including miscellaneous maintenance-energy requirements. The model predicted that, for the steady-state fermentations, there was significant net consumption of ATP due to SLP (-0.12 mol ATP/mol CO). However, this was offset by sufficient production of ATP via ETP (+0.14 mol ATP/mol CO) to result in a small net gain of ATP.

An unsteady-state approach had to be used for the oscillatory fermentations. The time derivative in Equation 12 was evaluated by graphically differentiating the C_i *vs.* time data. The net ATP yield predicted by the model was slightly negative throughout the experiment. Since the culture could not be sustained under a long-term ATP deficit, this result suggests that there is more ATP produced than is accounted for by the model. The assumed ratio of 2 protons ejected per acetyl-CoA may have been too conservative. Assuming 4 moles of protons ejected per mole of acetyl-CoA produced instead of 2 would give a $Y_{X/ATP}$ value of 10 g cells/mol ATP, which closely matches the accepted value for glucose (25). Alternatively, there is evidence (G. J. Shen and J. G. Zeikus, unpublished results) that the electron carrier for the reduction crotonyl-CoA to butyryl-CoA is membrane-bound and may thus participate in ETP.

The mechanism driving the oscillations is believed to be related to metabolic regulation, rather than the CO mass-transfer rate or the liquid flow rate. Both the rate of CO addition and the liquid flow rate were constant throughout the experiments. The mean residence time of the liquid (66 h) was much shorter than the period of the oscillations (about 250 h). Although there were oscillations in both the acetate and butyrate concentrations, these oscillations were out of phase, indicating that the carbon flux was being alternately regulated through the two-carbon and four-carbon pathway branches. The cells obtain twice as much ATP via SLP per carbon equivalent by producing acetate from acetyl-CoA than butyrate. However, acetate production eliminates fewer electrons per carbon equivalent. Thus, the oscillations may have arisen from the cells alternately responding to needs to eliminate electrons and generate ATP. Consistent with this hypothesis, the oscillations in the CO uptake rate are in phase with butyrate production and out of phase with acetate production. More CO has to be consumed when butyrate is produced to maintain an equivalent rate of ATP production via SLP.

Gas Mass-Transfer Issues in Synthesis-Gas Fermentations

Oxygen mass transfer from the gas to the liquid phase is commonly rate-limiting in commercial-scale, aerobic fermentations (*26*). For this reason, design of fermenters for aerobic applications centers around providing an adequate volumetric mass-transfer coefficient (K_La). In a commercial-scale synthesis-gas bioprocess, providing sufficient mass transfer would be expected to be even more challenging, for two reasons. First, about twice as many moles of gas must be transferred per electron equivalent in the substrate for fermentations based on synthesis gas than those based on glucose. Second, under mass-transfer limiting conditions, the volumetric mass-transfer rate is directly proportional to the gas solubility (*18*), and the molar solubilities of CO and H_2 are only 77% and 65% of that of oxygen, respectively (27).

Mass-transfer-limiting conditions are readily identified in synthesis-gas fermentations by applying an unsteady-state mass balance to the gas uptake rate data. This approach has been used to demonstrate mass-transfer limitations in a variety of bioreactor configurations, including batch (*28*), stirred tanks (*18*), airlift fermenters, and trickle-bed reactors (*16*). Under such conditions, the gas uptake rate is constant. Thus, increases in the concentration or intrinsic reaction rate of the cells will not translate into improved productivity unless comparable increases are made in the gas mass-transfer rate.

Traditionally, gas mass transfer has been enhanced by increasing the power input to the bioreactor, which reduces the average bubble size and hence increases the interfacial area. In the previously described continuous CO fermentations using cell recycle (*17*), efforts were made to increase interfacial area by rapidly recycling the gas from the headspace, through a frit, and back into the fermentation broth. A high impeller rate was also used to maintain small bubble size. Even then, the highest specific CO gas consumption rate observed was 0.0044 mol/h•g cell. Calculations based on batch data with this organism at lower cell densities (*20*) have yielded CO consumption rates as high as 0.02 mol/h•g cell for much lower cell densities. These results suggest that, even with a high agitation rate and gas recycle, CO mass transfer was still rate-limiting. Moreover, this approach would not be economically feasible at the commercial scale, because power consumption increases with the impeller diameter to the fifth power and the impeller rate to the third power (*29*).

Formation and Stability of Microbubble Dispersions. Microbubble aeration has recently been proposed as an energy-efficient approach to enhancing synthesis-gas mass transfer (*30*). *M*icrobubbles are surfactant-stabilized gas bubbles having radii on the order of 25 μm. The surfactant layer provides a surface charge that prevents bubble coalescence by electrical repulsion (*31*). Microbubble dispersions have colloidal properties and can be pumped, unlike conventional foams that collapse upon pumping. The formation and coalescence properties of microbubbles have been studied (*30*). The microbubble generator consisted of a 5-cm diameter, stainless-steel disk spinning at 7000 rpm in the vicinity of stationary baffles. The number-averaged diameter was 107 µm for room-temperature microbubbles generated using Triton X-100 at a concentration of twice the critical micelle concentration (*30*). Modifications of the microbubble generator later reduced the number-averaged bubble diameter to 56 µm.

The rate of drainage of the microbubble dispersion was measured as a function of surfactant concentration and type. This technique gives information on stability and initial gas void fraction. As the concentration of surfactant increased beyond the critical micelle concentration, the stability of the dispersion increased to an asymptotic value that varied with the surfactant used. The initial gas void fraction of the dispersion was virtually unaffected by surfactant concentration, surfactant type, or the addition of sodium chloride. The constant value of the initial gas void fraction approximated the theoretical packing limit for monosized spheres. These results indicated that salts commonly used in growth media should not interfere with microbubble formation and stability.

The power required to generate microbubbles was measured using a Lightnin Labmaster unit capable of simultaneously measuring the impeller rate and power input (Bredwell and Worden, manuscript in preparation). The Power Number of the microbubble generator was measured to be 0.036, and the projected power requirement for microbubble generation for commercial-scale *B. methylotrophicum* fermentations was calculated to be 0.0081 kW per m^3 of fermentation capacity. Compared to a nominal power input for commercial-scale fermentations of 1 kW/m^3 (*26*), these data indicate that power requirements for microbubble production should be low at the commercial scale. Moreover, minimal power input would be required for the bioreactor, because the mass-transfer rate from microbubbles is virtually independent of agitation rate (*33*). Some power input would be required for liquid-mixing requirements, but this input could be minimized by the use of advanced, axial-flow impellers or a pneumatically mixed bioreactor configuration (e.g., airlift) reactor.

Non-toxic Surfactants for Microbubble Formation. The surfactant used to form the microbubbles must be non-toxic to the biocatalysts. The effects of several anionic, cationic and nonionic surfactants on the growth and product formation by *B. methylotrophicum* were determined in batch culture on CO (Bredwell, et al., submitted). A phosphate-buffered-basal (PBB) medium was used with the addition of 1, 2, or 3 times the critical micelle concentration of the surfactant in the media. The ionic surfactants, cetyl pyridium chloride and sodium dodecyl sulfate, inhibited growth at concentrations lower than the critical micelle concentration. The non-ionic surfactants tested were polyoxyethylene alcohols (Brij surfactants) and polyoxyethylene sorbitan esters (Tween surfactants). These surfactants had little or no effect on the growth rate of the bacteria. Concentrations of Tween 20, Tween 40, and Tween 80 between 0 and 3 times the critical micelle concentration had a negligible effect on the growth rate. The longer chain length surfactants (Brij 56 and Brij 58) appeared to inhibit growth at higher concentrations. Product concentration was measured using gas chromatography to evaluate the effects of the surfactant on the fermentation products. Carbon and electron balances were used to compute the stoichiometric equations. These equations, listed in Table V, show little effect of the Tween surfactants on the stoichiometry. Combined with the growth data, these results suggest that non-ionic Tween surfactants are well suited for making microbubbles for synthesis-gas fermentations.

Table V. Effect of Surfactants on Fermentation Stoichiometries

pH	Fermentation Stoichiometry
Control	4CO ---> $2.17CO_2 + 0.40CH_3COOH + 0.065C_3H_7COOH + 0.076C_2H_5OH$ + 0.61 CELLS
Tween 20	4CO ---> $2.17CO_2 + 0.45CH_3COOH + 0.026C_3H_7COOH + 0.110C_2H_5OH$ + 0.60 CELLS
Tween 40	4CO ---> $2.18CO_2 + 0.37CH_3COOH + 0.082C_3H_7COOH + 0.068C_2H_5OH$ + 0.62 CELLS
Tween 80	4CO ---> $2.20CO_2 + 0.37CH_3COOH + 0.078C_3H_7COOH + 0.096C_2H_5OH$ + 0.55 CELLS

Mass-Transfer Properties of Microbubbles. The mass-transfer properties of microbubble dispersions were measured using oxygen as the transferred gas (Bredwell and Worden, manuscript in preparation). The experimental system consisted of a 60-cm long column that had four ports along its length. A stream of oxygen microbubbles was combined with a stream of degassed water in a small mixing zone at the bottom of the column. The resulting steady-state oxygen profile across the column was measured using an oxygen minielectrode. An aqueous solution of Tween 20 at twice the critical micelle concentration was used to prepare the oxygen microbubbles. The concentration of surfactant in the degassed water stream varied from 0 to 5 times the critical micelle concentration.

The overall average mass-transfer coefficients based on the liquid phase ($K_{L,av}$) calculated from the experimental data varied between 0.00002 and 0.0002 m/s. The largest values, which were obtained when the bulk liquid contained no surfactant, are about a factor of 2 greater than the value predicted by the well-known theoretical result that the Sherwood Number (Sh) = 2.0. The lowest values were obtained when the surfactant concentration in the bulk liquid was 5 times the critical micelle concentration. The high $K_{L,av}$ values, coupled with the extremely high interfacial areas provided by microbubbles, resulted in K_La values up to 1800 h^{-1}, even without mechanical agitation. By comparison, reported K_La values for synthesis-gas fermentations are 2.1 h^{-1} for a packed bubble column, 56 h^{-1} for a trickle-bed bioreactor, and a range of 28 - 101 h^{-1} for a stirred-tank bioreactor operated with a high impeller rate (*16*).

Analysis of the microbubble mass-transfer data indicated that a significant fraction of the transferred gas was lost from the microbubbles within seconds. Consequently, to mathematically model the mass-transfer process, an unsteady-state approach was needed that accounts for phenomena that can often be neglected when using conventional bubbles, including changes in the microbubble size, gas composition, and intrabubble pressure.

Mathematical Model of Microbubble Mass Transfer. An unsteady-state mathematical model has been developed to explore the dynamics of microbubble mass transfer (Worden et al., submitted). The unsteady-state mass balance on the transferred gas component dissolved in the liquid phase is

$$\frac{DC_L}{Dt} = \frac{\partial C_L}{\partial t} + v_R \frac{\partial C_L}{\partial r} = v_R \left[1 - \left(\frac{R_B}{r}\right)^2\right] \frac{\partial C_L}{\partial r} + \frac{\partial}{\partial r}\left(r^2 \frac{\partial C_L}{\partial r}\right) \tag{13}$$

This equation was derived in terms of a substantial derivative (DC_L/Dt) that follows the movement of the gas-liquid interface as the bubble shrinks at a velocity v_R. Two different sets of initial conditions for C_l were used for the simulations: a pseudo-steady-state profile (PSS) and a gas-free profile (GF). The PSS initial condition assumes the initial concentration profile surrounding the bubble is that given by the steady-state solution to Equation 13 for the initial bubble radius and gas composition. The GF initial condition assumes the the liquid surrounding the microbubble is initially devoid of the transferring gas. The GF profile resulted in much steeper initial concentration gradients and hence more rapid initial gas mass transfer. The unsteady-state mass balances on the transferred gas and total gas in the microbubble, along with the initial conditions, are given below

$$X \frac{dR_B}{dt}\left(P_0 + \frac{4\sigma}{3R_B}\right) + \frac{dX}{dt}\left(\frac{P_0 R_B}{3} + \frac{2\sigma}{3}\right) = D_l \left.\frac{\partial C_L}{\partial r}\right|_R R_g T \tag{14}$$

$$v_R = \frac{dR_B}{dt} = \frac{D_l \left.\dfrac{\partial C_L}{\partial r}\right|_R R_g T}{P_0 + \dfrac{4\sigma}{3R_B}} \tag{15}$$

$$t = 0 \qquad X = X_0 \qquad R_B = R_0 \tag{16}$$

where X is the mole fraction of the transferred gas in the gas phase, and D_l is the mass diffusivity in the liquid phase. The gas and liquid concentrations are related using Henry's Law. In some cases, an additional film resistance is added at the interface to account for the surfactant shell.

The model was used to calculate the instantaneous mass transfer coefficient (k) at each time, as well as an average value (k_{av}), defined below, for comparison with experimental $K_{L,av}$ values:

$$k_{av} = \frac{\int_0^{t_f} \left(-D_l \left. \frac{\partial C_L}{\partial r} \right|_R \right) (4\pi R^2) dt}{\int_0^{t_f} \left(C^* - C^\infty \right) (4\pi R^2) dt} \tag{17}$$

where C^* is the liquid-phase concentration of the dissolved gas in equilibrium with the gas phase, and C^∞ is the concentration in the bulk liquid.

The model, which contained no adjustable parameters, was used to predict the mass-transfer rate from a microbubble into an infinite pool of degassed water. The rate of bubble shrinkage, and the corresponding changes in the k and k_{av} values are shown in Figure 1 for the GF initial condition. The predicted lifetime of an pure-gas microbubble is on the order of seconds in degassed liquid. During bubble shrinkage, k changes significantly, but k_{av} is approximately constant except for very early in the transfer process. This result validates the use of average mass-transfer coefficients to characterize microbubble mass transfer in experimental systems. The range of k_{av} predicted by the model compares favorably to the steady-state k values predicted by the theoretical model of Waslo and Gal-Or (32). The experimental $K_{L,av}$ values measured in pure water were similar in magnitude to the predicted k_{av} values. However, $K_{L,av}$ values measured in liquid containing high surfactant concentrations were an order of magnitude less. These results suggest that the mass-transfer resistance of the surfactant shell may be controlled by manipulating the properties of the fermentation medium.

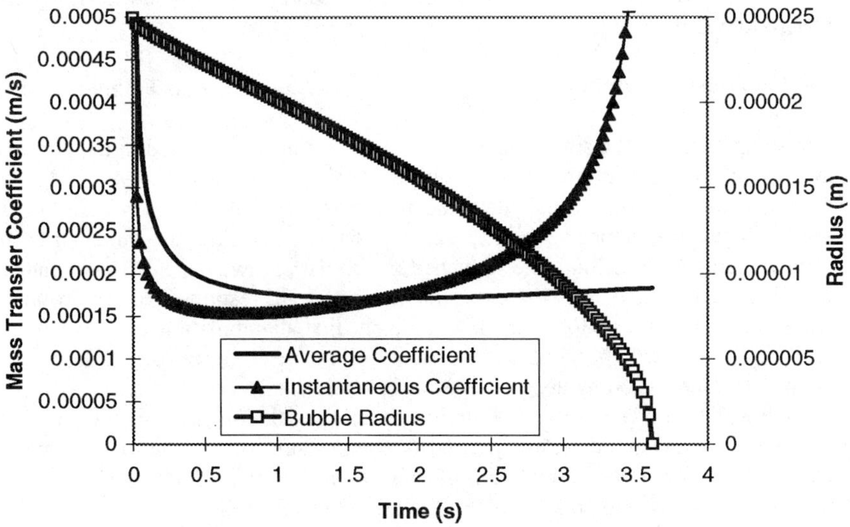

Figure 1: Predicted Rate of Change of Microbubble Radius and Mass Transfer Coefficient

As shown in Figure 2, the rate of microbubble mass transfer is predicted to decrease considerably as the initial concentration of transferred gas in the microbubble decreases. The average mass-transfer rate for a microbubble containing 100% transferable gas was predicted to be 14 times that for a microbubble containing only 20% transferable gas (e.g., air). Because synthesis gas consists of about 90% CO and H_2 (*2*), it is well suited for microbubble mass-transfer on this account.

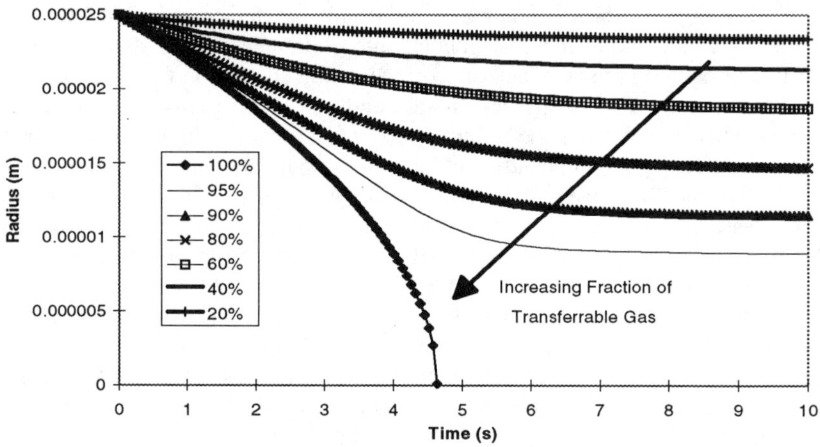

Figure 2. Effect of Concentration of Transferred Gas on Microbubble Shrinkage Rate

Implications for Commercial Development of Synthesis-Gas Fermentations

The significance of the studies summarized above can be discussed in the context of increasing bioreactor productivity in synthesis-gas fermentations. Metabolic modeling can be used to calculate the maximum theoretical yields possible from synthesis gas or combined with experimental data to calculate the fluxes of carbon, electron, and ATP through the branched metabolic pathways. Such information is useful in elucidating the patterns of regulation in response to environmental variables (e.g., pH). However, it is unlikely that manipulation of environmental variables alone will be sufficient to achieve the desired yields. Consequently, additional metabolic-engineering approaches will likely be necessary. Examples include isolation of mutants with altered genetic regulation patterns, elimination of enzyme(s) in unwanted pathways, and overexpression of rate-limiting enzymes in desired pathways. Implementation of these strategies will require development of recombinant tools (e.g., plasmids) for the microbes of interest.

Significant increases in productivity can be achieved via reactor engineering, as well, particularly through increasing cell concentration and increasing the rate of synthesis-gas mass transfer. The cell-immobilization studies indicated that *B. methylotrophicum* readily attaches to a variety of support materials that would be well-suited for industrial fermentations, and that these support materials do not inhibit cell growth or product formation. The cell recycle approach was highly successful. Both cell and product concentrations were increased several-fold over values obtained without cell recycle (17,19). Moreover, even during runs in excess of 1000 h, virtually no membrane fouling was observed.

The experimental and modeling results to date suggest that microbubble dispersions are well-suited for enhancing synthesis-gas fermentations. Extremely high K_La values have been measured for microbubbles without mechanical agitation. When the surfactant concentration in the bulk liquid was low, these coefficients approached the theoretical values, suggesting that the mass-transfer resistance of the surfactant shell can be maintained at low levels. The power-consumption rate to produce the microbubbles is projected to be quite low. Because the microbubbles would be produced in a relatively small vessel and pumped to the bioreactor, only the minimal amount of power input required to maintain sufficient mixing would be required in the bioreactor. Consequently, energy-efficient, pneumatically mixed configurations, such as the airlift could be used. Several surfactants have been identified that do not interfere with the growth and product formation yet form high-quality microbubbles. The dynamic microbubble model has been used to evaluate the influence of bubble shrinkage, surfactant-shell resistance, and changes in gas pressure and composition on the mass-transfer efficiency of microbubbles and to help interpret experimental results. Experiments are currently underway in our laboratory to evaluate the suitability of microbubble mass transfer in long-term synthesis-gas fermentations.

Although this paper has focused primarily on issues related to bioreactor productivity, there are also important separations issues related to synthesis-gas fermentations. First, the products are currently produced in relatively low concentrations, so cost-effective methods to separate them from dilute fermentation broths are needed. Second, the acids and alcohols produced in synthesis-gas fermentations become inhibitory as they accumulate. The effects of these products on the growth and stationary-phase product formation in *B. methylotrophicum* have been measured (*17*). Cell growth was found to be inhibited at alchohol concentrations on the order of 5 g/L, even though the stationary-phase CO metabolism was unaffected by such levels. Simultaneous fermentation and separation approaches would be expected to be useful in this situation, such as pervaporation membranes, which are selective for alcohols in their transport properties. Flux through the membrane is facilitated by the use of a vacuum, whereby components that diffuse through the membrane are immediately removed by evaporation. The resulting vapor, which is enriched in the alcohols, would then be condensed prior to further purification steps (e.g., distillation). Third, is will likely be necessary to recover and reuse the surfactant when microbubbles are used. However, it may be possible to rely on the surfactants that are naturally produced in the fermentation to form the microbubbles. This approach has been demonstrated in bench-scale yeast fermentations (*33*).

This paper has identified several engineering issues that currently limit the commercial prospects of synthesis-gas fermentations and has summarized recent research that addresses these issues. Such research, combined with complementary biocatalyst-development efforts, may make bioconversion of biomass-derived synthesis gas into fuels and chemicals a commercial reality.

Acknowledgments

This material is based on work supported by the National Science Foundation under grant #BCS 92 20396.

Literature Cited

1. Parkinson, G. and K Fouhy (1996) *Chem. Eng., March*, 1996, 37-43.
2. Najewicz, D.J. and A.H. Furman (1993) Pilot Scale Testing of Biomass Feedstocks for use in Gasification, *Proceedings of First Biomass Conference of the Americas*, NREL, Golden, CO, 498-512.
3. TRW Report, *Development Status of Key Emerging Gasification Systems*, September, 1980.
4. Wilson, J.S., J. Halow, and M.R. Ghate (1988) *Chemtech, February*, 123-128.
5. Overend, R.P. and Rivard, C.J. (1993) Thermal and Biological Gasification, *Proceedings of First Biomass Conference of the Americas*, NREL, Golden, CO, 470-497.
6. Mills, G.A., (1993) *Fuel, 73*, 1243-1279.
7. Zeikus, J.G., R. Kerby, and J.A. Krzycki (1985) *Science, 227*, 1167-1173.
8. Williams, D., J. Colby, C.M. Lyons and J. Bell, (1986) *Biotechnology and Genetic Engineering Reviews, 4,* 169-211.
9. Morsdorf, G., K. Frunzke, D Gadkari, and O. Meyer, (1992) *Biodegradation, 3*, 61-82.
10. Diekert, G., and G. Wohlfarth, (1994), *Antonie van Leeuwenhoek, 66*, 209-221.
11. Grethlein, A.J., B.K. Soni, R.M. Worden, and M.K. Jain (1992) *Appl. Biochem. Biotechnol., 34/35*, 233-246.
12. Vega, J.L., K.T. Klasson, D.E. Kimmel, E.C. Clausen, and J.L. Gaddy (1990) *Appl. Biochem. Biotechnol., 24/25,* 329-340.
13. Rostrup Nielsen, J.R. (1971) *J. Catalysis, 21,* p. 171.
14. Grethlein, A.J., and M.K. Jain, (1992) *Trends in Biotechnol., 10*(12), 418-423.
15. Worden, R.M., A.J. Grethlein, M.K. Jain, and R. Datta, (1991) *Fuel, 70*, 616-619.
16. Klasson, K.T., M.D. Ackerson, E.C. Clausen, and J.L. Gaddy, (1992) *Enz. Microb. Technol., 14*, 602-608.

17. Grethlein, A.J., (1991), *Metabolic Engineering of Product Formation During Carbon Monoxide Fermentations by Butyribacterium methylotrophicum*, PhD. Dissertation, Michigan State University, East Lansing, MI.
18. Worden, R.M., A.J. Grethlein, J.G. Zeikus, and M.K. Jain (1989) *Appl. Biochem. Biotechnol., 20/21,* 687-698.
19. Grethlein, A.J., R.M. Worden, M.K. Jain, and R. Datta (1990) *Appl. Biochem. Biotechnol., 24/25,* 875-884.
20. Lynd, L., R. Kerby, and J.G. Zeikus, (1982), 'Carbon Monoxide Metabolism of the Methylotrophic Acidogen *Butyribacterium methylotrophicum*', *J. Bacteriology, 149*(1), 255-263.
21. Grethlein, A.J., R.M. Worden, M.K. Jain, and R. Datta, (1991), 'Evidence for Production of n-Butanol from Carbon Monoxide by *Butyribacterium methylotrophicum, J. Ferm. Bioeng.*, *72*(1), 58-60.
22. Papoutsakis, E.T. (1984) *Biotechnol. and Bioeng., 26,* 174-187.
23. Roels, J.A. (1983) *Energetics and Kinetics in Biotechnology,* Elsevier, New York, NY, 41.
24. Niedhardt, F.C. (1987) *E. coli and Slamonella typhimurium, Cellular and Molecular Biology*, ASM Press, Washington, DC.
25. Payne, W.J. (1970) *Ann. Rev. Microbiol, 24,* 17.
26. Van't Riet, K., (1979), *Ind. Eng. Chem. Process Des. Dev., 18*(3), 357-364.
27. National Research Council (1928) *International Critical Tables,* Vol 3, McGraw-Hill, New York, NY, 254.
28. Phillips, J.R., E.C. Clausen, and J.L. Gaddy (1994), *Appl. Biochem. Biotechnol., 45/46,* 145-157.
29. McCabe, W.L., J.C. Smith, and P. Harriott (1993) *Unit Operations of Chemical Engineering, Fifth Ed.*, McGraw-Hill, New York, NY, 250.
30. Bredwell, M.D., M.D. Telgenhoff, and R.M. Worden, (1995) *App. Biochem. Biotechnol., 51/52, 501-509.*
31. Sebba, F. (1987) *Foams and Biliquid Foams—Aphrons*, Wiley, New York, NY., 47-78.
32. Waslo, S. and Gal-Or, B., (1971), *Chem. Eng. Sci., 26,* 829-838.
33. Kaster, J.A., D.L. Michelsen, and W.H. Velander (1990) *Appl. Biochem. Biotechnol., 24/25,* 469-484.
34. Chatterjee, S., A.J. Grethlein, R.M. Worden, and M.K. (1996) *Journal of Fermentation and Bioengineering, 81,* 158-162.

Chapter 19

Solution of Environmental Problems Through Biomass Conversion Using Microbial Technology

Ayaaki Ishizaki

Department of Food Science and Technology, Faculty of Agriculture, Kyushu University, Hokozaki, Higashi-ku, Fukuoka 812–81, Japan

1. Introduction

Today, we are faced with serious global environmental problems. Among them, counter-measure against the elevation of carbon dioxide in the atmosphere must be an urgent tasks. Microbial technology can contribute to the solution of this problem. First , we must note that the composition of the atmosphere over the globe has undergone drastic changes in the past 4.5 billion years of the earth's history which has resulted in the development and diversification of living organisms and changes in the biota of the earth. However, these changes have been occurred in a very limited part of the earth i.e. biosphere in which the atmosphere and oceans are included. These changes have been accompanied by changes in the flora and fauna as well. The ecosystem of the globe and the environment influence each other in the recycling of atoms in the biosphere to renew the composition of the air and to establish new equilibrium from time to time. However, the global problem of the environment today is a new subject which never arose in the past history of the earth and is caused by new technologies developed to control nature and to ensure a comfortable life. The speed of these developments have been very rapid indeed. Therefore, there has not been enough time for humans to adapt the environment brought by these technologies to suit. Such drastic change has never occurred in the past history of earth and this is the first experience for all living organisms, thus the problem is becoming critical and difficult to solve.

The amount of the elemental carbon existing in the crust of the earth is great but this carbon is inactive since they are contained in the crust such that no chemical reactions occur resulting in no emission of carbon dioxide from the crust into the atmosphere. Thus, carbon dioxide which influences environment is restricted to the elemental carbon recycling in the biosphere. Since there is about 300 ppm of carbon dioxide in the air so the total amount of elemental carbon containing in the atmosphere is about 700 Gt (gigatons). On the land, heterotrophic organisms including humans release about 50 Gt of elemental carbon (in the form of carbon dioxide) per year into air. If the same amount of elemental carbon

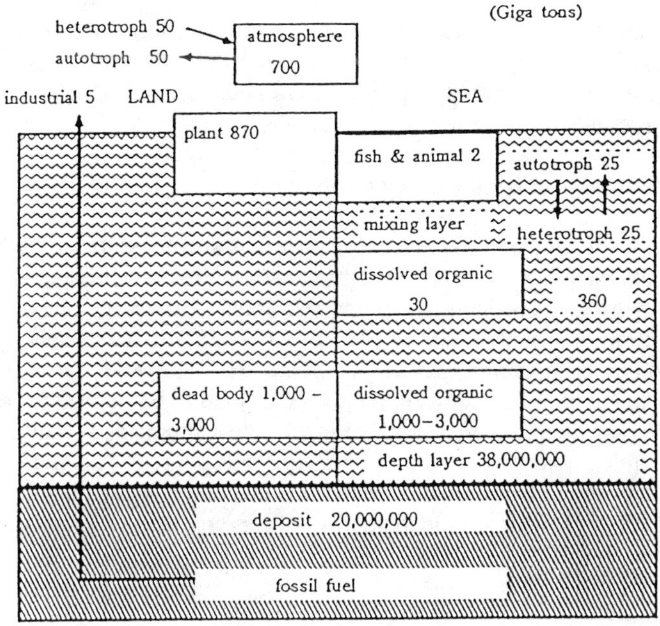

Fig. 1. Elemental carbon recycling in the biosphere

was fixed into organic materials by autotrophic organisms, a constant concentration of carbon dioxide would be maintained in the atmosphere. In the oceans, balance would be maintained by exchange elemental carbon between autotrophs such as sea weeds and heterotrophs such as fish, resulting in a constant concentration of dissolved carbon dioxide and bicarbonate in the mixing layer of ocean. The amount of elemental carbon thus recycled between inorganic (carbon dioxide) and organic matter both on the land and in the sea is a very small amount compared to the total amount of elemental carbon existing in the globe. However, the amount of elemental carbon recycling in the biosphere, 50 Gt, is big enough to influence the carbon dioxide concentration in air because the size of carbon dioxide pool in biosphere is only 700 Gt .

The approximately 5 Gt of elemental carbon released from the combustion of fossil fuels is considerably large when compared to the amount of elemental carbon recycling in the biosphere, 50 Gt. It is therefore reasonable to attribute elevating levels of carbon dioxide concentration in the air to carbon dioxide emissions from combustion of fossil fuels. In order to prevent such increase of carbon dioxide concentration in the air, numbers of autotrophic organisms must increase so as to fix 5 Gt of elemental carbon into organic materials.

In the case of nitrogen, the effect of increased artificial nitrogen fixation (synthetic ammonia and urea) to the environment has not been observed. To date, the amount of nitrogen recycling through artificial fixation has reached 30 Gt per year which is almost equivalent to the amount of biologically fixed nitrogen (by nitrogen fixing bacteria such as *Rhizobium*). The luck of influence of the rapid growth of artificially fixed nitrogen on the environment of the globe is due to the huge nitrogen pool size in the biosphere, where about 3,800,000 Gt of elemental nitrogen is presented.

2. Photosynthesis and High Energy Crops

Carbon dioxide emission in the biosphere is attributed to two sources, one biological and the other industrial. Industrial carbon dioxide emission is mainly from the exhaust gases of industrial processes. Where carbon dioxide occurs in high concentration in such processes, the carbon dioxide in the exhaust gas can be trapped and then recovered for uses as a raw material. However, carbon dioxide released from automobiles, for example, is too diluted to be trapped for recovery. Photosynthesis is the most successful means for the trapping and recovery of carbon dioxide released to the air from diluted emission sources. A counter-measure against the elevation of carbon dioxide in the atmosphere is to strengthen the path of carbon dioxide recycling through photosynthesis and then efficiently utilize the plant biomass formed as a result.

2-1 Oil Palm

In terms of photosynthesis, tropical areas offer great advantages over temperate zones due to high solar radiation. High energy crops native to South East Asia (and Africa and South America) are therefore desirable tools for the counter-measure to carbon dioxide elevation.

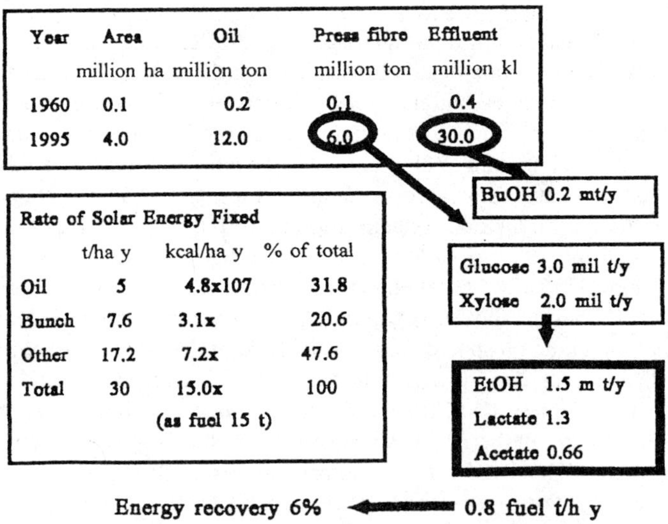

Fig.2. Total biomass production and waste materials in palm oil industry.

Oil palm is one crop which has a very high energy fixation per land area, i. e. 5 t/ha year of oil ($4.5x10^7$ kcal/ha year of energy) can be harvested. However, as shown in Fig.2 , oil is only 1/3 of the total biomass of the tree. Hence the total energy fixed by biomass is estimated to be $13.5x10^7$ kcal/ha year, this being equivalent to 15 tons of fuel. In the world as a whole, the plantation area for oil palm trees is estimated to be about 4,000,000 ha (equivalent to 12,000,000 t oil/year). Thus, the total biomass fixed in oil palm trees is equivalent to about 30,000,000 t of fuel. As shown in Fig. 2, 2/3 of the biomass in the oil palm industry is agricultural waste. A number of valuable resources are wasted and dumped in the plantation and oil mills. Among them, potential resources are press fiber and waste liquor. The chemical composition of the carbohydrate component of delignified palm fiber is 56.4% glucose, 36.0% xylose, 5.9% arabinose, and 1.7% mannose (1). Our experiments showed that delignified palm fiber can be digested to form glucose and xylose by commercial cellulases such as Meicellase and Onozuka and the hydrolysate can be subjected to fermentation to produce biofuel and chemicals (1). Our study estimates that 3,000,000 t of glucose and 2,000,000 t of xylose could be recovered from this waste per year.

Another big resource in the palm oil industry is the waste liquor from the oil mill. Two kinds of waste from the crude oil separation process are discharged, sterilized condensate and separator sludge. The mixture contains about 30,000 ppm of BOD and it is very hard to reduce BOD to the regulation limits. Our technology proved that the waste which digested by commercial cellulase can be fermented well by *Cl. saccharoperbutylacetonicum* N1-4 to produce acetone, butanol and ethanol without any additional medium supplementation (2-3). Extraction of butanol by fatty acid methylester stimulated fermentation rates by release of end product inhibition (4) and the extract was similar to diesel fuel. We are confident that a process using crude palm oil (CPO, the product from the oil mill) methylester will stimulate the fermentation rate to produce biodiesel efficiently. Thus, about 200,000 t of biodiesel could be recovered from the waste by this process (Fig.3).

2-2 Sago Palm

Gregarious sago palm is limited to tropical areas such as Malaysia (Sarawak), Indonesia, and Papua New Guinea (5). The starch productivity of sago palm is several times greater than that of tapioca (cassava), wheat, and rice, i.e. sago palm has a high efficiency of photosynthesis. Recently, Sarawak, a state of Malaysia, has successfully cultivated a sago plantation on 7,000 ha of land and has developed 85,000 ha of land for planting. The productivity of starch is reported to be 25 t/ha y. As shown in Fig. 4, one tree is planted per 100 m^2 (10 m x 10 m) land and one shoot from the mother tree will be remained every year. Thus, one grown tree can be harvested from 100 m^2 land every year after ten years ini-

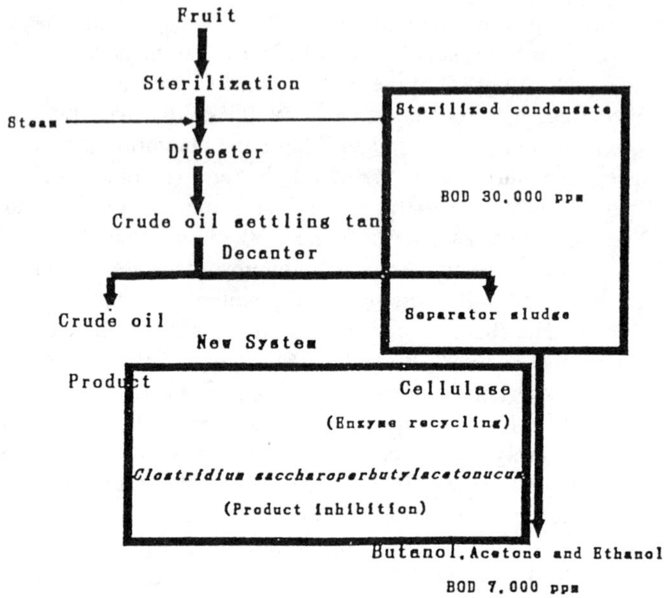

Fig. 3. Proposed process for bioconversion to reform palm oil waste liquor.

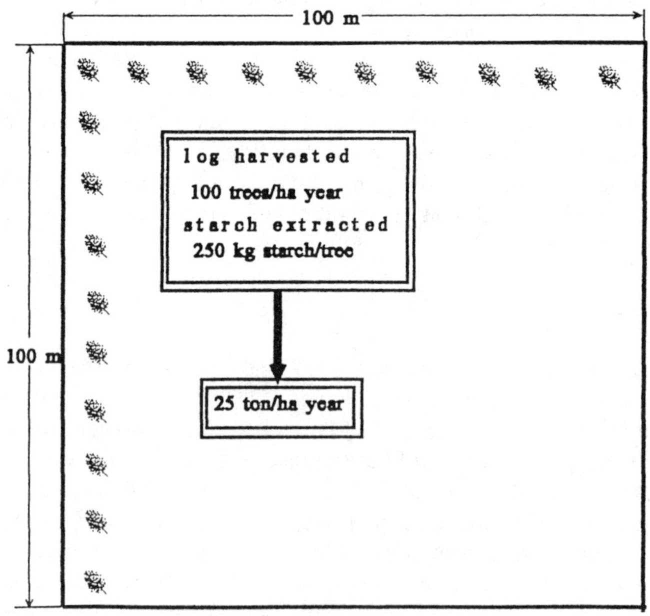

Fig. 4. Starch productivity of sago palm plantation.

tial propagation. In this plantation, 25 tons of starch per hectare can be recovered and this is almost 3 times higher than that of rice. Approximately 1,500,000 ton of starch will be harvested from this plantation per year and the starch could be used for bioconversion. Either 750,000 ton of ethanol or 1,350,000 of lactic acid could be produced by anaerobic bioconversion from this plantation.

2-3 Natural Rubber Waste

Natural rubber is also one of biomass made by recycling carbon dioxide by photosynthesis. Cost saving and increase productivity of natural rubber give a chance to win the competition of natural rubber against synthetic rubber made by petrochemical industry. Recently, our laboratory has developed NRSP (Natural Rubber Serum Powder) which is the spray dried product of the waste from natural rubber industry. The waste is discharged from the centrifugal machine to separate latex. The waste is subjected to digestion by papain, a proteinase, prior to spray drying. It is noted that NRSP contains strong growth promoters for various kind of microorganisms particularly anaerobes such as lactic acid bacteria (6), wide range of *Bifidobacterium* (7), and *Zymomonas* (8). Usually, anaerobic microorganisms require complex nutritious factors. *Bifidobacterium* is important intestinal microflora and there is a great interest in the possibility of increasing number of this microorganism in intestine by the special growth promoter. It is expected that NRSP can increase the number of *Bifidobacterium* in intestine to enhance the health of human.

Zymomonas mobilis is capable of producing ethanol with a higher productivity than yeast, but this microorganism requires complex nutritional factors. This microorganism thus can not grew in simple and economical medium such as that used in industrial ethanol production. By our study, the medium prepared with NRSP with soy bean hydrolysate as a source of complex nutritional factors exhibited the same fermentation capability with *Z. mobilis* as medium prepared using yeast extract (8).

3. Anaerobic Fermentation

To develop any chemical or biochemical technology for recycling carbon dioxide to serve as a counter-measure against carbon dioxide elevation in the atmosphere, it is obvious that the process should not use fossil fuel. In this regard, it is also better to avoid aerobic bioprocesses which are accompanied by carbon dioxide emission. Aerobic bioprocesses involve the oxidation of organic substrates by oxygen and this is the same principle of fossil fuel combustion. It is therefore preferable that anaerobic bioprocess should be adapted for bioconversion which allows less carbon dioxide emission. Some anaerobic reactions do not involve the production of carbon dioxide, some result in small quantities being produced, but others produce comparatively large amounts during the entire course of the fermentation. For example, fermentation involving the metabolic pathway in which pyrvate is converted into acetyl-CoA, with the release of carbon dioxide, inherently produce one mole of carbon dioxide for each mole of substrate consumed. Hence, anaerobic fermentation yielding ethanol, butanol, acetone, acetoin, and butanediol are accompanied by the production of carbon dioxide, while lactate formation from pyrvate is not accompanied by the release of carbon dioxide. Thus, lactate fermentation has advantages over other anaerobic fermentation such as ethanol fermentation from the point of view of carbon dioxide release during the biochemical reaction.

In general, anaerobic fermentation has advantages as 1. energy gaining metabolism, 2. less carbon dioxide, 3. accompanying metabolites as electron acceptor, while, it has disadvantages as 1. low cell density, 2. product inhibition, and 3. complex nutrition requirement. These disadvantages are sometimes bottleneck to attain efficient fermentation.

4. Lactate Industry

Lactic acid fermentation has great advantage over other bioconversions. The stoichiometry for homofermentation from hexose can be expressed,

$$C_6H_{12}O_6 \longrightarrow 2C_3H_6O_3$$

whilst lactate and acetate are formed from pentose without the release of carbon dioxide.

$$C_5H_{10}O_5 \longrightarrow C_3H_6O_3 + C_2H_4O_2$$

As indicated in the above stoichiometry, neither reaction loses any atoms of carbohydrate during bioconversion. Thus, bioconversion using lactic acid fermentation does not produce carbon dioxide and does not lose any material. Thus, this is an excellent counter-measure against the global problem of carbon dioxide accumulation. Processes by which glucose and xylose are first formed from lignocellulosic agricultural waste and next converted into lactic acid (from hexose) /or lactic acid and acetic acid (from pentose) are an anaerobic bioconversion; and resulting organic acid is transformed into commodity chemicals such as polylactate, biodegradable plastics, by synthetic methods. Recent technology involved in polylactate production can blend the various forms of polyester such as PHA to improve the characteristics of the plastics. Our laboratory has developed the fermentation technology to produce PHB from lactic acid and acetic acid (9,10) and this PHB could be used for preparing mixtures of polylactate and PHA whose characteristics is better than that of lactate homopolymer.

Lactococcus lactis IO-1 isolated in our laboratory (11) is capable of efficiently fermenting xylose up to 50 g/l initial concentration into acid (12,13). However, the rate of fermentation in xylose medium is slower than in glucose medium but a high dilution rate culture with cell recycling could be applied to stimulate xylose consumption rate overcoming lactate inhibition (14). We have observed an interesting phenomenon that NRSP, the product from natural rubber

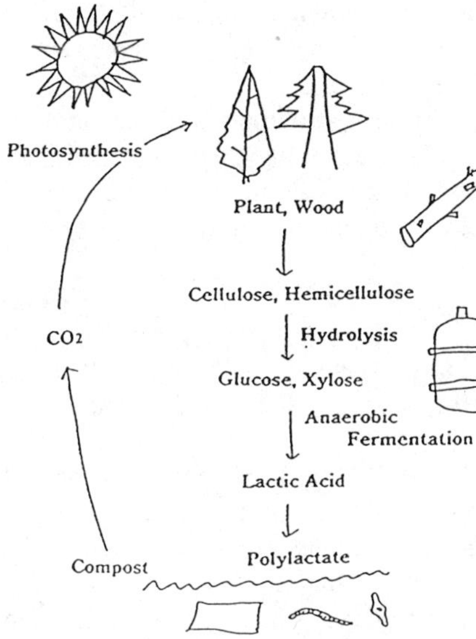

Fig. 5 The concept of "**Lactate Industry**" for postpetrochemistry.

waste, changed the molar ratio of the two products (lactic acid/ acetic acid) of lactic acid fermentation using xylose to increase lactic acid production (unpublished).

Bioconversion through lactic acid fermentation can therefore contribute to the reduction of the carbon dioxide concentration in atmosphere. We propose a new concept of biochemical industry named "Lactate Industry" (15,16) in which biomass is converted into lactic acid without carbon dioxide release and no elemental loss. The resulting lactic acid could be used to produce commodity chemicals such as biodegradable plastics. Used plastics would be digested into carbon dioxide by composting and recycled back to biomass by photosynthesis (Fig. 5).

5. Conclusion

Elevated levels of carbon dioxide in the atmosphere is a serious problem that will interfere with life of all organisms on the planet. To solve this crisis and to allow our continued survival, a favorable ecosystem in which all elements are recycled in the biosphere from the organic state to the inorganic state has to be maintained. Excess carbon dioxide is released into atmosphere by the consumption of fossil fuel to support our comfortable way of life. Carbon dioxide emission from non biological action can not yet be balanced by the recycling of elemental carbon in the biosphere. However, new technology introduced as a counter-measure to this environmental problem will inevitably results in other changes to our ecosystem. Such change is sometimes serious problem that all living organisms in the earth never have.

To tackle the global task of reduction of carbon dioxide accumulation through the application of biotechnology, we must assess the other risks that the introduction of the new technology could cause. We have to avoid these dangers we would be suffered by the new technology.

Literature Cited

1. Koba, Y. and Ishizaki, A.: *Agric. Biol. Chem.*, **1990**, 54, pp 1183-1187.
2. Lee, T.M., Ishizaki, A., Yoshino, S., and Furukawa, K.: *Biotechnol. Letters*, **1993**, 17, pp 649-654.
3. Ishizaki, A., Lee, T.M., and Todd, J.I.: Proceeding for 1996 PORIM International Palm Oil Congress. PORIM, Kuala Lumpur, **1996**, pp 114-121.
4. Grobben, N.G., Eggink, G., Cuperus, F.P., and Huizing H.J.: *Appl. Microbiol. Biotechnol.*, **1993**, 39, pp 494-498.
5. Stanton, W.R.: Proceeding of the Third International Sago Symposium. Yamada, N and Kainuma, K. ed., Tokyo, **1986**, pp 50-64.
6. Ishizaki, A.: Microbial Utilization of Renewable Resource Vol. 6. Osaka Univ., **1989**, pp 253-241.
7. Ishizaki, A.: *Biosci. Biotech. Biochem.*, **1995**, 59, 1150-1151.
8. Tripetchkul, S., Tonokawa, M., and Ishizaki, A.: *J. Ferment. Bioeng.*, **1992**, 74, pp 384-388.
9. Tanaka, K., Katamune, K., and Ishizaki, A.: *Biotechnol. Letters*, **1993**, 15, pp 1217-1222.
10. Tanaka, K., Katamune, K., and Ishizaki, A.: *Can J. Microbiol.*, **1995**, 41 (Suppl.), pp 254-257.
11. Ishizaki, A., Osajima, K., Nakamura, K., Kimura, K., Hara, T., and Ezaki, T.: *J. Gen. Appl. Microbiol.*, **1990**, 36, pp 1-6.
12. Ishizaki, A., Ueda, T., Tanaka, K., and Stanbury, P. F.: *Biotechnol. Letters*, **1992**,14, pp 599-604.
13. Ishizaki, A. and Ueda, T.: *J. Ferment. Bioeng.*, **1995**, 80, pp 287-290.
14. Ishizaki, A, and Vonktaveesuk, P.: *Biotechnol. Letters*, **1996**, 18, (in press, Dec. issue)
15. Ishizaki, A.: *Kagaku to Seibutsu* (Japanese), **1996**, 34, pp 232-237.
16. Ishizaki, A.: *Seibutsu Kogaku Kaishi* (Japanese), **1996**, 74, pp 289-291.

INDEXES

Author Index

Affiliation Index

Subject Index